The Biotechnology Revolution?

The Biotechnology Revolution?

Edited by
Martin Fransman, Gerd Junne
and Annemieke Roobeek

BLACKWELL
Oxford UK & Cambridge USA

Copyright © Basil Blackwell Ltd 1995
First published 1995

Blackwell Publishers
108 Cowley Road
Oxford OX4 1JF
UK

238 Main Street
Cambridge, Massachusetts 02142
USA

British Library Cataloguing in Publication Data

A CIP catalogue record for this book is available from the British Library.

Library of Congress Cataloging-in-Publication Data

The Biotechnology revolution / edited by Martin Fransman, Gerd Junne
and Annemieke Roobeek.
p.   cm.
Includes bibliographical references and index.
ISBN 0–631–19596–3 (pbk)
1. Biotechnology. I. Fransman, Martin. II. Junne, Gerd.
III. Roobeek, Annemieke J. M., 1958–.

TP248.2.B573   1995
660'.6—dc20   94–6313
                                                                          CIP

Typeset in 10 on 12 pt Palatino
by Pure Tech Corporation, Pondicherry, India
Printed in Great Britain by T.J. Press Ltd, Padstow, Cornwall.

This book is printed on acid-free paper

Contents

Contributors

Iftikhar Ahmed Technology and Employment Branch, Employment and Development Department, International Labour Organization (ILO), Geneva, Switzerland

H. David Banta Consultant, TNO/WHO Programme in Health Care Technology Assessment, Professor of Technology Assessment, University of Limburg, The Netherlands

W. Jos Bijman Research Fellow, Landbouw-Economisch Instituut (LEI-DLO), The Hague, The Netherlands

Frederick H. Buttel Professor of Rural Sociology, University of Wisconsin-Madison, U.S.A.

Mark Cantley Head of the Biotechnology Unit, Directorate for Science, Technology and Industry, OECD, Paris, France

Jack Doyle Director of the Agriculture and Biotechnology Project for the Environmental Policy Institute in Washington, D.C., U.S.A.

Martin Fransman Director, The Institute for Japanese–European Technology Studies, University of Edinburgh, U.K.

Christopher Freeman Emeritus Professor of Science Policy at the University of Sussex, U.K.

Phyllis Freeman Associate Professor and Director of Clinical Programs of the Law Center of the College of Public and Community Service at the University of Massachusetts, Boston, U.S.A.

Rod N. Greenshields Director, GB Biotechnology Ltd, Swansea, Wales, U.K.

Wendy Harcourt Editor of *Development*, Society for International Development (SID), Rome, Italy

John Hodgson Editor, *Bio/Technology*, London, U.K.

Rogier A. H. G. Holla Center for Technology, Policy and Industrial Development, Massachusetts Institute of Technology (MIT), Cambridge, Mass., U.S.A.

Jaap Jelsma Associate Professor, Centre for Studies of Science, Technology and Society, University of Twente, The Netherlands

Gerd Junne Professor in International Relations at the University of Amsterdam, The Netherlands

Martin Kenney Professor, Department of Applied Behavioral Sciences, University of California, Davis, U.S.A.

Sheldon Krimsky Professor, Department of Urban and Environmental Policy, Tufts University, Medford, Mass., U.S.A.

Gisbert A. M. van Marrewijk Department of Plant Breeding, Agricultural University, Wageningen, The Netherlands

D. de Nettancourt Biotechnology Division, Directorate-General for Science, Research and Development (DG XII), Commission of the European Communities, Brussels, Belgium

Gerardo Otero Assistant Professor of Latin American Studies, Simon Frazer University, Burnaby, B.C., Canada

Anthony Robbins Professor of Public Health at the Boston University School of Medicine, Boston; Head of the National Vaccine Program, U.S.A.

Clare Robinson Editor, *Trends in Biotechnology*, Cambridge, U.K.

Annemieke Roobeek Cornelis Verolme Professor of Technology and Economy at Nijenrode University – The Netherlands Business School; Senior Research Fellow of the Royal Netherlands Academy of Sciences (KNAW), Faculty of Economics and Econometrics, University of Amsterdam, The Netherlands.

Harry Rothman The Centre for Science and Technology Policy, Bristol Business School, Bristol Polytechnic, Bristol, U.K.

Lindsay Sawyer Department of Biochemistry, University of Edinburgh Medical School, Edinburgh, U.K.

Francisco C. Sercovich Industrial Development Officer, Regional and Country Studies Branch, United Nations Industrial Development Organization (UNIDO), Vienna, Austria

K. Schügerl Professor, Institut für Technische Chemie, University of Hanover, Germany

Margaret Sharp Senior Research Fellow, Science Policy Research Unit, University of Sussex, U.K.

Joachim H. Spangenberg Deutscher Umwelttag, Frankfurt, Germany

Shoko Tanaka Ph.D. Programme, Department of Government, Cornell University, Ithaca, U.S.A.

Z. Towalski The Biotechnology Training Unit, Blackpool and the Fylde College, Blackpool, U.K.

Indra K. Vasil Graduate research professor in the laboratory of plant cell and molecular biology, Department of vegetable crops, University of Florida, Gainesville, Florida, U.S.A.

Alyson Warhurst Research Fellow, Science Policy Research Unit, University of Sussex, U.K.

Raymond A. Zilinskas Research Associate Professor, Maryland Biotechnology Institute, College Park, Maryland, U.S.A.

Acknowledgements

The following chapters were first published elsewhere (see below). The editors gratefully acknowledge permission to reprint these contributions in the original or a slightly adapted form.

Chapter 2, Christopher Freeman, The Diffusion of Biotechnology Through the Economy: The Time Scale, chapter III of *Biotechnology. Economic and Wider Impacts*. Paris: OECD, 1989.

Chapter 4, Plenum Publishing Corporation and Geraldo Otero, The Coming Revolution of Biotechnology: A Critique of Buttel, *Sociological Forum*.

Chapter 13, Elsevier Trends Journals and Clare Robinson, The genome race is on the road, *Trends in Biotechnology*, vol. 10 (January/February 1992).

Chapter 15, Office of Technology Assessment, The Biotechniques in Agricultural Research, *New Developments in Biotechnology: 4. U.S. Investment in Biotechnology*, Washington, July 1988, pp. 194–219.

Chapter 16, Indra K. Vasil, The Realities and Challenges of Plant Biotechnology, *Bio/Technology*, vol. 8 (April 1990).

Chapter 25, The Genetics and Politics of Frost Control, in *Biotechnics and Society*. Westport, CT: Greenwood Publishing Group.

Chapter 27, John Hodgson, Biotechnology: Feeding the World?, *Bio/Technology*, vol. 10 (January 1992).

Chapter 28, Gerd Junne, The Impact of Biotechnology on International Commodity Trade, in E. J. DaSilva et al. (eds), *Biotechnology:*

Economic and Social Aspects. Issues for Developing Countries. Cambridge: Cambridge University Press, 1992.

Chapter 35, Francisco C. Sercovich, *Industrial Biotechnology Policy: Guidelines for Semi-Industrial Countries*, United Nations Industrial Development Organization, Report prepared by the Regional and Country Studies Branch, Industrial Policy and Perspective Division, Report PPD.194, 22 April 1991.

Chapter 36, Raymond A. Zilinskas, Biotechnology and the Third World: the missing link between research and applications, *Genome*, vol. 31 (1989).

Every effort has been made to trace all the copyright holders, but if any have been inadvertently overlooked the publishers will be pleased to make the necessary arrangement at the first opportunity.

Martin Fransman would like to acknowledge the patience and support of his family–Tammy, Judy, Karen and Jonathan.

Gerd Junne and Annemieke Roobeek would also like to acknowledge the patience and good spirits of their children, Fabian, Rosanne and Thorsten.

1

Editors' Introduction

Biotechnology has provided humankind with substantial, indeed incredible, powers over nature. For a long time it has been assumed that evolution occurs through the mutation of genes which in turn generates the variety which, together with the process of selection, drives the evolutionary process. Biotechnology, by providing the means to instantly combine the genetic material of different organisms, has given humankind the awesome power to override 'natural' evolution, thereby achieving an unprecedented degree of control.

These powers are remarkable and there is justified wonder regarding the scientific enterprise that has made such control possible. However, 'The Revolution in Biotechnology'[1] in scientific terms is not the main focus of this book, although some of the main achievements of biotechnology are described here as background material to understand and evaluate statements on the possible impact of these developments. The scientific advances do not in themselves constitute, or imply the necessity of, a biotechnology-led socio-economic revolution. It is this wider impact on economy and society with which this book mainly deals. This raises the question: what is meant by a 'revolution' in this context?

Christopher Freeman has suggested five criteria which a new technological system would have to satisfy to have major effects on economy and society (see Chapter 2):

1. A new range of products accompanied by an improvement in the technical characteristics of many products and processes.
2. A reduction in costs of many products and services.
3. Social and political acceptability.
4. Environmental acceptability.
5. Pervasive effects throughout the economic system.

If these criteria are fulfilled in the case of biotechnology, we may speak of a 'Biotechnology Revolution'.

Our purpose in putting together this collection of readings is to provide an overview of the emergent biotechnologies and to analyse many of the impacts and implications of the use of these technologies. Taken together, the different contributions in this volume will allow the reader to answer the question whether the above mentioned criteria are indeed fulfilled in the case of biotechnology.

The structure of the book therefore follows more or less these criteria. In the four contributions to the introductory section, the question of whether or not we can talk of a 'Biotechnology Revolution' is dealt with in more detail.

Part II provides a description of old and new biotechnologies. It points to the new range of products and to the improvement in the technical characteristics of many products and processes which result from modern applications of biotechnology. Modern biotechnology combines old and new biotechnologies. In our view, no sharp dividing line should be drawn between the two. The term biotechnology as used in this volume is broader than just 'genetic engineering'. Our interest in the socio-economic implications of biotechnology is better served by a broader definition. To make use of the results of genetic engineering, after all, the more traditional technologies such as tissue culture, fermentation and enzyme technology, have to be applied. The new possibilities created by the breakthrough of genetic engineering broaden the application of these more traditional technologies to such an extent that they too change qualitatively.

Part III is the largest part. It provides a survey of the areas of application of biotechnology. About half of all biotechnology research takes place in the health sector. The social and economic repercussions, however, will probably be at least as great in the field of agriculture. Applications to agriculture may develop just in time in order to help world agriculture to adapt to long-term global climatic change. But the capacity to use the new technologies will vary from one country to another, aggravating the existing large inter-country development gap rather than reducing it. Though there are many applications of biotechnology in other industries, biotechnology may cause less of a revolution in these fields.

The diffusion of biotechnology is the topic of Part IV. Before the impact of any technology can be discussed, its diffusion has to be analysed. Impediments to the further diffusion of biotechnology can be embedded in the technology itself, in industrial structures, regulatory rules, and in public attitudes. To cause far-reaching changes, however, it is not necessary that the diffusion is really pervasive.

Many changes in investment patterns, trade flows, employment opportunities etc. are the result of *partial* and *unequal* patterns of diffusion – between sectors, geographical areas, large and small companies etc. Different forms of regulation can either impede or speed up the diffusion of biotechnology, which is probably, beside nuclear energy, the most heavily regulated technology field of all.

Part V concentrates on the impact of biotechnology. Biotechnology will probably cause fundamental changes in the world food system. It will affect trade in many commodities. It will have an important impact on demographic structures of the population in many countries. It will affect employment to an extent which can be compared to the 'Green Revolution'. And it will have an impact on the protection of the environment, although the direction of this impact is not yet certain, since dangers for the environment and new means to prevent or clear up pollution may counterbalance each other.

In Part VI, finally, a number of policy issues will be discussed. What are the options for public policies at different levels of economic development? What will be the corporate structures and strategies in the different sectors affected? What are the implications of different policies and industrial structures in the United States, Europe, Japan, and other countries? What role is and can be played by international organizations in stimulating the use of modern biotechnology and in coordinating national policies?

This book in a way complements the three volumes on 'The Microelectronics Revolution' (1980), 'The Information Technology Revolution' (1985) and 'The Materials Revolution' (1988), all edited by Tom Forester. The selection of these 'new technologies' – microelectronics and information technology, materials, and biotechnology – reflect the fact that these are at the very core of the 'new technological paradigm'.[2] Microelectronics demonstrated first how far-reaching the repercussions of these changes are. Developments in microelectronics and information technology have fundamentally changed industries and services and directly affected almost everybody's life and work. Expectations regarding the importance of biotechnology and the pervasiveness of its applications have no doubt been influenced by the experience with microelectronics. The development of modern biotechnology, however, is of somewhat more recent origin, and it will take some time until the full impact is felt.

The different 'new' technologies, nevertheless, should not be seen in isolation from each other. They not only have much in common, they also affect each other. Modern biotechnology could hardly advance without the support of information technology; information technology needs new materials for further breakthroughs (e.g. superconductivity), and biotechnology has started to contribute new tailor-made materials, and ultimately may lead to a totally new generation of information processing (bioinformatics). Rather than a series of discrete revolutions, caused by different technological breakthroughs, the world is more likely to experience rapid change as a result of the interaction of these different technologies with each other in the decades to come.

NOTES

[1] Cf. Jean L. Marx (ed.), *A Revolution in Biotechnology* (Cambridge: Cambridge University Press, 1989).

[2] Giovani Dosi, *Technical Change and Industrial Transformation. The theory and an application to the semiconductor industry* (London: Macmillan, 1984); Luigi Orsenigo, *The Emergence of Biotechnology. Institutions and Markets in Industrial Innovation* (London: Pinter Publishers, 1989); Annemieke J. M. Roobeek, *Beyond the Technology Race. An Analysis of Technology Policy in Seven Industrial Countries* (Amsterdam: Elsevier Science Publishers, 1990).

Part I

The Debate

2

Technological Revolutions: Historical Analogies

Christopher Freeman

Introduction

Modern biotechnology is opening up innumerable exciting new possibilities which may dramatically affect society over the next century. Some of these new developments are already in the arena of commercial exploitation, especially in the fields of health care and of agriculture. But the results of interviews with nearly one hundred companies suggest that we are still only in the early stages of the full-scale application of this revolutionary new technology. In this chapter we discuss the problems of diffusion of the technology through the economy as a whole and consider the probable time scale of this diffusion process.

We shall make some comparisons with other pervasive technologies which have had very widespread economic consequences in the past, such as the introduction of electric power, and more recently computer technology and microelectronics. It is always dangerous of course to make analogies of this kind and it is extremely important to take account of the *differences*, as well as the *similarities* between the various generic technologies which have so deeply transformed industrial societies over the past century.

However, although the unique features of modern biotechnology and its distinct areas of application must always be kept in mind, there are useful lessons which can be learnt from earlier waves of technical change which had very widespread economic and social consequences. It is evident, for example, that a transformation of the technologies in use in many sectors of the economy must lead on the one hand to large-scale investment in new types of plant and equipment and on the other hand to a change in the skill profile of the labour force. It must also lead to changes in company organisation and industrial structure.

We know from the past experience of the introduction and diffusion of such revolutionary new technologies that these changes in capital stock, in the skill profile and in organisational structures of industry cannot possibly take place in a short period. They are a matter of decades rather than years or months. The recognition of the relatively long time scale involved in diffusion is extremely important as it can avert two dangers which might otherwise have adverse policy consequences.

First is the danger of 'technological super optimism', which tends to ignore the hard economic realities of relative costs, profitability and size and consumer acceptance of entirely new products. Second is the danger of 'technological conservatism' which fails to recognise the enormous long-term potential of generic technologies for the ultimate development of an entirely new range of products and services. The first can lead to serious under-estimation of the time scale involved in diffusion processes; the second to equally serious errors of under-estimation of the potential of long-term transformation. We shall illustrate these points from the history of computer technology and electric power and we shall then consider the specific features of the new biotechnology and its diffusion in the economic and social system.[1]

The Analogy of the Electronic Computer:
Microelectronics and the 'Automatic Factory'

We shall first take the example of electronic computer technology to illustrate the general problem of estimating the probable scale and timing. We take this example because (at least until the advent of biotechnology) it is probably the best known example of a pervasive technology in the second half of the twentieth century. Moreover it is one which is well documented and which is generally agreed to be of extraordinary importance for all OECD Member countries.

With the first application of the electronic computer during and just after the Second World War it was realised that this new technology had an enormous potential for transforming industrial processes, office systems and records and communication systems. However, opinions differed sharply on the probable time scale and extent of these developments. Some scientists and engineers anticipated very rapid and large-scale applications with immense social consequences (including large-scale unemployment) already in the 1950s and 1960s. On the other hand it is well established that such a well informed industrial leader as Thomas J. Watson (Senior) of IBM did not believe, even in the early 1950s, that there would be any big commercial market for electronic computers.[2]

He felt that the one SSEC machine which was on display at IBM's New York offices could solve all the scientific problems in the world involving scientific calculations. He saw no commercial possibilities. This view, moreover, persisted even though some private firms that were potential users of computers – the major life insurance companies, telecommunications providers, aircraft manufacturers and others were reasonably informed about the emerging technology. A broad business need was not apparent.

It was not until the Korean War that IBM was persuaded to undertake production of a small batch of electronic computers and even then it was only with a change of management that they entered the commercial market. As against this conservative view of a very limited market for computers, imaginative scientists like Norbert Wiener (1949) envisaged a huge scale of applications and forecast large-scale unemployment as a result.

A much more balanced view of the probable time scale and social consequences of the diffusion of the electronic computer was taken by John Diebold (1952),[3] one of the most authoritative and imaginative consultants in this field. In his book *The Advent of the Automatic Factory* he showed remarkable foresight and depth of understanding of the problems involved. Whilst recognising the enormous potential of the electronic computer for the transformation of all industrial and office processes, he saw quite clearly that this would be a matter of several decades and not of a few years. Indeed most of the 'factory automation', which is today described as 'FMS' (Flexible Manufacturing Systems) or 'CIM' (Computer Integrated Manufacturing) did not show a really rapid take-off until the 1980s, even though most of the technical innovations which come under this heading were clearly foreseen by John Diebold in 1952. The 'automation' of the 1950s was really a kind of advanced mechanisation (mainly in the automobile industry) rather than computerisation.

Diebold stressed several reasons for believing that the diffusion process would be much slower than many computer enthusiasts imagined at that time. The most important were these:

(i) True computerised 'automation' would involve the redesign of all industrial processes and products. It would be quite impossible to achieve this in a short period. The simple availability of computers was only the first step. An enormous amount of R&D, design and new investment in machinery and instruments would be needed in every branch of industry.

(ii)   Such a process of redesign would affect *both* products and processes. Diebold gave examples to show that this could only occur if there was a change in the structure and organisation of firms, as well as in the attitude of management. This change would involve much closer integration of R&D, design, production engineering and marketing – a horizontal rather than a vertical flow of communication and information within firms.

(iii) Not only would computer-based automation change the configuration and organisation of every factory, it would also involve a big change in the skill composition of the work force. Diebold rejected the idea of mass unemployment arising from automation and also the idea of 'deskilling' the work force. On the contrary he stressed the *new* skills that would be required, especially in design and maintenance, and saw automation as a means of overcoming the fragmentation and dehumanisation of work. But he also saw that it would take a long time before the new skills were available and people were retrained.

(iv)   Diebold recognised the importance of the *economic* aspects of diffusion. Computers would diffuse not only because they were technically advantageous. They had also to be *cheap*. It was only with the advent of microelectronics in the 1960s and the microprocessor in the early 1970s that computerisation took off in small and medium-sized firms (SMEs), as well as in large firms, and in batch production as well as in flow process industries such as chemicals. Moreover (and this is the most important point when we are looking at economy-wide effects), computer technology could only realise its potential outside a few 'leading-edge' industries when computerised *systems* became relatively cheap and accessible.

Events since 1952 have fully confirmed Diebold's analysis. Even though the computer industry itself was growing at an extremely rapid rate for the next thirty years it took a whole series of complementary radical and incremental innovations such as Computer-Aided Design (CAD), Computer Numerical Control (CNC), Large-Scale Integration (LSI) and big developments in software engineering and process instrumentation before computer-based automation could diffuse to most industrial and service sectors. Even now the economic advantages are by no means always clear-cut and there are often considerable teething problems when firms attempt to introduce FMS, or other forms of computerisation.

The Analogy with Electric Power

A similarly long time scale was necessary for the diffusion of electric power and its innumerable applications, from the time of its first appearance in the 1880s, and for similar reasons. Not only did it take two or three decades before generating and transmission systems made the new energy source universally available in the industrialised countries; it took even longer to redesign machinery and equipment in other industries to take advantage of electricity, and to make the necessary skills available.

Warren Devine[4] has given an illuminating account of the debates which took place at the end of the 19th century and the early part of this century on the implications of electric power for the future of factory processes.

Replacing a steam engine with one or more electric motors, leaving the power distribution system unchanged, appears to have been the usual juxtaposition of a new technology upon the framework of an old one . . . . Shaft and belt power distribution systems were in place, and manufacturers were familiar with their problems. Turning line shafts with motors was an improvement that required modifying only the front end of the system . . . . As long as the electric motors were simply used in place of steam engines to turn long line shafts, the shortcomings of mechanical power distribution systems remained.

It was not until after 1900 that manufacturers generally began to realise that the indirect benefits of using unit electric drives were far greater than the direct energy-saving benefits. Unit drive gave far greater flexibility in factory layout, as machines were no longer placed in line with shafts, making possible big capital savings in floor space. For example, the US Government Printing Office was able to add 40 presses in the same floor space. Unit drive meant that trolleys and overhead cranes could be used on a large scale, unobstructed by shafts, countershafts and belts. Portable power tools increased even further the flexibility and adaptability of production systems. Factories could be made much cleaner and lighter, which was very important in industries such as textiles and printing, both for working conditions and for product quality and process efficiency. Production capacity could be expanded much more easily.

Table 2.1 Chronology of electrification of industry[1]

1870	1875	1880	1885	1890	1895	1900	1905	1910	1915	1920	1925	1930

a   Direct Drive

    Line shaft drive _ _ _ _ _ _

       Group drive _ _ _ _ _ _ _

       Unit drive _ _ _ _ _ _ _

    DC Transmission _

b   DC motors in manufacturing

    'Battle of the currents'

    AC transmission _ _ _ _

    AC motors in manufacturing _

c

1870:	Steam 52%	Water 48%	
1890:	Steam 78%	Water 21%	Electricity <1%
1900:	Steam 81%	Water 13%	Electricity 5%
1910:	Steam 65%	Electricity 25%	Water 7%
1915:	Electricity 53%	Steam 39%	Water 3%
1925:	Electricity 78%	Steam 16%	Water 1%

d

1870   DC electric generator (hand-driven)

1873   Motor driven by a generator

1878   Electricity generated using steam engine

1879   Practical incandescent light

1882   Electricity marketed as a commodity

1883   Motors used in manufacturing

1884   Steam turbine developed

1886   Westinghouse introduces AC for lighting

1888   Tesla develops AC motor

1891   AC power transmission to industrial use

1892   Westinghouse markets AC polyphase induction motor; General Electric Company formed by merger

1893   Samuel Insull becomes President of Chicago Edison Company

1895   AC generation at Niagara Falls

1900   Central Station steam turbine and AC generator

1907   State-regulated territorial monopolies

1917   Primary motors predominate; capacity and generation of utilities exceeds that of industrial establishments

1. a.   Methods of driving machinery;
   b.   Rise of alternating current;
   c.   Share of power for mechanical drive provided by steam, water and electricity;
   d.   Key technical and entrepreneurial developments.

Source: W. Devine: From Shafts to Wires: A Historical Perspective, *Journal of Economic History*, Vol. 43, pp. 347–373.

The full expansionary benefits of electric power to the economy depended, therefore, not only on a few key innovations in the 1880s, but on the development of a new paradigm or production and design philosophy. This involved the redesign of machine tools and much other production equipment. It also involved the relocation of many plants and industries, based on the new freedom conferred by electric power transmission and local generating capacity. Finally, the revolution affected not only capital goods but a whole range of consumer goods, as a series of radical innovations led to the universal availability of a wide range of electric domestic appliances going far beyond the original domestic lighting systems of the 1880s. Ultimately, therefore, the impetus to economic development from electricity affected almost the entire range of goods and services.

But this complex diffusion process took about half a century and it was not actually until the 1920s that electricity overtook steam as the main source of industrial power in the United States (table 2.1). It was not until the 1950s and 1960s that widespread ownership of electric consumer durables became the norm in Europe and Japan.

From these historical analogies it is evident that there is a major difference between the diffusion process for a single product and the diffusion process for a generic technology with numerous potential applications in a variety of different industrial sectors. Once it is on the market in an acceptable form, a single product may be adopted by more than half the population of potential adopters within a decade. This occurred for example in many OECD countries for a variety of consumer durables such as television, in the 1950s and 1960s. Mansfield's studies[5] also showed that this rate of adoption occurred for some types of industrial and transport equipment, such as the diesel locomotive and the continuous strip mill. However, there are other case studies of both agricultural and industrial innovations (such as Metcalfe's 1970 study of the diffusion of the size box in the cotton industry) which show many 'laggards' and non-adopters even when the economic and technical advantages are clear.[6] When we come to consider whole clusters of related innovations with new generations of products, a much longer time scale is involved.

The New Biotechnology as a Change of Technology System

General criteria for major economic effects of new technologies

Most diffusion research in the post-war period has concentrated on the diffusion of individual products and processes, and on incremental types of innovation. Schumpeter[7] was almost alone among leading

20th century economists in looking at 'creative waves of destruction' – the effect of major new technologies as they pervade the economic system. More recently a number of economists have made further contributions to this Schumpeterian approach.

Nelson and Winter[8] used the expression 'generalised natural trajectories' to describe cumulative clusters of innovations, as for example, those associated with electric power. Dosi[9] used the expression 'technological paradigm' by analogy with Kuhn's[10] scientific paradigms. In these terms 'incremental innovations' within an established paradigm may be compared with Kuhn's 'normal science'. Carlota Perez[11] has developed the concept of 'techno-economic paradigms' to describe those changes in technology which pervade the entire economy and provide the new 'common sense' for a whole generation of engineers and managers.

Clearly biotechnology is already a new paradigm in Dosi's sense and a new 'natural trajectory' in Nelson and Winter's sense for the development of products and processes. Whether it is such an important trajectory that it will ultimately come to affect management decision-making in most branches of the economy remains an open question. The new biotechnology has undoubtedly led to enormous excitment in the research community and many new companies were established with venture capital to pursue R&D. This 'research explosion' was without parallel. But the pervasiveness of a new trajectory or technology system depends on the range of *profitable* opportunities for exploitation. Until recently, despite its undoubted importance for the future, biotechnology has led to *profitable* innovations in only a relatively small number of applications in a few sectors and in a few countries.

In Schumpeter's model, the profits realised by innovators are the decisive impulse to surges of growth, acting as a signal to the swarms of imitators. But this 'swarming' behaviour, generating a great deal of new investment and employment, depends on falling costs of adoption and very clear-cut advantages and/or competitive pressures. Later on, of course, after a period of profitable fast growth, profitability may decline. Schumpeter stressed that changing profit expectations during the growth of an industry are a major determinant for the sigmoidal pattern of growth. As new capacity is expanded, at some point (varying with the industries in question), growth will begin to slow down. Exceptionally, this process of maturation may take only a few years, but more typically it will take several decades and sometimes longer still. Biotechnology is a very long way from this mature stage and the main interest is in when it will enter the 'swarming' phase and on what scale.

For a new technological system to have major effects on the economy as a whole it should satisfy the following conditions:

(i) *A new range of products accompanied by an improvement in the technical characteristics of many products and processes,* in terms of improved reliability, new properties, better quality, accuracy, speed or other performance characteristics. This leads to the opening up of many new markets, with high and rapid growth potential and the rise of new industries, based on these products.

(ii) *A reduction in costs of many products and services.* In some areas this may be an order of magnitude reduction; in others, much less. But it provides another essential condition for Schumpeterian 'swarming', i.e. widespread perceived opportunities for new profitable investment. The major revolutions such as electric power and computing were both labour-saving *and* capital-saving, but also offered a reduction in the cost of other major inputs, such as energy.

(iii) *Social and political acceptability.* Although many economists and technologists tend to think narrowly in terms of the first two characteristics, this third criterion is also important. Whereas the first two advantages are fairly quickly perceived, there may be long delays in *social* acceptance of revolutionary new technologies, especially in areas of application far removed from the initial introduction. Legislative, educational and regulatory changes may be involved, as well as changes in management and labour attitudes and procedures. Changes in taste, especially in sensitive areas such as food and drink, are often unpredictable.

(iv) *Environmental acceptability.* This may be regarded as a subset of (iii) above, but, especially in recent times, it has become important in its own right. It is of particular significance in relation to the *Limits to Growth* debate and the debate over nuclear power. It finds expression in the development of a regulatory framework of safety legislation, and procedural norms which accompany the diffusion of any major technology. Particularly hazardous technologies or those which are extremely expensive to control are severely handicapped, even if they do have some economic and technical advantages, as in the case of nuclear power.

(v) *Pervasive effects throughout the economic system.* Some new technologies, as for example the float-glass process, have revolutionary effects and are socially acceptable, but are confined in their range of applications to one or a very few branches of the

economy. It follows from (i) and (ii) above that for a new technology to be capable of affecting the behaviour of the economy
as a whole, in the Perez sense, it must clearly have effects on
technical change and investment decisions in many or all important sectors.

The case of new biotechnology

Using these five criteria it is relatively easy to see why nuclear technology does not qualify as a change of techno-economic paradigm
since it fails on almost every one of them. Electric power or the
microelectronic, computer-based information technology by contrast
satisfy all five criteria.

Clearly, the new biotechnology is likely to satisfy the first of the
above criteria. It is beginning to give rise to a range of new products
and processes in health care, in medical diagnostics, in water treatment, in veterinary applications, in agriculture and forestry, in the
food industry, in services and in mineral extraction and processing.
New varieties of plants with drought-resistant, pest-resistant and
other specifically engineered attributes are one of the most active
areas with enormous potential. This is confirmed for example by the
structural change in the activity of chemical firms entering the seed
and bulb industries on a large scale. The future potential is even
greater, extending to a broad range of chemical and food products
and processes and perhaps ultimately to an even wider spectrum. If
the hopes of 'biochips' are realised they could extend to the whole
range of microelectronics.

When we come to the second criterion the picture is less clear-cut.
Perhaps the best analogy is with the first two generations of electronic
computers before the advent of the integrated circuit and the microprocessor. At this stage, in the 1950s, computers certainly found cost-
reducing applications in such areas as pay-roll and invoicing, but the
cost of computers was still relatively high so that the range of adoption was limited. In the case of the new biotechnology there are
indications that rDNA-derived and other new processes will be less
costly than traditional manufacturing processes in certain health-care
and agricultural sectors, but it is also evident that in other important
areas, such as animal feeding stuffs, mineral extraction, energy and
chemical feed stocks, the lack of cost competitiveness has slowed
down or prevented more widespread application. This is an important limitation on the speed of diffusion in key industrial sectors.

However, these limitations may be overcome as a result of further research and development, or as a result of rising costs and

diminishing supplies of non-renewable materials or both. In his analysis of the 'Economic Potential of Biotechnologies' Rehm[12] pointed out that very much further research and development was needed not only in the field of chemical feedstocks but also in relation to biomass, metal leaching and oil recovery. His point is still valid. Relative prices and estimates of probable profitability have not yet led to a major expansion of these fields of R&D.

In the NAS Report on *New Frontiers in Biotechnology*, Cooney[13] pointed out in 1984 that:

Products from the biochemical process industry . . . take advantage of the same economies of scale experienced in commodity chemicals production. . . . Most biochemicals are made using inexpensive raw materials, such as sugar, and they offer good potential value added. The profit margins depend on the efficiency in transforming these raw materials into products. It is this biochemical problem that needs to be translated into a biochemical process. At this point one begins to see the need for integrating improved conversion yields, better metabolic pathways and new reactor mechanisms. This requires integrating biochemistry, microbiology and chemical process technology.

However, this integration of disciplines and skills is by no means easy to achieve as it requires new forms of organisation and structure in firms (and in universities). It is a problem comparable to that identified by Diebold in the case of factory automation. Diebold realised that an enormous amount of design and development work was necessary for each specific application of computers to industrial processes and that the skills were often not available for this work. Nor were firms organised to achieve results. In the case of biotechnology similar points are valid but the extent of redesign may be less far-reaching. Postgate[14] may be right when he says that one major disadvantage of biotechnological processes need not prevent them from becoming cost effective (i.e. that the product usually has to be concentrated from relatively dilute solutions). He is right too that they have the advantage of not requiring high temperatures and pressures. However, the experience so far with the scaling up of biotechnological processes to meet the requirements for large-scale production of *bulk commodities* is not encouraging in relation to comparative costs.

Costs remain high and it is an open question whether biotechnological processes will replace the present processes for bulk chemicals in the next 20 years. Applications in the copper industry appear more promising, as Warhurst has shown.[15]

Links upstream to more fundamental research have been a central feature of the new biotechnology and exceptionally important for chemical and drug firms.[16] This will continue to be extremely important in relation to most new products. At present, the dependence of biotechnology on information technology is considerable. 'Supercomputers' and advanced information systems are essential for advanced research, development and design work in molecular biology. Biosensors also demand integration with electronic technology. The two technologies are likely to become increasingly interdependent in future generations of 'intelligent' computer systems and process control systems. But again the present experience of '5th Generation' computers and 'artificial intelligence' suggests that these developments will also extend well into the next century and are surrounded by great uncertainty.

The discussion so far has concentrated mainly on the problems of process technology in relation to the potential large-scale future applications of biotechnology outside the present rather limited area.

This is because the economy-wide effects of the new biotechnology depend upon the resolution of these problems (our second criterion). It is already clear that biotechnology is having effects in the pharmaceutical industry, medical care and agriculture. Whether these effects extend to the whole of the chemical industry, oil recovery, the energy industries, the food industry and ultimately an even wider range of manufacturing and service industries, will depend upon the progress of research, development and design over the next 10–20 years. This in turn relates to social and organisational problems in the 'national system of innovation' – the management and scale of R&D, the interfaces between different parts of the system, the availability of skills, the encouragement (or lack of it) to the experimental application of new processes and so forth. Finally, the incentives to conduct R&D and to introduce new processes depend on the development of relative costs in alternative processes.

Whether or not the new biotechnology becomes a new 'techno-economic paradigm' dominating future economic development in the next century, depends also upon whether it satisfies other criteria. So far we have discussed almost exclusively the first two criteria relating to technical and economic performance.

Continued acceptance cannot be taken for granted, although the initial public debate and the regulatory mechanisms already introduced offer a favourable basis for diffusion in most countries. Various retarding factors in such sensitive areas as agriculture, food and public health are however present.

Clearly, new social and institutional problems will arise as the scale of application extends. These could be very great, involving for example, the restructuring of agricultural and health services on a worldwide scale. Kristensen[17] has suggested that there may even be a 'role reversal' between the present group of industrialised countries and the present group of underdeveloped countries. Because of their land area, demographic pressures and environmentalist pressures as well as because of new technological developments, the present industrial (mainly OECD) countries may become the main suppliers of agricultural products (indeed Kristensen suggests this is already happening, even without biotechnology. Some of the present Third World countries on the other hand, especially the NICs, may become the main centres of manufacturing production and exports during the 21st century.

This speculative 'futurology' is intended *not* as a precise forecast of future events, which is in any case impossible, but to illustrate the type and magnitude of the structural changes and the social and institutional adjustments which may occur as biotechnology begins to have really widespread effects. Big changes in the internal structure of agricultural production within each country are also probable. One small example of this may illustrate the point. The UK is now exporting date palms on a significant scale to the Middle East. This business has been pioneered by what was once a small horticultural enterprise on traditional lines but is now a medium-sized firm with R&D facilities and hundreds of employees. Classical (and neo-classical) trade theory would probably have considered the idea of the UK being an exporter of date palms as absurd on grounds of comparative advantage. But new technologies change many parameters and this certainly applies to biotechnology.

The pervasiveness of new biotechnology

Following this brief discussion on the four criteria so far considered, we may now turn to the final (fifth) criterion to be taken into account in assessing the macro-economic consequences of the new biotechnology: pervasiveness.

The new biotechnology is clearly more pervasive than more narrowly focused technologies, such as nuclear power. It has already found applications in primary industries (agriculture, forestry and mining), in secondary industries (chemicals, drugs, food) and in tertiary industries (health care, education, research, advisory services). However, the *actual* range of applications is still far narrower than the potential, and the 'known potential' (a concept which may be compared to 'proven reserves' in the oil industry) is still much narrower than in the case of computer applications.

In fact, biotechnology has often been compared to information technology, whose influence can be felt in all economic sectors. However, it is necessary to emphasize some fundamental differences at least for the time being. First of all, the fact that biotechnology operates through living organisms (or parts thereof) limits the field of activity to materials that can be biologically manipulated. Numerous industrial sectors would then be excluded from the *direct* influence of biotechnology (e.g. the metal and steel industry, mechanical industry, telecommunications and so on), although an indirect or mediated influence cannot be excluded.

Information technology, on the other hand, operates through substitution or change of a given production factor (e.g. labour) and has been able to penetrate almost all products and processes of human activity.

There is no process that has not been or cannot be modified by the use of information technology. From this feature derives the functional pervasiveness of information technology: informatics and telecommunications are employed not only by technical personnel (e.g. production, engineering, R&D), but also by non-technical personnel (e.g. administration, financing, marketing, sales). This does presently not hold true for biotechnology, which, in this respect, is more akin to chemical technology.

In the longer term, a linking of biological and information technologies might materialise in specific devices (biochips, neurocomputers, biorobotics) endowed with much higher capacities for storage and processing. The merging of information technology with the power of the new biotechnology would give the latter all the pervasiveness-aspects of information technologies and would influence human activity in ways which are presently difficult to imagine.

As we have seen, the future path of diffusion is still surrounded by considerable uncertainties, associated both with future technological developments and with socio-political developments. For this reason, figure 2.1 illustrates simplistically four possible 'scenarios'.

Scenario 1 would represent an accelerated diffusion of the new biotechnology into many new industrial sectors and applications, including the rapid development of many new industrial processes as well as products. It would represent a more rapid advance than that which occurred historically with earlier waves of new technology. For reasons which have been discussed (time and scale of R&D, education, training, capital investment, social and structural change) this scenario seems highly improbable. Only rapid changes in relative costs, prices and profitability might induce such a development.

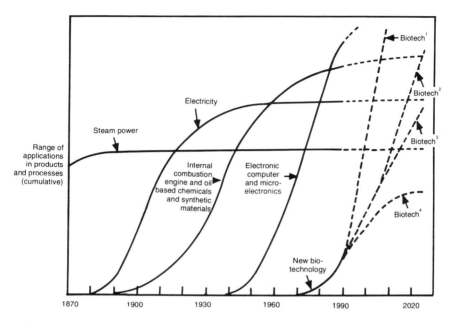

Figure 2.1 A simplified illustrative representation of the diffusion of 'mega-technologies'
Source : C. Freeman

Scenario 4 would represent a much slower rate of diffusion than now seems likely from the identifiable potential of the technology. It would mean not only that 'biochips' but also many other of the hoped for applications of biotechnology in the chemical, food, agricultural, energy, medical, environmental, mineral and other industries remain still in the future even after another 20 years of R&D. This also seems a rather unlikely scenario in the light of all that is known about the current pace of R&D activity and the technological potential.

Something between Scenario 2 and Scenario 3 therefore is identified as the most probable, with biotechnology beginning to be *a major basis* for new investment and the growth of the economy in the second or third decade of the 21st century.

Of course, the level of aggregation of this Scenario is very high. A more precise analysis would have to break down biotechnology into various fields of application in order to assess both promoting and retarding diffusion factors in each of them. Diffusion rates vary greatly between sectors. Health applications of biotechnology are likely to diffuse much faster than agricultural applications, and within the health sector, diagnostic products are diffusing faster than therapeutic products. Past experience in agriculture indicates that new technologies often take 10–20 years before they are adopted, but these rules may not always apply to the new innovations coming from biotechnology. If the introduction of an agricultural innovation has clear economic advantages, and if there is a fast flow of information, no supply bottlenecks and no regulatory delays, diffusion times in agriculture could be shorter than in the past. The new biotechnology appears to satisfy at least the first three of these four conditions. But even then, social and structural adjustments have to accompany and support technical change.

The fact that the new biotechnology is not likely to become the predominant technology for most industries and services in the 20th century should be no cause for surprise. Nor does it mean that it is unimportant for economic growth and international competitiveness. On the contrary, it is clear that already it will be at the heart of a rapidly growing cluster of new industries and an essential element of competitive survival in an increasing number of established industries. Moreover, one reason for the intense research interest is that the unexpected can always happen in such a fast-moving area.

Finally, it is essential to remember that the discussion so far has been mainly in terms of the quantitative contribution of the new biotechnology to economic performance. The qualitative and social changes associated with its diffusion are likely to be far more important and cannot be measured in such categories as GNP or industrial and agricultural production, as the Introduction has indicated. Its effects on health care are already profound and its influence on the entire culture and social fabric of OECD countries may be even more profound. These aspects of biotechnology raise fundamental philosophical and ethical problems which are beyond the scope of this chapter. Suffice it to mention here the extraordinary implications of 'genetic fingerprinting' for crime detection, for personal medical records and prognosis and for the insurance industry. Although the economic effects of this new activity may be limited and it is likely to develop first as a 'service' provided by specialised firms to government and other industries, it raises very fundamental issues of personal and social behaviour as well as legal issues. These qualitative implications of biotechnology are likely to be one of the biggest challenges over the next decade.

NOTES

[1] On these and related issues, see also: Freeman, C., The Challenge of New Technologies, OECD 25th Anniversary Symposium on Opportunities and

Risks for the World Economy: The Challenge of Increasing Complexity (Paris: OECD, 1986); Freeman, C., *Technology Policy and Economic Performance: Lessons from Japan* (London: Frances Pinter, 1987); Freeman, C. and Perez, C., Innovazione, Diffusione, e Nuovi Modelli Tecno-Economici, *L'Impresa*, No. 2, Milan (1986), pp. 7–14.

2
Katz, B. G. and Phillips, A., Government, Technological Opportunities and the Emergence of the Computer Industry, in Giersch, H. (ed.), *Emerging Technologies: Consequences for Economic Growth, Structural Change and Employment* (Tübingen: J. C. B. Mohr, 1982).

3
Diebold, J., *Automation: The Advent of the Automatic Factory* (New York: Van Nostrand, 1952).

4
Devine, W., From Shafts to Wires: Historical Perspectives, *Journal of Economic History*, Vol. 43 (1983), pp. 347–373.

5
Mansfield, E., Technical Change and the Rate of Imitation, *Econometrica*, Vol. 29, No. 4 (1961), pp. 741–766.

6
Metcalfe, J. S., The Diffusion of Innovation in the Lancashire Textile Industry, *Manchester School*, No. 2 (1970), pp. 145–162.

7
Schumpeter, J., *Business Cycles*, 2 volumes (New York: McGraw Hill, 1939).

8
Nelson, R. R. and Winter, S. G., In Search of a Useful Theory of Innovation, *Research Policy*. Vol. 6, No. 1 (1987), pp. 36–76.

9
Dosi, G., Technological Paradigms and Technological Trajectories, *Research Policy*, Vol. 11 (1982), p. 147–63.

10
Kuhn, T. S., *The Structure of Scientific Revolutions* (Chicago University Press, 1982).

11
Perez, C., Structural Change and the Assimilation of New Technologies in the Economic and Social System, *Futures*, Vol. 15, No. 4 (1983), pp. 357–375; Perez, C., Microelectronics, Long Waves and World Structural Change: New Perspectives for Developing Countries, *World Development*, Vol. 13, No. 3 (1985), pp. 441–463.

12
Rehm, H. J., The Economic Potential of Biotechnologies, Kiel Conference on Emerging Technology, Kiel Institut für Weltwirtschaft, published in Giersch, H. (ed.), *Emerging Technologies: Consequences for Economic Growth, Structural Change and Employment* (1982).

13
Cooney, C. L., Biochemical Engineering Solutions to Biotechnological Problems, in *New Frontiers of Biotechnology* (Washington, D.C.: NAC, 1984).

14
Postgate, J., Microbes, Microbiology and the Future of Man, *The Microbe* (Society for General Microbiology, Cambridge University Press, 1984), pp. 319–333.

15
Warhurst, A., The Potential of Biotechnology for Mining in Developing Countries: the Case of the Andean Pact Copper Project, D.Phil. thesis, University of Sussex, 1986.

16
Faulkner, W., Linkage between Academic and Industrial Research: the Case of Biotechnological Research in the Pharmaceutical Industry, D.Phil. thesis, University of Sussex, 1986.

[17] Kristensen, R., *Biotechnology and the Future Economic Development* (Copenhagen: Institute for Future Studies, 1986); Hollander, S. G., *The Sources of Increased Efficiency: A Study of Du Pont Rayon Plants* (MIT Press, 1965).

3

Biotechnology: An Epoch-Making Technology?

Frederick H. Buttel

Introduction

All social science is, in a sense, partisan in its explicit or implicit belief that the subject matter under study is important, and therefore worthy of inquiry. Thus, an elementary sociology of knowledge on 'high-technology' surely must begin with the observation that the social scientists who explore such a topic do so, at least in part, out of a conviction that current patterns of technological innovation are, or are destined to be, crucial in the restructuring of local, regional, national, and global economies and social structures. There can be little doubt that science is becoming increasingly important as a productive force in modern capitalist–industrial societies, and that social scientific analysis of science and technology is becoming increasingly central in all the major social science disciplines. The growing adherence to the primacy of 'high-technology', however, is by no means confined to those who specialize in science and technology topics. Indeed, there has been a broad, and largely unrecognized, social scientific acceptance of relatively casual reasoning about the future import of some of the more visible emergent or embryonic technologies, including but not limited to biotechnology.

'High-technology' is a buzzword coined in the late 1970s by those seeking to emphasize – and, more often than not, to promote if not to profit from – a cluster of emerging technical forms such as microelectronics and biotechnology (see, for example, Dickson, 1984, for a useful discussion of the social forces that have led to high-technology 'hype' in corporate, governmental, scientific, media, and social science circles). Since the late 1970s the expression 'high-technology' has slipped into the social science lexicon and has implicitly been elevated to the level of a concept with little scrutiny. Most germane for present purposes, the notion has increasingly been incorporated into the major social science disciplines in a manner reminiscent of the imagery formulated by Schumpeter, especially in his classic work, *Business Cycles* (1939) (see also Schumpeter, 1912, 1942).

A wide range of implicit and explicit theorization of the dynamics and tendencies of 'high-technology' and social change can be said to be neo-Schumpeterian[1] in nature. My concern in this chapter will be primarily with analyses that lie toward the self-conscious pole of neo-Schumpeterianism. Nonetheless, I will suggest as well that the inferred neo-Schumpeterian logic of high-technology is a matter of considerable importance throughout the social sciences.

While contemporary observers of high-technologies more often than not omit any direct reference to the work of Schumpeter, there are nonetheless a large number of claims in the contemporary literature on science and technology that bear surprising similarity to his classic works on technology, business cycles, and socioeconomic transformation. Among the most important neo-Schumpeterian premises are the following:

(1)   The origins of technical change are thought to lie in the 'heroic entrepreneur' who takes the high risks of implementing a set of innovations in return for high profits; the late twentieth-century version of this notion is the much-glamorized, small, 'venture' or 'start-up' high-technology firm. Such 'locomotive technologies' (see, e.g., Sharp, 1986: 275) originate in innovative bursts, rather than in the prevailing routines of dominant institutions.

(2)   Technological change is regarded as discontinuous in nature – just as Schumpeter (1942: 84ff.) saw the origin of early twentieth-century innovation to be, for example, neither in price cutting nor incremental improvements in harness making, but rather in mass production of automobiles. Similarly, contemporary discussions see new high-technology forms as decisive breaks with the past.

(3)   Technological change occurs initially through the clustering of major innovations in one or a limited number of sectors; a 'locomotive technology' subsequently becomes generalized by a swarm of secondary innovators who are attracted by, and ultimately compete away, the monopoly profits of the early innovator-entrepreneurs.

(4)   Technological change is a process of 'perennial gales of creative destruction' (Schumpeter, 1942) in which innovations in an ascendant sector lead to the obsolescence and decline of other technologies and/or productive spheres.

(5)   These gales of creative destruction create giant discontinuities with or disruptions of past social arrangements.

(6)   Current high-technologies will be 'revolutionary' or epoch-making technologies. It is common, in fact, for the word 'revolution' to appear in the titles of books and articles on high-technologies (e.g., Molnar and Kinnucan, 1989; see Elster, 1983: ch. 5, and Rosenberg, 1982: ch. 1, for overviews of Schumpeter's theory).

Among the components of contemporary 'neo-Schumpeterian' views of high-technology is the premise that these new technical forms will, if effectively promoted and implemented, become the essential basis of a revitalized U.S. and world economy.[2] Indeed, while there recently has been sustained economic expansion in Japan, the newly-industrialized countries of Asia (Korea, Hong Kong, Taiwan, and Singapore), and Germany, and an apparent partial 'economic recovery' in the U.S., the U.K., and several other advanced industrial countries, the world economy as a whole has been in a state of chronic stagnation since the early 1970s. For example, while aggregate GNP in the Organization for Economic Cooperation and Development (OECD) countries increased by 5.1 per cent annually in 1960–1968 and 4.7 per cent annually in 1968–1973, the rate of GNP growth declined to 2.6 per cent in 1973–1979 and was about 2.8 per cent in 1986–1987 (Greenhouse, 1987). Current projections for the immediate future suggest that little improvement is on the horizon. U.S. trends have closely paralleled those for the OECD countries as a whole. Across a wide range, social scientists and policymakers have come to believe that innovation in high-technology– a new round of creative destruction – is a fundamental prerequisite for restoring the economic vitality of the OECD and other world nations.

Such neo-Schumpeterian premises, however, seldom appear without embellishment from quite different theoretical traditions – if not with some definitive criticism of Schumpeter (see, e.g., Roobeek, 1987a). Among the more prominent combinations of theoretical argumentation are analyses that in one way or another combine Schumpeterian logic about high-technology and social change with a more global or world-economy logic of 'long-wave' theory (see especially Sharp, 1986, and, I might add, Buttel et al., 1984; see also Mandel, 1978; Gordon, Edwards and Reich, 1982; Freeman, 1984; and Wallerstein, 1984, for more general discussions of the connections between Schumpeter's work and Kondratieff's, 1979, increasingly influential 'long-wave' theory). A second prominent variant is the blending of neo-Schumpeterianism with 'regulation' theory (e.g., Roobeek, 1987a, 1987b, 1990; Cowan and Buttel, 1988).

A further aspect of the ubiquity of neo-Schumpeterianism is the fact that this perspective on high-technology and social change has adherents from scholars who otherwise embrace a wide range of political–ideological commitments. Schumpeter's primary works involved an eclectic blend of positions that appeal to virtually the full range of the theoretical spectrum and which could be incorporated into theories as diverse as neoclassical economics and Marxism. For example, stress on the imperative of new technical forms undergirding a new cycle of economic expansion and on their revolutionary or epoch-making nature is understandably common among neoclassical economists and non-Marxist sociologists and political scientists (see, for example, OTA, 1984; Molnar and Kinnucan, 1989). But this position has strong allegiance among Marxists as well (see Markusen, Hill and Glasmeier, 1986; Castells, 1985; Castells and Henderson, 1987). Also note that both proponents (e.g., National Research Council, 1987) and opponents (Doyle, 1985; Yoxen, 1986) of these technologies[3] tend to portray them in quite similar – that is, revolutionary or epoch-making – terms, as have Western politicians from essentially the full range from left to right (Roobeek, 1987b: 140; 1990). Social scientists typically are more measured in their discussions of the revolutionary implications of high-technologies than are most corporate, government, and scientific promoters. Yet there is an emerging consensus among many social scientists that a set of new 'high-technologies' – biotechnology, microelectronics, micro- and super-computers, robotics, and other new information technologies, and new materials such as fibre optics and superconducting ceramics – will greatly change the social organization of all societies around the world over the next two or so decades (Roobeek, 1987b; and some of my own work, e.g., Buttel et al., 1984).

For example, Roobeek (1987a, 1987b) has argued that the long phase of economic decline in the OECD countries has been set in motion by a number of 'control problems' of the preceding socio-technical regime, which she calls 'Fordism' following Aglietta (1979) and Lipietz (1987) (see also Jessop, 1988). Among the control problems of Fordism identified by Roobeek are maturation of the product life cycles of previous locomotive technologies, overcapacity and market saturation, the increased importance of diversified consumer demand at the same time that production systems remain based on mass production, excessive dependence on nonrenewable raw materials, and the divergence between rising wages and declining productivity growth. She suggests that microelectronics, biotechnology, and new materials provide solutions to these control problems. While cautious about suggesting that new technologies will provide automatic solutions, she argues that a new 'social structure of accumulation' is being shaped by the nature of these new technologies, by corporate strategies to increase profitability, and by the role of the OECD states in promoting new high-technologies. For example, microelectronics and biotechnology are said to open up vast new possibilities for computer-integrated production of specialty products for highly-differentiated consumer lifestyles. Accordingly, state intervention in the promotion of high-technology, in her view, is laying the ground-work for the transformation of corporatist welfare states into new systems of more accumulation-centred corporatist arrangements among corporations, governments, and universities.

This chapter will assess critically the notion that one particular high-technology, biotechnology, will be as 'revolutionary' or epoch-making in its impacts as many observers have proclaimed. By biotechnology I mean the application of a set of relatively new processes – particularly recombinant DNA ('genetic engineering'), monoclonal antibodies, protoplast fusion, cell and tissue culture, and bioprocessing – to com-mercial use (see OTA, 1984). These new technologies received their principal scientific impetus in the early 1970s by the development of recombinant DNA, which has enabled researchers to insert genetic material from one organism into another of a different species. This is turn has permitted scientists to direct which genes are used by cells and to exercise far greater control over the production of biological molecules than was possible with traditional technologies.

The argument that follows – that biotechnology is not likely to be a revolutionary technology – will strike some as heretical. While there are not yet, to my knowledge, any published critiques of this thesis, this conclusion has generated considerable debate in private com-munication. In this extension of my original paper I will therefore address more directly some of the possible objections to the case I wish to make.

To anticipate one of the most important possible objections, I wish to make clear at the outset that a claim of the nonrevolutionary nature of biotechnology should not be seen as equivalent to arguing that biotechnology (or other contemporary high-technologies) will play an unimportant role in social change, or that biotechnology will fail to have major social effects. Indeed, biotechnology will almost certainly have some major distributional impacts on consumers, far-mers, Third World peoples, and other social groups; will substantially affect the social relations of scientific research; and will alter indus-trial structures and world markets. Rather, I wish to suggest why the neo-Schumpeterian imagery of biotechnology as a first-order, epoch-making causal force in social change may be misleading. I will thus suggest that while social scientists who study science and technology have yet to develop frameworks for distinguishing between those that induce and are induced by broader social relations, this is an urgent task.

Nonetheless, while I will essentially come down on the side of claiming that the revolutionary character of biotechnology has been exaggerated, my analysis has several purposes that go beyond establishing the thesis of the nonrevolutionary nature of biotechnology. These corollary purposes are as follows. First, the very exercise of attempting to assess the degree to which biotechnology will be a first-order, epoch-making causal force in social change over the next two or three decades should prompt a recognition that there is currently lacking a systematic body of empirical evidence that can be brought to bear on the rival hypotheses of the revolutionary versus nonrevolutionary character of biotechnology, or any other technology. Biotechnology, in particular, is an embryonic technology; relatively few biotechnology products have reached the market, and it is inherently problematic to project patterns of research and development in such an incipient technological area many years into the future. But even in more mature areas of high-technology, such as the 'information technologies', there are neither unambiguous data nor available frameworks for assessing the degree to which a technology is revolutionary or epoch-making with any acceptable degree of precision. In part, this difficulty reflects as much on the lack of systematic, theoretically-guided historical research on the relationships between technological and macrosocial change as it does on the lack of contemporary data. Thus, second, this chapter is intended to advance a provisional framework, or set of criteria, for assessing the degree to which a technology is revolutionary or epoch-making. As this framework is admittedly very preliminary, the conclusions that are drawn should be regarded as tentative and as constituting a research agenda for both sociologists of science and technology and macrosociologists more generally. Third, as noted earlier, my reasons for exploring the logic of claims for the revolutionary nature of biotechnology go beyond this particular issue, however important it might be regarded in its own right. Instead, this chapter is intended to provoke a recognition among sociologists and other social scientists of the pitfalls of what is now a widespread pattern of casual social scientific thinking about the new technologies that are currently under development in the advanced societies. As we enter an era in which scientific and technological forces become increasingly important in shaping modern social structures, social scientists will need to be more precise in their treatment of categories and concepts relating to science and technology.

Putting Biotechnology into Perspective

For several centuries social scientists have been well aware that technological changes have been closely implicated in broader processes of social change. Some of the classical theorists – most notably Marx – saw technology playing a relatively autonomous role in shaping new social and institutional arrangements. Yet there has been relatively little systematic attention given to *ex post*, let alone *ex ante*, assessment of how fundamental or 'revolutionary' a particular technology has been or will be. This difficult conceptual and methodological issue unfortunately cannot be dealt with to complete satisfaction in a chapter of the scope set forth here. Instead, I will elaborate a preliminary framework for assessing the degree to which a technical form has been or will be central in large-scale socioeconomic change while recognizing that there is ample room for disagreement with its premises.

The framework adopted here is essentially Schumpeterian. This framework has been selected for two major reasons. First, neo-Schumpeterian theory is arguably the most advanced in dealing with issues of the type of concern in this chapter (see, for example, Sharp, 1986). Second, since much social scientific reasoning about high-technologies takes on an implicitly neo-Schumpeterian cast, the criteria I advance are appropriate for examining received arguments on their own terms.

A new technical form can be said to be revolutionary or epoch-making in the neo-Schumpeterian sense if it meets several criteria. First, of course, the technology must have relatively wide applicability, which biotechnology surely does. Second, a technology will be a particularly likely candidate for revolutionary or epoch-making status if it is applicable in both the sphere of production (to reduce production costs) and of consumption or circulation (to create large new categories of consumer and producer goods, thereby deepening or reinforcing the dynamic of capital accumulation). Third, a revolutionary technology should be applicable to the leading or ascendant economic sector(s), or to sectors likely to be leading ones in the future. While the new information technologies may well meet all of these criteria (see Roobeek, 1987b, 1990; Sharp, 1986, for similar assessments), the analysis that follows suggests that biotechnology will not. Hence, we predict that while biotechnology will be of substantial significance, it will be a subordinate technology, whose pace and dynamics are determined largely by the other, more fundamental social factors and technical forms.

Unfortunately, as noted earlier, the current state of data on biotechnological innovation is far too thin to permit a confident analysis of the revolutionary vs nonrevolutionary character of biotechnology, even if the criteria for such an assessment were unambiguous. Thus, the bulk of the evidence to be drawn upon in the analysis that follows consists of a series of illustrative examples. These examples, however, have been selected from a wide range of applications of this new technology.

Biotechnology as substitutionism

In one of the three or four most insightful book-length treatments of biotechnology,[4] Goodman et al. (1987) call two essential points to our attention. First, they demonstrate the long historical continuity of biotechnology with previous instances of 'substitutionism' in agriculture and food (e.g., the substitution of synthetic fibres for cotton, synthetic dyes for indigo, reconstituted foods for 'whole, natural' foods). Second, Goodman and colleagues stress the role that biotechnology will play in creating substitutes for *existing* products and discuss how substitutionism may well become the dominant aspect of agricultural biotechnology. While Goodman and colleagues seemingly do embrace (implicitly) the notion of biotechnology being a revolutionary technical form, their essential arguments – the predominance of substitutionism and of biotechnology being a new tool of 'appropriationism'[5] – suggest a far different image.

Much like Goodman and colleagues, I hypothesize that the development and application of biotechnology will serve primarily to cheapen or otherwise improve the production of *existing* products or to provide substitutes for existing products or services. This is the key component of the overall hypothesis that biotechnology will be evolutionary in its applications and impacts. A quick glance at the major contemporary applications of biotechnology (either planned or under development) provides considerable support for this contention.

For example, *Biotechnology Newswatch* (1988: 6–7), a trade periodical, published a list of what are considered to be the eighty-one most important pharmaceuticals currently under research and development. Virtually all of these new drugs, only nine of which had at the time of publication been approved by the (U.S.A.) Food and Drug Administration, are either direct substitutes for old products (e.g., Eli Lilly's *Humalin* insulin, which is being employed to substitute for porcine-derived insulin) or new products developed to deal with diseases for which existing pharmaceuticals are expensive and/or ineffective (e.g., Cetus's interleukin-2 [*Proleukin*], which is being promoted as an improvement on existing chemotherapy drugs). In industrial microbiology in the pharmaceutical and chemical industries, biotechnology makes possible major improvements in longstanding methods of fermentation, cell culture, continuous-process biochemical engineering, and so on. Biotechnology, to be sure, is leading to some 'new' drugs such as human growth hormone (Genentech's *Protropin*) and to some new products and services such as human gene therapy and the numerous diagnostic kits that have been developed through monoclonal antibody technology. But in the main, it appears that biotechnology will be deployed in nonfarm industry mainly in the 'substitutionism' mode as discussed by Goodman, Sorj and Wilkinson (1987) for agriculture and food.

While many observers, ranging from Goodman and his colleagues (1987) to Kenney (1986) and Kloppenburg (1988), correctly point to the myriad of ways that biotechnology may affect food and agriculture, a closer examination suggests more continuity than revolutionary change. Biotechnology in its 'appropriationist' applications reflects, virtually by definition, continuity with the past. That is, insofar as biotechnology will be applied to improve crops and livestock species, the new processes or materials are basically improvements on, and thus are being substituted for, conventional or traditional methods. Biotechnology thus represents a potentially powerful new tool – undoubtedly a powerful one – for increasing the efficiency of agricultural research. In addition, biotechnology opens up opportunities for product substitutions. As Vergopoulos (1985: 297) has put the matter, 'agriculture will not be industrialized, as some had long been anticipating, but will be replaced by industry'. But it should be kept in mind that substitutionism is just that – in other words, primarily cost-reducing alternatives to products and services that are already widely generalized in the advanced countries.

Further, the principal applications to agriculture are not, in the main, revolutionary in the sense that the petrochemical/'green' revolution was a major break – either with the agricultural techniques pioneered in Northwest Europe in the eighteenth century or with the 'traditional' food production practices of Third World peasants. A cursory glance at the major 'appropriationist' applications of biotechnology to agriculture show that, in the main, they serve to 'patch up' the problems of Western agriculture or otherwise obviate the productivity plateaux of conventional, petrochemical-based systems. Good examples are herbicide-tolerant crop varieties (which are being developed to rationalize pesticide usage and to stimulate sales of pre-biotechnology agrochemicals), salt-tolerant crop varieties (to mitigate salinization problems caused by irrigation), and 'ice-minus' bacteria (that enable frost tolerance in high-value horticultural crops – an increasingly attractive possibility as this production sector has become more capital-intensive and globally integrated). Moreover, the research goals that might enable biotechnology to become a bona fide fifth world agricultural revolution (see Lipton with Longhurst, 1985, for a brief overview of the four previous agricultural revolutions) are not yet (or are no longer) being actively researched in either public or private-sector laboratories. Autosufficient biological nitrogen fixation in nonleguminous crops and increased photosynthetic efficiency, while potentially capable of major impacts on agricultural production systems, have been all but abandoned as commercial biotechnology research goals, because of the perception that their achievement lies several decades into the future.

Thus, there seems to be relatively little revolutionary significance in the applications of biotechnology to agriculture, food, chemicals, or pharmaceuticals. Again, as stressed earlier, this conclusion is not incompatible with the notion that biotechnology will have major socioeconomic impacts. Biotechnology has already contributed to the transformation of public research institutions – chiefly in the direction of privatization of biological information (Kenney, 1986; Kloppenburg, 1988) – and surely will result in significant structural changes such as the decline of Third World agricultural export sectors in crops such as sugar, vanilla, palm oils, cocoa, and so on (Junne, 1987; Busch and Lacy, 1986). But these transformations, no matter how central they might be to those who experience them, are not the first-order causal forces that the electric motor was earlier in this century or microelectronics and the new information technologies may well be in the future. The electric motor, for example, permitted mass, assembly-line manufacturing, which contributed to the formation of the industrial working class, and led to whole categories of new products (especially consumer durables such as refrigerators, vacuum cleaners, and washing machines) and, most importantly, to an enormous vector of profitable capital accumulation. Microelectronics and the new information technologies may portend comparable changes, but biotechnology, which is principally a substitution technology, probably will not.

The preceding argument, of course, stands or falls on whether we can make a meaningful distinction between 'new' and 'substitute' products, particularly when virtually all goods and services have long since undergone commodification. There is admittedly a certain arbitrariness to declaring, for example, that recombinant insulin is a substitute for porcine-derived insulin, while a microcomputer and spreadsheet software are not mere substitutes for adding machines and ledger books. As implied in the preceding, the most important criterion for assessing the novelty vs substitutionality of products is that of whether a major new vector of capital accumulation is made possible. Indeed, the 1980s literature on biotechnology was filled with startling estimates of the projected sales volume of biotechnology products within one to two decades. It was virtually never estimated, however, whether these sales volumes would be several orders of magnitude greater than current ones (e.g., whether biotechnology-related agrochemical sales would be significantly greater in real terms than agrochemical sales in the 1970s or 1980s). One suspects, in fact, that sales volume estimates for biotechnology products were little more than computations of the sales of products that biotechnologies were estimated to replace! Thus, if we accept the criterion of whether a new technical form makes possible a vibrant new vector of capital accumulation, a provisional index of which would be several-fold increases in sales volumes, there is arguably no evidence for the epoch-making stature of biotechnology.

Sectoral patterns of biotechnology applications

The electric motor and synthetic organic chemical technology placed such a distinctive stamp on global social organization earlier in the twentieth century because they were applied in the 'leading' or 'dynamic' sectors of the economy. Both were industrial-manufacturing technologies in eras when the 'secondary sector' and its social relations were predominant in shaping social organization and social change (see, for example, Roobeek, 1990). But our hypothesis here is that because biotechnology is mainly being applied to *declining* sectors of the national and global economy it will tend to be a subsidiary technical form.

This component of the overall argument is closely related to the first, i.e., it would logically be the case that novel, epoch-making, accumulation-facilitating production processes and products would be implicated in an ascendant sector of the economy. In addition, I will stress that the role of biotechnology in financial and producer services is not likely to be significant.

This argument is not meant to deny that 'manufacturing matters' (to borrow the title of Cohen and Zysman's [1987] book), and I recognize the dangers of reifying the notion of economic sectors. As Cohen and Zysman stress, manufacturing is crucial to the economic viability of 'post-industrial' nations, even in the highly-routinized sectors such as textiles or bulk chemicals that face declining shares of the national product. Biotechnology will undoubtedly be significant in rationalizing several declining sectors of the economy, and will in all probability do so with a fair amount of Schumpeterian-like 'creative destruction'.

Likewise, the major ongoing transformations of the national and global economies render the concept of 'sectors' increasingly problematic. There is no longer any rigid isolation between clearly-defined macro-sectors such as agriculture, forestry, 'heavy' and 'light' manufacturing, services, and so on or specific industrial sectors such as electronics, chemicals, and autos that exist relatively independently from one another. The traditional conception of sectors is essentially a category of *national economy*, which is of decreasing relevance as the internationalization of production and circulation proceeds apace (see Friedmann and McMichael, 1988). The fast-food industry is a particularly dramatic example of this blurring of boundaries. That industry is nominally considered to be part of the service sector. Yet the fast-food industry has become increasingly vertically integrated with agriculture on an international scale (e.g., in potato and especially beef production) and with manufacturing (as in the tendency for some fast-food franchising firms to manufacture their own plates, cups, and so on).

This blurring and restructuring of sectors will no doubt continue apace over the next several decades. Thus, there is a certain arbitrariness in the delineation of 'rising' and 'declining' sectors. Nonetheless, this usage will still serve to illustrate the basic argument about the evolutionary character of biotechnology.

The industrial sectors to which biotechnology is being applied include many of the longstanding or newly declining industries in the national and world economies: agriculture, food, forestry, fisheries, bulk chemicals, energy, and mining. Each of these sectors has for one decade, if not several, exhibited declining shares of national product in the world's advanced industrial countries. For a new technical form to be revolutionary in the neo-Schumpeterian sense, it cannot be rooted primarily in declining economic sectors.

There is, to be sure, one dynamic sector to which biotechnology is currently being applied – that of pharmaceuticals. Biotechnology will in all probability assist the pharmaceutical sector in remaining relatively profitable and dynamic in the national and world economies (though it is unlikely to account for a growing share of the national products in the advanced countries). The vast majority of new biotechnology products already commercialized or nearing commercialization are proprietary drugs, which testifies to the importance of this new technology to the industry. There are several reasons for this: first, new biotechnology-derived drugs are based on a relatively simple technology – genetically-modified bacteria, which are 'engineered' to produce the desired proteins. The higher plants and animals are much more difficult to modify genetically in commercially-relevant ways. Second, pharmaceuticals have relatively large markets. Third, some small biotechnology firms will be able to market lowvolume, high-value drugs directly to physicians, which enables them to circumvent one of the major problems faced by small start-up companies: their lack of a marketing infrastructure.

While the importance of biotechnology in pharmaceuticals, and the role of pharmaceuticals and health care in services, is clear, it should be stressed as well that the long-term future of the pharmaceutical industry will by no means be a linear extrapolation of previous trends (e.g., of pharmaceutical sales or of the value of health-care services in real terms). A significant share of pharmaceutical sales currently is based on the legacy of the post-war welfare state – i.e., national health-care plans in most of the advanced countries and more limited forms of health insurance, such as in the U.S. – and of the state-underwritten 'social wage' in general. Thus, the degree to which pharmaceuticals will remain a vibrant sector depends crucially on the persistence of health-insurance schemes in the advanced countries, many of which have been under assault since the 1980s. It is typically held, for example, that the erosion of the welfare state is likely to be a cornerstone of the emergent regime(s) of accumulation (e.g., Gordon et al., 1982). A significant decline in state health-insurance funding in the advanced countries, resulting in declining health-service availability for the poor, and thus to declining proprietary drug sales, could relegate the pharmaceutical sector to a declining role.

Put somewhat differently, among the reasons that microelectronics technologies may have more fundamental socioeconomic effects than biotechnology is that the former are applicable to certain parts of the 'service sector' while the latter is not. Of course, the conventional definition of 'the' service sector embraces such a congeries of activities – from fast food to finance, from waiting at tables to engaging in leveraged buy-outs – that it has little internal coherence. For present purposes, the conception of the service sector is a more limited one, mainly confined to financial services, insurance, and related activities integrally linked to major industrial production sectors (as in Cohen and Zysman, 1987) and to the state sector (as depicted by O'Connor, 1973). Nonetheless, while microelectronics and associated information technologies will be important in increasing productivity and extending the scope of the dynamic components of the service sector, biotechnology will have few applications in this sphere. Biotechnology's applications to services will probably be primarily confined to health services.

Premature, speculative investment in biotechnology

Finally, two additional factors favouring the hypothesis of the evolutionary nature of biotechnology are found in its financial–industrial structure. First, biotechnology was prematurely commercialized, and its major applications lie decades in the future. Second, the historical – and, to a considerable extent, contemporary – patterns of investment in biotechnology have been highly speculative and often 'defensive' in nature.

We can now see in retrospect that biotechnology was prematurely commercialized.[6] It became an object of substantial private investments at a point at which major product revenues were several years into the future. The initial private investments in biotechnology were made by venture-capital firms in the United States in the late 1970s and early 1980s, and by 1984 the basic industrial structure of U.S. biotechnology research and development – the coexistence of about three hundred small start-up companies with approximately fifty large multinational chemical, pharmaceutical, energy, and food firms with in-house biotechnology research programmes (OTA, 1984; Kenney, 1986) – was in place.

It is useful to note that these 'venture-capital' investments began even before the basic intellectual property-rights structure had been defined (the Diamond v. Chakrabarty U.S. Supreme Court decision in 1980, which permitted the patenting of genetically-modified life forms). This rush of investment at such an early point seems to have been due more to exaggerated promotion and generous federal research-and-development tax subsidies (see Krimsky, 1982, and Wright, 1986, for useful historical analyses) than to sound financial-productive reasoning or even to Schumpeterian-style heroic entrepreneurialism. Also, while it was not entirely clear in the late 1970s and early 1980s that the technical and regulatory problems of translating biotechnology research into sales revenues would be so formidable, in a sense this was not a particularly important investment consideration at the time. Early venture-capital investors had no intention to own, much less operate, companies; they were merely trying to carry some small firms through to a stage at which they would have a few patents, the ability to raise funds through public stock offerings, and the assets to buy out the venture capitalist's share of the business at a high profit. Venture capitalists, in general, maintained their start-up company investments only long enough to get their money out at a handsome rate of return. Thus investments in biotechnology start-up companies have tended to be based on speculation about small firms' abilities to obtain patents and to appeal to institutional and corporate investors on the stock market. Somewhat later, a number of large multinational firms invested in biotechnology, often for defensive reasons – worrying that *if* the small start-up firms came up with important innovations, the large firms lacking a 'presence' in biotechnology might be left with obsolete product lines (Kenney, 1986; Wright, 1986).

Three recent developments provide evidence of the speculative or defensive nature of biotechnology investments. First, with the ending of the generous research-and-development tax subsidies of the Economic Recovery Tax Act of 1981 in the 1986 federal tax reform, new investment in additional start-up companies and expansion of existing companies has slowed to a trickle. Second, during the 1987 stock market crash, biotechnology start-up company stocks were harder hit than virtually any other category of high-technology or non-high-technology industrial stocks. Many biotechnology stocks declined by more than 50 per cent the first day of the crash alone, suggesting the extent to which the financial basis of the industry had become premised on the general climate of speculative stock appreciation of 1986 and 1987. Agricultural biotechnology stocks, in particular, have plummeted in value now that more realistic information about the timing and size of new product lines has become available (Pollack, 1989). Third, many of the major multinational biotechnology firms have reduced their research and development programmes and dramatically adjusted their planning horizons. For example, the Monsanto Corporation, one of the major agricultural input and specialty chemical firms in the U.S., has laid off biotechnology research scientists and phased out whole areas of research and development (such as plant growth regulators, which only a few years earlier were touted as portending the obsolescence of synthetic–organic–chemical pesticides). Monsanto officials now acknowledge that they underestimated the time-to-market for biotechnology products and are bracing themselves for continued losses in their biotechnology product lines for up to a decade (Andre d'Olme, 1988).

In sum, patterns of investment in biotechnology have been substantially speculative (in the case of start-up firms) or defensive (on the part of large multinational firms). Total private investment in biotechnology in the United States, which stood at about $6 billion in 1988, stagnated for more than two years. This modest level of investment has been attracted more out of hope and intuition than by the more short-term financial planning that tends to predominate in most other firms. This type of financial structure seems ill-suited to be a bearer of revolutionary social and technical forces. These investment patterns, to be sure, have a certain similarity with Schumpeter's early emphasis on the heroic entrepreneur (e.g., Schumpeter, 1912). But this entre-preneurship has been more financial than productive, geared as it is to producing profits from financing arrangements and stock market transactions.

Conclusion

My thesis that biotechnology will not be a revolutionary sociotechnical form, at least in the foreseeable future, can now be complemented with two further hypotheses. First, the future importance of biotech-nology will lie in *rationalization* of a number of primary and traditional manufacturing sectors in tandem with the transition to a new world economy based in considerable part on new microelectronics and information technologies. Second, biotechnology will be a subordi-nate technology, subsidiary to or derivative from the social relations of the predominant information technologies (see Sharp, 1986, for a parallel argument).

Several caveats are in order with regard to the foregoing assessment of biotechnology. First, since sociology and its kindred social sciences have only begun to address questions such as those of concern in this chapter, our concepts and methods for doing so are severely limited. Thus, the conclusions I have drawn must be regarded as tentative. Second, the criterion used here for assessing the future of biotechnol-ogy – its potential as an epoch-making technology – does not imply a technological determinist or 'autonomous technology' (see Winner, 1977) perspective. An epoch-making technology is not one in which technology is *itself* the major force that determines social organiza-tion. Rather, it is one that is a part of the predominant forces of social change, which are necessarily social in nature. Epoch-making technologies do not literally make epochs. Rather, such locomotive technologies provide profitable investment outlets when and if extant social conditions – a practicable set of social relations, or regime of accumulation – are at hand. Interestingly, this position is close to that of Schumpeter himself; as Walker (1985: 244) has sug-gested, Schumpeter's analysis was far more subtle and variegated than are many contemporary varieties of neo-Schumpeterian reason-ing which tend toward technological determinism.

Third, it is quite possible that we may just be viewing the 'prehistory' of biotechnology at this point. That is, we may be so many decades away from the major applications of biotechnology (e.g., 'bioinformatics', as discussed by Roobeek, 1987a, 1987b) that we cannot readily foresee how this technology may take its place along with the more tangibly significant information technologies. Nonetheless, even if these major applications do come about, they will occur at a point sufficiently far in the future that they are unlikely to affect fundamentally the social and technical relations of the emerging world economy of the next two or three decades.

Uncritical acceptance of the notion that the high-technologies will be revolutionary, epoch-making technical forms is by no means the only high-technology myth to have been appropriated by the social sciences in an uncritical way. Others, for example, are that Silicon Valleys and Route 128s can be created in virtually any location proximate to a major research university (Cowan and Buttel, 1988) or that universities can readily generate vast additional sums of private research funding and royalty income by engaging in joint research and development with industry (Dickson, 1984). It is also useful, in this era of neo-Schumpeterian celebration of the start-up company entrepreneur, to recall that later in his career (in his classic *Business Cycles* [1939]) Schumpeter began to revise his conception of the historical bearers of new innovations to incorporate an endogenous element – the routinization of research and development activities within large firms as a self-reinforcing dynamic of technical innovation – into his theory. It is also germane that small high-technology start-up companies, once almost universally revered for their technological and marketing innovativeness, are now beginning to fall out of favour in some mainstream business school circles. Critics of 'chronic entrepreneurialism' (the proliferation of small start-up companies) point to the tendency to duplicative, fragmented investment, to selling technology to foreign firms, and to being unable to effectively market new innovations (Pollack, 1988).

Social scientists must be sceptical about the many unsubstantiated generalizations about the high-technologies that have become increasingly prevalent in recent years. Indeed, one of the purposes of this chapter has been to demonstrate that as we enter the so-called high-technology world economy of the future, casual reasoning about technology and social change will become increasingly unacceptable. It is hoped that this chapter is a step in the direction of critically assessing conventional wisdom about the new high-technologies and constructing an empirically grounded sociology of contemporary technological and social change.

NOTES

[1] I refer to many contemporary works on high-technology and social change as being neo-Schumpeterian because 'Schumpeterian-like' premises are typically invoked alongside ones that are quite unlike those of Schumpeter. It is especially interesting to contemplate the fact that most contemporary radical scholarship that draws explicitly or implicitly on Schumpeter omits mention of his expectation that capitalism would inevitably yield to some type of socialist order.

[2] Note, however, that this premise can be set forth with both pejorative and nonpejorative overtones. The former posture is typically expressed by those who see high-technology innovation as a relatively straightforward response to economic stagnation, while the latter involves reservations that the problems of modern capitalist economies will be addressed at the expense of the disadvantaged.

[3] It goes without saying, of course, that it is sometimes difficult to disaggregate opposition to a particular technology from opposition to the institutional structures or arrangements within which technology is being deployed, which in turn would give rise to technological outcomes that are perceived as being adverse or undesirable. For example, it is unclear whether Doyle's (1985) position on biotechnology falls into the former or latter category.

[4] It should be stressed that their *From Farming to Biotechnology* is almost exclusively focused on agriculture, food, and fibre, and their arguments cannot therefore be extended to other realms of production in a straightforward way. Nonetheless, integral to their analysis is the notion that non-farm substitutionist biotechnologies (i.e., those deployed in the 'downstream' part of the food system) involve an inevitable blurring of the boundaries between food and fibre production on one hand, and the chemical industry on the other. Thus, the applicability of their argument to nonfood sectors is greater than might be apparent.

[5] By appropriationism, Goodman and colleagues mean the response of industrial capital to the constraints of nature as a natural production process by taking over discrete elements of this process, transforming them into industrial activities, and re-incorporating them into agriculture as purchased inputs. A useful example would be the manufacture and use of synthetic fertilizers. Fertilizer manufacturers thereby reduce the importance of 'nature' in agricultural production.

[6] I have argued elsewhere, for example, that the prematurity of investment in biotechnology had much to do with 1970s regulatory politics in the U.S. Many prominent molecular biologists, faced with the possibility of strict federal regulation of recombinant DNA research, began to argue against 'restrictive' regulation on the grounds that such strictures would slow the development and commercialization of wonder drugs, miracle crop varieties, and so on. Shortly thereafter there emerged a flood of new start-up

firms which typically combined the skills of a senior, university-based molecular biologist with the financial resources of a venture capital house. It was only after there were several dozen start-up firms in existence that multinationals began to invest (Buttel, 1989).

REFERENCES

Aglietta, M. (1979). *A Theory of Capitalist Regulation*. London: New Left Books.
Biotechnology Newswatch. (1988). PMA releases 'most comprehensive' list of biotechnology products. *Biotechnology Newswatch*, **8**, 6–7.
Busch, L. and Lacy, W. B. (1986). Biotechnology and the restructuring of the world food order. Paper presented at the XI World Congress of Sociology, New Delhi, India, August.
Buttel, F. H., Cowan, J. T., Kenney, M. and Kloppenburg, J., Jr. (1984). Biotechnology in agriculture: The political economy of agribusiness reorganization and industry–university relationships. *Research in Rural Sociology and Development*, **1**, 315–343.
Buttel, F. H. (1989). Theoretical issues in the regulation of genetically engineered organisms: a commentary. *Politics and the Life Sciences*, **7** (February), 135–139.
Castells, M. (1985). High technology, economic restructuring, and the urban-regional process in the United States. In M. Castells (ed.), *High Technology, Space, and Society*, 11–40. Beverly Hills, CA: Sage Publications.
Castells, M. and Henderson, J. (1987). Techno-economic restructuring socio-political processes and spatial transformation: a global perspective. In J. Henderson and M. Castells (eds), *Global Restructuring and Territorial Development*, 1–17. Beverly Hills, CA: Sage Publications.
Cohen, S. S. and Zysman, J. (1987). *Manufacturing Matters*. New York: Basic Books.
Cowan, J. T. and Buttel, F. H. (1988). Subnational corporatist policymaking: the organization of state and regional high-technology development. *Research in Politics and Society*, **3**: 241–268.
Dickson, D. (1984). *The New Politics of Science*. New York: Pantheon.
d'Olme, Andre. (1988). Personal Communication, Manager, Plant Science, Europe and Africa, Monsanto Corporation, 17 July.
Doyle, J. (1985). *Altered Harvest*. New York: Viking.
Elster, J. (1983). *Explaining Technical Change*. New York: Cambridge University Press.
Freeman, C. (1984). Long Waves in the World Economy. London: Frances Pinter.
Friedmann, H. and McMichael, P. (1988). The world-historical development of agriculture: Western agriculture in comparative perspective. Paper presented at the World Congress for Rural Sociology, Bologna, June.
Goodman, D., Sorj, B. and Wilkinson, J. (1987). *From Farming To Biotechnology*. Oxford: Basil Blackwell.

Gordon, D. M., Edwards, R. and Reich, M. (1982). *Segmented Work, Divided Workers*. New York: Cambridge University Press.

Greenhouse, S. (1987). When the world's growth slows. *New York Times*, 27 December (Section 3), 1, 24.

Jessop, B. (1988). Regulation theory, post-Fordism and the state: more than a reply to Werner Bonefield. *Capital and Class*, **34**, 147–163.

Junne, G. (1987). Avenues of future social science research on implications of biotechnology. *Development*, **4**, 86–90.

Kenney, M. (1986). *Biotechnology: The University–Industrial Complex*. New Haven: Yale University Press.

Kloppenburg, J., Jr. (1988). *First the Seed*. New York: Cambridge University Press.

Kondratieff, N. D. (1979). The long waves in economic life. *Review*, **2**, 517–562.

Krimsky, S. (1982). *Genetic Alchemy*. Cambridge, MA: MIT Press.

Lipietz, A. (1987). *Mirages and Miracles*. London: New Left Books.

Lipton, M. with Longhurst, R. (1985). *Modern Varieties, International Agricultural Research and the Poor*. Washington, DC: World Bank.

Mandel, E. (1978). *Late Capitalism*. London: New Left Books.

Markusen, A., Hall, P. and Glasmeier, A. (1986). *High Tech America*. Boston: Allen & Unwin.

Molnar, J. J. and Kinnucan, H. (1989). Introduction: the biotechnology revolution. In J. J. Molnar and H. Kinnucan (eds), *Biotechnology and the New Agricultural Revolution*, 1–18. Boulder, CO: Westview Press.

National Research Council. (1987). *Agricultural Biotechnology*. Washington, DC: National Academy Press.

O'Connor, J. (1973). *The Fiscal Crisis of the State*. New York: St Martin's Press.

Office of Technology Assessment (OTA). (1984). *Commercial Biotechnology*. Washington, DC: OTA.

Pollack, A. (1988). A look at entrepreneurs: doubt on American ideal. *New York Times*, 14 June, A1, D6.

Pollack, A. (1989). Farm gene makers' money woes. *New York Times*, 24 April, D1, D6.

Roobeek, A. J. M. (1987a). The crisis in fordism and the rise of a new technological paradigm. *Futures*, **19** (April), 129–154.

Roobeek, A. J. M. (1987b). *Governments Locked in the Technology Race: The Political and Economical Background to the Technology Race Between Industrialised Countries*. Amsterdam: Faculty of Economics, Department of Business Economics, University of Amsterdam.

Roobeek, A. J. M. (1990). *Beyond the Technology Race. An Analysis of Technology Policy in Seven Industrial Countries*. Amsterdam: Elsevier.

Rosenberg, N. (1982). *Inside the Black Box*. New York: Cambridge University Press.

Schumpeter, J. (1912). *Theorie der Wirtschaftlichen Entwicklung*. Leipzig: Duncker and Humbolt.

——.1939. *Business Cycles*. 2 vols. New York: McGraw-Hill.

——.1942. *Capitalism, Socialism, and Democracy*. New York: Harper & Row.

Sharp, M. (1986). Conclusions: technology gap or management gap? In

M. Sharp (ed.), *Europe and the New Technologies,* 263–297. Ithaca: Cornell University Press.

Vergopoulos, K. (1985). The end of agribusiness or the emergence of biotechnology. *International Social Science Journal, 37,* 285–299.

Walker, R. A. (1985). Technological determination and determinism: industrial growth and location. In M. Castells (ed.), *High Technology, Space, and Society,* 226–264. Beverly Hills, CA: Sage.

Wallerstein, I. (1984). *The Politics of the World Economy.* New York: Cambridge University Press.

Winner, L. (1977). *Autonomous Technology.* Cambridge, MA: MIT Press.

Wright, S. (1986). Recombinant DNA technology and its social transformation, 1972–1982. *Osiris* (2nd series), *2,* 303–360.

Yoxen, E. (1986). *The Gene Business.* New York: Oxford University Press.

4

The Coming Revolution of Biotechnology: A Critique of Buttel

Gerardo Otero

For several years, social scientists, policymakers, and biological scientists have regarded biotechnology as one of the new technologies that would revolutionize production in medicine and agriculture. Now however, one of the very pioneers in the study of socioeconomic impacts of biotechnology has turned the argument around (Buttel, 1989a). Rather than revolutionary, Frederick Buttel has proposed that biotechnology is more a 'substitutionist' technology to be applied in declining sectors of the economy. For these reasons, he has argued that biotechnology should no longer be considered an 'epoch-making' technology as are electronics and informatics.

The first section of this chapter is dedicated to a presentation of Buttel's arguments. In the second section I provide both external and internal critiques of Buttel's position based on a different perspective, which uses the concept of the 'third technological revolution'. One of my central contentions in this regard is that the appropriate way of dealing with new technologies is by looking at their impact as a global and interrelated phenomenon, and not on an individual, case by case, basis. Finally, the concluding section suggests the necessity to bring into the analysis the majority of the world's people: those living in the Third World. I present some research questions and hypotheses on the potential regional implications of biotechnology in these societies. Whether directly or indirectly, Third World populations are bound to be the most affected by the deployment of agricultural biotechnologies.

Is Biotechnology Revolutionary? Buttel's Argument

Although the very definition of 'biotechnology' has been the subject of controversy, I base my discussion on one provided by the General

Accounting Office (GAO) of the U.S. Congress: 'Today, biotechnology is generally considered to be a component of high technology, and the "new biotechnologies" are those resulting from recently developed, sophisticated research techniques, including plant cell and protoplast culture, plant regeneration, somatic hybridization, embryo transfer, and recombinant DNA methods' (GAO, 1986: 10). In my view, biotechnology is possibly the most important technical force that will shape world agriculture over the coming decades. But biotechnology is a two-edged sword. Which way it cuts will depend largely upon who wields it and how (Kloppenburg, 1988; Kloppenburg et al., 1988).

Buttel, however, has become impatient with biotechnology. In the few years since he and his colleagues began to examine the potential impacts of new technologies (Buttel et al., 1985), there have not been any major changes in the agrarian social structure or productivity due to biotechnology. Therefore, Buttel (1989a, 1989b) has declared that biotechnology is not a revolutionary or epoch-making technology.

It was interesting to read what I thought was largely a 'devil's advocate' paper, arguing against positions that Buttel himself had advanced in the past. The initial argument for toning down the extent of biotech's impact was Kenney and Hibino's paper presented at the 1987 meetings of the Rural Sociological Society. Both papers can be seen as self-criticism of previous work, an activity that is usually healthy. And yet Buttel's paper overcompensates. Because my own work on biotech has been influenced by the critical scholarship represented by Buttel's work, I cannot agree with his new position.

Buttel's strategy to declare that biotechnology is not a revolutionary or 'epoch-making' technological form is *ad hoc*. First he sets three criteria, all of which have to be fulfilled by any given technology if it is to be labelled 'revolutionary'. He admits that it is a fairly arbitrary set of criteria, but posits it as a framework to provoke discussion. The first criterion is that the technology must have a wide applicability. The second is that it should be applicable to production, as a cost reducing technology, as well as to consumption, in creating large new categories of consumer and producer goods. Third, applicability should be directed to leading or ascendant sectors of the economy.

Once these criteria have been set, Buttel goes on to establish why, in his view, biotechnology does not fulfil them adequately. First, Buttel argues that while biotechnology will have wide applicability, it will only 'patch up' problems of Western agriculture. An example of this would be the creation of herbicide-resistant plant varieties. But this general assessment of biotechnology is done on the basis of regarding it as a 'substitutionist' technology, following Goodman and colleagues (1987). Buttel provides no reasons why substitutionism will not create major changes in productivity and social structure.

A second disqualifying factor against biotechnology is that, in Buttel's view, it will be applied to declining sectors (agriculture and manufacturing), rather than to the leading or ascendant one (services). Services is regarded as the leading sector in 'postindustrial' societies, although this characterization of present capitalism is not discussed. Because of this, biotechnology is regarded as a 'subsidiary technical form'. Unlike microelectronics and associated information technologies, biotechnology's 'application to services will probably be confined to health services', says Buttel (1989a: 254). Buttel is right when he says that medicine or health services will be affected by biotechnology. And if we take into account health care's current proportion in the gross domestic product (GDP), and the fact that it is increasing faster than any other category, there is no doubt that biotechnology will have a substantial market. In fact, one of the recent concerns of 'corporate America' is that the costs of medical care are soaring – and are now over 10 per cent of the GDP (Garland et al., 1989). Thus, even if biotechnology were to be 'confined' to health services, it would have great potential. And pharmaceuticals are only a small piece of future products in human genetics.

Buttel gives two other reasons against considering biotechnology as revolutionary. In his view, biotechnology was 'prematurely commercialized, and its major applications lie decades ahead into the future'. Moreover, he argues, 'the historical . . . patterns of investment in biotechnology have been highly speculative and often "defensive" in nature' (1989a: 256). Based on these observations, Buttel considers that the biotechnology industry faces not transformation – but rather, imminent decline.

External and Internal Critiques

As an external critique of Buttel's analysis I question the very premises of his formulation of the research problem. My task in the internal critique is to assess Buttel's internal consistency in his use of concepts, and whether his substantive propositions are actually backed up by available data.

On the external level, I find Buttel's 'problem' rather problematic. It is a false problem to ask whether biotechnology is 'revolutionary' or not. It would be quite difficult to establish consensus on basic criteria by which to judge what is revolutionary or epoch-making, and what is not. More importantly, we must ask first whether there is a third technological revolution within a long wave movement in effect (Mandel, 1978, 1980). Is this new revolution – based on electronics, informatics, new materials, and biotechnology – actually leading to a new phase of capitalism? If we can answer this affirmatively, then we could ask what is the place of biotechnology in such revolution and in reformed capitalism. In the context of these questions it would be misleading to ask whether one of those new technologies, individually, is 'epoch making' or not. In sum, we should keep the question of the 'third technological revolution' conceptually as a single phenomenon, rather than as one made up of several juxtaposed revolutions based on each of the new technologies. While it is clear that the new technologies have different rhythms of development, with electronics in the leading position, there are at least two reasons why we should consider them as a conceptually single phenomenon.

First, the combined use of two or more of those technologies in new productive processes generates tremendous 'synergies'. The successful use of certain innovations may actually pose the technical requirement to combine the use of two or more of the new technologies. For instance, the most sophisticated dairy operations may use embryo transfer technologies to improve the herd, while using computers to maintain exact yield records per cow and to monitor their feed requirements (Sun, 1986: 151). Second, capitalism as a world economy entered a period of profound crisis in the early 1970s (Mandel, 1978). Its development had been based on heavy industry, which is indeed declining or being profoundly restructured (Piore and Sabel, 1984). The current restructuring of the world economy is predicated on productivity increases, which in turn depend on new technologies (Hatsopoulos et al., 1988; Young, 1988; Florida and Kenney, 1990). Mandel has convincingly demonstrated the existence of 'long waves of capitalist development' (1980) and predicted an upswing of capitalist growth in the early 1990s. Therefore, engaging in a partial analysis of whether one or the other new technology is revolutionary can only be misleading. The task at hand, for development sociology at least, seems to be in decoding the implications of the 'reformation of capitalism' (Sklair, 1989) for the new international division of labour.

An internal critique of Buttel's analysis will further highlight the need for a different perspective. Out of the three criteria that he proposes, Buttel concludes that number 3 is the one biotech least fulfils: it is not applicable to leading or dynamic sectors, but to declining ones. It is mainly this limitation of biotechnology that Buttel sets out to establish in his paper. His claim that biotechnology does not meet the third criterion is based on three arguments: (a) it is a 'substitutionist' technology both in agriculture and in the pharmaceutical industry, so that it will only 'patch up' the problems of Western agriculture and medicine; (b) the initial research may have been oriented toward revolutionary goals (e.g., the nitrogen fixation agenda), but these have been abandoned in favour of short-term commercial and profit interests; and (c) biotechnology is applicable to declining sectors (manufacture and agriculture), and the biotechnology industry itself is declining, evidenced by disproportionately poor performance of biotech stocks in the 'Black Monday' crack of 1987, and the reduction of in-house biotechnology research in large chemical and pharmaceutical firms.

Substitutionism

Buttel tends to minimize the importance of biotech on the grounds that it will merely *substitute* for existing products. I think, however, that this may be just the beginning of the application of new techniques, but they offer virtually endless possibilities. Even confined to a substitutionist role, biotechnology could have profound implications for both productivity and the international division of labour, which are bound to generate major social changes. In fact, some have already occurred with the introduction of high fructose corn syrup in the United States. From 1978 to 1987, 42% of sugar used in the United States was substituted by the new sweetener based on corn and produced with new enzymatic techniques. This change was profoundly damaging for several Caribbean countries and the Philippines, for a large part of their foreign exchange came from sugar exports to the United States (Ahmed, 1988; Otero, 1989a). Current research in Germany is developing a substitute for coffee (Quintero, 1989). How will that affect the economies of several Central and South American countries for which coffee is a major export crop? Will they continue to shift production from basic to illegal crops? To be sure, mere 'substitutionism' can have profoundly damaging effects on primary goods exporting countries. On the other hand, scientists can currently do more with biotechnology than their sciences allow them to understand. In other words, the technology is more advanced than science in this case, as far as knowledge and understanding of certain processes goes. The new techniques may eventually serve creative minds to develop more revolutionary products than mere substitutes for presently existing ones. Now, is it a question of time? Sure, but Buttel's criteria for defining what is revolutionary do not establish time limits, which would be quite arbitrary anyway. Thus, whether through substitutionism or new kinds of products and processes, biotechnology holds a tremendous potential for promoting major changes in production structures.

Research agendas

Although Buttel does not define 'revolutionary research agendas', one might infer that they result in new products that go beyond substitutionism. These would include new categories of goods and processes that will create equally new markets for means of consumption and means of production.

With this definition in mind, whether or not 'revolutionary' research agendas have been abandoned is more of an empirical question. Buttel asserts that such agendas have been abandoned, at least in the industry labs. While this might be true for the most part, there are many university labs that may be doing very significant 'basic' research that may sooner or later become 'applied', as has happened in the past in various biotechnology-related disciplines. Mexico has a whole research centre dedicated to nitrogen fixation research at the National Autonomous University of Mexico, and its researchers are in the frontier of knowledge, publishing in respected international journals and beginning to establish links with industry (Otero, 1989b). Nitrogen fixation is one of the areas of research that Buttel explicitly regards as revolutionary.

The whole issue of research agendas is one that may change rapidly with the intervention of the U.S. government. During the Reagan administration there were two areas of legislation that were reoriented to enhance the commodification of science: P.L. 96–517 (enabling universities to patent federally funded research results) and the new possibilities of patenting plants, microorganisms, and animals. There is a current debate in which a strong (perhaps official) position is questioning the Mertonian conception of science, discipline-based research, and the peer-review method of allocating funds, favouring a more pragmatic approach that talks about establishing 'strategic priorities' and interdisciplinary research. In fact, there have always been priorities in allocating federal research funds in the United States, with the Department of Defense usually charged with allocating the largest piece of the pie (66% in 1987). But non-defence federal research funds have been largely allocated according to the peer-review process. If this changes, biotech research could go in either of

two ways: orienting research agendas toward human needs or toward
the interests of transnational corporations (TNCs) and the state's interest in making the U.S. economy more competitive at all costs. The
National Science Foundation (NSF) has taken a step in the latter
direction by creating interdisciplinary research centres. While I ignore
the particulars about how they are being established, these centres get
funds outside of the usual peer-review process, at least to get them
started. Also, I suspect that they are very clearly 'mission oriented'
with 'strategic priorities' in mind. The latter may well include research agendas that Buttel would call 'revolutionary' for biotechnology, even if most are geared to the perceived need to take knowledge
to the marketplace as quickly as possible.

Declining sectors

Macroeconomic theory has used the concept of sectors to categorize
the various parts of the economy into more or less homogeneous
groups: the primary goods producing sector (agriculture, mining, oil,
etc.), the secondary sector (or manufacturing or industrial in a strict
sense), and the tertiary sector (services; this is perhaps the most hete-
rogeneous). In terms of employment, the tertiary sector has been the
most dynamic in the United States, but in productivity growth the
manufacturing sector has been more dynamic. Baumol has effectively
questioned the 'deindustrialization of America' thesis. The fact that
manufacturing absorbs a lesser proportion of the labour force
does not indicate that it is declining. To the contrary, it means
that its productivity is quite dynamic: 'throughout the industrial
world . . . productivity in manufacturing has grown considerably
faster than it has in a large group of services. . . . This means that,
though manufacturing outputs have grown, less and less of each
nation's labor force has been needed to produce them' (Baumol, 1989:
612).

One can certainly break down the analysis of sectors, and this is
done by 'industry' – and here, even agriculture would be considered
an industry. In Buttel's article, 'industry' and 'sector' are confused
when he mistakenly refers to the pharmaceutical 'sector' as a potentially dynamic one. Strictly speaking, pharmaceuticals are an industry. It would have been much better for Buttel's purposes to talk about
'industries' in his third criterion, rather than about sectors, because
the latter concept is much too general and may contain very heteroge-
neous industries in terms of their dynamism.

Now, is agriculture a declining industry? Buttel answers 'yes',
against the empirical evidence. Whether we call agriculture a sector or

an industry, its productivity has been historically quite dynamic, indeed surpassing the rest of the economy in its productivity (measured as output per unit of labour) rate of growth (National Research Council, 1989: 33). What has happened though, as in manufacturing, is that a far smaller number of people are able to produce enough food to feed the United States *and* to have an exportable surplus. In fact, in the last few years when the U.S. economy has been losing competitiveness in world markets, manifested in a growing trade deficit, agriculture has been the only sector that has accounted for a significant trade surplus of about $13 billion (National Research Council, 1989: 30–32).[1]

Thus, how do we measure 'dynamism'? By production and productivity increases, or by the proportion of the labour force absorbed by an industry or a sector? If the latter is taken as a measure, which is what Buttel seems to have done implicitly, then indeed 'services' would be the most dynamic sector, with 66 per cent of the U.S. labour force in 1980 (Baumol, 1989). But this is problematic for several reasons: it would be a major inconsistency with his designation of the 'Green Revolution' as a package of revolutionary technologies, and it would be unclear why electronics and informatics should be considered epoch-making technologies by Buttel's account.

'Green Revolution' (GR) was the name given to the technological package behind the modernization of U.S. agriculture which was exported to Third World countries in the 1960s and 1970s. But its application in the United States began in the 1930s, with the introduction of improved corn varieties, namely hybrid corn. Later on came other improved varieties, chemical fertilizers, pesticides, herbicides, and increasingly sophisticated agricultural machinery. As a package, these technological innovations not only gave a tremendous boost to productivity; they also displaced large numbers of farmers from their occupations (Cochrane, 1979; Kloppenburg, 1988). If anything, the so-called 'Green Revolution' accounted for a major leap forward in agricultural productivity. But it is also one of the major factors in the dramatic reduction of the labour force dedicated to farming, now a mere 2.2 per cent in the United States.

In Third World countries where the GR was adopted, the negative socioeconomic consequences were much graver. It was regionally polarizing, favouring zones with irrigated agriculture, and it entailed profound social differentiation processes, by which the better-off farmers benefited disproportionately and poor farmers went out of business, due to the scale-bias of GR technologies toward large farms (Cleaver, 1972; Hewitt de Alcántara, 1978; Pearse, 1980). Environmentally, there have been many problems of ground-water pollution, soil

erosion, loss of genetic diversity, and increased crop vulnerability
resulting from cultivation of increasingly homogeneous plant varieties (Kloppenburg, 1988; Otero, 1991). All of these problems appeared in the United States as well, but Third World countries were in
a much more vulnerable situation: they were not undergoing the
robust process of industrialization that the United States was
experiencing after the 1930s. Hence the results were more dramatic in
the underdeveloped world: there were massive migrations to the
cities where too few employment opportunities could be found. Then
the tertiary sector began to grow rapidly in underdeveloped countries. But this was certainly not a sign of dynamism of the sector;
rather, it masked severe unemployment and underemployment of
people who had been displaced from agriculture. I doubt that Buttel
wants to refer to this type of 'dynamism', but his arguments lend
themselves to this interpretation.

Is the biotechnology industry declining too? As may be expected
from the foregoing, Buttel would also answer 'yes' to this question in
a twofold sense: the biotech industry is declining both in the stock
market and with regard to the amount of Research and Development
conducted 'in-house'.

It is true that many biotech firms suffered disproportionately in the
Wall Street crack of 1987. While the overall decline in stock prices for
the 400 largest companies in the United States between the 12th and
the 28th of October was 28%, the dip in stock prices for 60 prominent
biotechnology firms was 44 per cent (Crawford, 1987). But should not
these financial difficulties be attributed to the infancy of the new
industry, and to the fact that it has taken longer than expected to
come up with commercial products? If anything, such difficulties
will accelerate the concentration trends in biotech. In fact,
Genentech, the flag-ship firm of the industry, has been taken over
by Roche Holding Ltd, the Swiss parent of giant drugmaker
F. Hoffman-La Roche & Co. (Hamilton et al., 1990). This could set off
a shopping frenzy by the giant TNCs for small biotech firms in the
chemical and pharmaceutical industries (Peterson and Armstrong,
1990).

Jumping to the conclusion that biotechnology is a declining industry
due to a conjunctural crisis resembles the 'short-termist' mentality of
which U.S. corporate executives are accused, when compared with
their Japanese counterparts. A recent survey of 480 biotechnology
firms (almost half of the approximately 1,100 in the United States)
found a healthy industry. Sales in 1988 were 33 per cent higher than
in the previous years, and total assets expanded by 6 per cent. The
biggest concern expressed by executives was financial: whether they

would have the ability to finance growth and technology development. Therefore, 66 per cent of the biotech companies expect to be acquired by larger firms at some point in the 1990s (*Genetic Engineering News*, 1989: 11). This financial mechanism clearly reinforces the concentration trends.

With regard to research, reducing 'in-house' efforts does not amount to withdrawing from research altogether. To the contrary, it probably means making a more efficient use of R&D funds, by contracting with universities. For example, Monsanto corporation, a giant in the agrichemical industry with substantial interests in biotechnology, fired over 100 biotechnology researchers in 1986; but it simultaneously gave a $60 million grant to Washington University for research in the same field. Such increased university–industry links are amply documented and represent a new type of relationship in the biological sciences (Blumenthal et al., 1986; Kenney, 1986; Kleinman and Kloppenburg, 1988; Otero, 1989c). In sum, we cannot jump to the conclusion that the biotechnology industry is declining and withdrawing from research solely on the basis of a conjunctural business downswing and a shift in research policy.

Conclusions: Bringing the Majority of the People In

My main conclusion from the foregoing analysis is that we gain a deeper understanding of complex events by considering biotechnology as part of the 'third technological revolution'. To the extent that new technologies are at the core of the world economic restructuring, biotechnology will play a major role in transforming agriculture. Just as the substitution of the horse for the tractor was revolutionary by any standard, biotechnology's 'substitutionism' will bring about major changes in production and social structures. To be sure, these will not come overnight, for we are dealing with a *technological* revolution. Besides substitutionism, genetic engineering is leading scientists and industrialists to reconceptualize *what life is*. We are only beginning to see its potential. Like the Copernican revolution, biotechnology will lead to unforeseen developments.[2]

Moreover, the 'biotechnology industry' itself has to be reconceptualized. Taken as individual firms, biotechnology companies have seen an expansion in sales and investments after recovering from 'Black Monday'. Due to short-term financial problems, however, they are losing their independence and being integrated into industries that will use their technology. Thus, rather than an industry, genetic engineering will become an enabling technology for the already existing chemical and pharmaceutical industries. In fact, giants in these industries have been particularly aggressive in absorbing former biotechnology 'start-ups' – and both of those industries present highly concentrated profiles. This emerging market structure will undoubtedly have a major bearing on the ways biotechnology products are disseminated in the world. The contrasts in this regard with the 'Green Revolution', which was promoted by public and semi-public institutions, will be profound.

Although the bulk of biotechnology research and development is taking place in advanced societies, deployment of its fruits will have implications for the world economy as a whole. As Iftikhar Ahmed has recently suggested, the 'application of biotechnologies to agriculture would automatically affect 60 per cent of the Third World population who depend on agriculture alone for their livelihood' (1989: 553). There have been important reflections on the implications of current changes for advanced capitalist societies, such as that by Piore and Sabel in *The Second Industrial Divide*, but few address the place of Third World countries in the new international division of labour. Given the vast heterogeneity existing among Third World societies, they will be affected differentially by the third technological revolution, depending on the profile of their socioeconomic structures. Some have recently become industrialized, precisely on the basis of new technologies. The larger countries have a certain potential to jump on the bandwagon of the technological revolution; others might be integrated to the world economy simply as producers of cheap labour-power, while still others could be marginalized from the main economic trends. Thus, one thing that should be looked at more closely is the new stratification of underdeveloped societies that is bound to emerge. An initial formulation has been provided by Castells (1986).

Castells proposes that a first stratum of Third World countries is made up of the so-called Newly Industrialized Countries (NICs), currently Singapore, Taiwan, Korea, and Hong Kong. Their export-led industrialization has been predicated mostly on local capital, which accounts for up to 75 per cent of total exports. Another stratum would be made up of those countries characteristically in the 'new international division of labor' (Fröbel et al., 1980). The countries included in this stratum are becoming mostly export platforms for the most labour-intensive parts of global labour processes within transnational corporations. Included in this category are: Thailand, Philippines, Malaysia, some Caribbean islands, and parts of Mexico, China, and Brazil. A third stratum is made up of large Third World countries such as Brazil, Mexico, Argentina, China, and India, with some industrial potential to join the 'third technological revolution' with local resources. The main danger for these countries is the development of a disarticulation in their economies and societies between integrated and disintegrated sectors to the world economy. The latter would be largely marginalized from the benefits of new development. The major OPEC countries make up another stratum. They have been unable to industrialize, despite heavy capital influxes for several years during the oil boom. This may be explained by a number of factors: being trapped in political games of the superpowers, having corrupt bureaucracies, attempting to develop heavy industries with just as heavy technological dependence at the time that these industries were declining in the world economy. Finally, Castells sees the majority of Third World countries as being condemned to economic obsolescence, unemployment, misery, hunger, illness, and individual violence in their large urban centres. Countries in the latter stratum would either be marginalized from the world economy or would experience a 'perverse integration' (Castells and Laserna, 1989) through the production and export of illegal crops.

Finally, I suggest four areas of future research to assess the socioeconomic impact of biotechnology in developing countries. First, we should further clarify the new stratification of countries in the international division of labour, and determine which forms of integration to the world economy are most promising for the majority of the people in terms of the distribution of benefits from development. This would also involve the study of 'structural processes of technological innovation', which should combine structural analysis with that of the protagonist actors in the economic dynamic of developing countries. I suggest at least the following actors for close scrutiny: transnational corporations, governments, local entrepreneurs, and international agencies. What will 'reformed capitalism' involve for the changing relationships among these actors and how will they distribute the benefits of development?

Second, we should evaluate the potential of the systems of science and technology in developing countries and their existing links with industry. What is the extent of technology transfer and how could this be further promoted? Where scientific capacity or such links are nonexistent, technological dependence or marginalization are inevitable.

Third, what is the character of legal structures in regard to intellectual and industrial property? Do they promote or hinder the development of a local biotechnology industry?

Fourth, we need an analysis of the various industries that will be affected by products of biotechnology and of the industries or institutions that will be charged with their dissemination. In contrast to the Green Revolution, which was promoted mostly by public and semi-public institutions, private industry will probably be the main promoter of biotechnology. Given its unequivocal interest in profit-maximization, it is likely that the impact of this new institutional framework will be even more socially polarizing than the case of the Green Revolution.

ACKNOWLEDGEMENTS

The author was a Professor–Researcher at the University of Guadalajara (Mexico) and Visiting Senior Lecturer at the University of Wisconsin-Madison when he conceived and drafted this chapter. He wishes to acknowledge the helpful comments by Martin Kenney, Daniel Kleinman, and three anonymous reviewers for *Sociological Forum*.

NOTES

[1] 'Agriculture' is used here in a restricted sense, as equivalent to farming. A broad definition of agriculture would include the agricultural inputs industry and the food processing industry. Biotechnology will have an impact on both in important ways. For instance, new plant varieties with herbicide resistance will expand the sale of inputs, and new tomatoes with greater solid contents will increase the profitability of food makers. On the other hand, the environmental implications of the former trend may be largely undesirable (Otero, 1990).

[2] I thank Martin Kenney for this idea in his comments to a previous version of this chapter.

REFERENCES

Ahmed, Iftikhar. (1988). The bio-revolution in agriculture: Key to poverty alleviation in the Third World? *International Labour Review*, **127**.
——. 1989. Advanced agricultural biotechnologies: Some empirical findings on their social impact. *International Labour Review*, **128**.
Baumol, William J. (1989). Is there a U.S. productivity crisis? *Science*, **243**, 611–615.
Blumenthal, David, Gluck, Michael, Louis, Karen Seashore and Wise, David. (1986). Industrial support of university research in biotechnology. *Science*, **231**, 242–246.

Buttel, Frederick H. (1989a). How epoch making are high technologies? The case of biotechnology. *Sociological Forum*, **4**, 247–260.
——. 1989b. Social science research on biotechnology and agriculture: A critique. *The Rural Sociologist*, 5–15.
Buttel, Frederick, Kenney, Martin and Kloppenburg, Jack. (1985). From Green Revolution to biorevolution: Some observations on the changing technological bases of economic transformation in the Third World. *Economic Development and Cultural Change*, **34**, 31–51.
Castells, Manuel. (1986). High technology, economic policies and world development. Working paper, Berkeley Roundtable on the International Economy, University of California, Berkeley.
Castells, Manuel and Laserna Roberto. (1989). The new dependency: Technological change and socio-economic restructuring in Latin America. *Sociological Forum*, **4**, 535–560.
Cleaver, Harry. (1972). Contradictions of the Green Revolution. *Monthly Review*, **2**, 80–111.
Cochrane, Willard. (1979). *The Development of American Agriculture: A Historical Analysis*. Minneapolis, MN: University of Minnesota Press.
Crawford, Mark. (1987). Biotechnology's stock market blues. *Science*, **238**, 1503–1504.
Florida, Richard and Kenney, Martin. (1990). *The Breakthrough Illusion: Corporate America's Failure to Move From Innovation to Mass Production*. New York: Basic Books.
Fröbel, F., Henricks, J. and Kreye, O. (1980). *The New International Division of Labor*. Cambridge: Cambridge University Press.
GAO. (1986). Biotechnology: Agriculture's regulatory system needs clarification. Report to the Chairman, Committee on Science and Technology, USA House of Representatives, GAO/RCED 86–59, March.
Garland, Susan B., Hoerr, John, Galen, Michelle and others. (1989). Ouch: The squeeze on your health benefits. *Business Week*, **3134**, 110–118.
Genetic Engineering News. (1989). Ernst & Young survey finds healthy industry. *Genetic Engineering News*, **9**, 11.
Goodman, David, Sorj, Bernardo and Wilkinson, J. (1987). *From Farming to Biotechnology*. Oxford: Basil Blackwell.
Hamilton, Joan O'C., Jereski, Laura and Weber, Joseph. (1990). Why Genentech ditched the dream of independence. *Business Week*, **3146**, 36–37.
Hatsopoulos, George N., Krugman, Paul R. and Summers, Laurence H. (1988). U.S. competitiveness: Beyond the trade deficit. *Science*, **241**, 299–307.
Hewitt de Alcántara, Cynthia. (1978). *Modernización de la Agricultura Mexicana*. Mexico: Siglo XXI Editores.
Kenney, Martin F. (1986). *Biotechnology: The University–Industrial Complex*. New Haven, CT: Yale University Press.
Kenney, Martin F. and Hibino Barbara. (1987). Biotechnology in Mexican

agricultural research: Development of a research agenda. Paper presented in the Meetings of the Rural Sociological Society, in Madison, Wisconsin, 12–15 August.

Kleinman, Daniel L. and Kloppenburg, Jack R., Jr. (1988). Biotechnology and university–industrial relations: Policy issues in research and the ownership of intellectual property at a land grant university. *Policy Studies Journal*, **17**, 83–96.

Kloppenburg, Jack R., Jr. (1988). *First the Seed: The Political Economy of Plant Biotechnology, 1492–2000*. New York: Cambridge University Press.

Kloppenburg, Jack R., Jr, Kleinman, Daniel L. and Otero, Gerardo. (1988). La biotechnología en Estados Unidos y el Tercer Mundo. *Revista Mexicana de Sociología*, Vol. L, No. 1, 97–120.

Mandel, Ernest. (1978). *Late Capitalism*. London: New Left Books.

——. 1980. *Long Waves of Capitalist Development: The Marxist Interpretation*. Cambridge and Paris: Cambridge University Press and Editions de la Maison des Sciences de l'Homme.

National Research Council. (1989). *Alternative Agriculture*. Washington, DC: National Academy Press.

Otero, Gerardo. (1991). Biotechnology and economic restructuring: Toward a new technological paradigm in agriculture? In Albert Sasson (ed.), *Economic and Socio-Cultural Implications of Biotechnology for the Third World*. Paris: UNESCO (forthcoming).

——. 1990. Biotechnology, agriculture and the environment. Paper presented at the annual meetings of the Rural Sociological Society, Norfolk, VA, 7–12 August.

——. 1989a. Commodification of science: Biotechnology in the United States and Mexico. Paper presented in the meetings of the Society for Social Studies of Science, Irvine, California, 15–18 November.

——. 1989b. Ciencia, nuevas technologías y universidades. *Ciencia y Desarrollo* (Mexico City), Vol. xv, 49–59.

——. 1989c. Industry–university relations in biotechnology and the sugar and dairy industries: Contrasts between Mexico and the United States. World Employment Programme research working paper. Geneva: International Labour Organization.

Pearse, Andrew. (1980). *Seeds of Plenty, Seeds of Want: Social and Economic Implications of the Green Revolution*. Oxford: Oxford University Press.

Peterson, Thane and Armstrong, Larry. (1990). Roche's big buy may set off a shopping frenzy. *Business Week*, **3146**, 38.

Piore, Michael J. and Sabel, Charles F. (1984). *The Second Industrial Divide: Possibilities for Prosperity*. New York: Basic Books.

Quintero, Rodolfo. (1989). Prospectiva de las Agrobiotecnologías en América Latina y el Caribe. Paper presented in the Seminar on General Perspectives and Impacts of Agribiotechnologies in Latin America and the Caribbean: Political Implications and Strategies, Paipa, Colombia, 14–17 August.

Sklair, Leslie. (1989). *Assembling for Development: The Maquila Industry in Mexico and the United States*. Boston: Unwin Hyman.

Sun, Marjorie. (1986). Will growth hormone swell milk surplus? *Science*, **233**.

Young, John A. (1988). Technology and competitiveness: A key to the economic future of the United States. *Science*, **241**, 313–316.

5

Biotechnology: A Core Technology in a New Techno-Economic Paradigm

Annemieke J. M. Roobeek

In Search of an Integrated Innovation Concept

In recent years more attention is being focused on formulating an integrated innovation concept. Rather than centring an innovation in the isolated environment of an individual enterprise, the integrated approach studies at a higher aggregated level the clustering of innovations around one or more core technologies within the complex of international competitive relations, interindustrial relations, macroeconomic conditions, national and regional political structures, and socio-institutional conditions. Instead of looking for a monocausal or branch-specific cause, the integrated approach seeks the origins of innovation in various endogenous and exogenous developments.[1]

Recent theories about the relationship between technological innovation and economic development highlight five central themes.[2] First, technological progress is a major driving force behind the dynamics of international economic development. Second, technological innovation occurs in clusters. Third, these clusters bring about long-term change. Fourth, the traditional instruments of economic science are incapable of giving a realistic representation of the relationship between technology and economy. Fifth, and last, the institutional and social conditions greatly influence the invention and diffusion of innovations.

The search for a more integrated approach to the innovation process springs from the theoretical and practical need to gain greater insight into the processes of structural change sparked by innovation. The research into the dynamics of the development of innovation around core technologies sheds new light not only on the – potentially – pervasive economic effects, but also on the bottlenecks that may occur in the diffusion process and the industrial and institutional adjustments that may be required.

One of the most important strands elaborated by the group of innovation economists around Christopher Freeman of SPRU (Sussex) and MERIT (Maastricht) is the long-term dynamics based on the occurrence of unequal 'technological revolutions' in industrial development. They take this notion a few steps further than their precursor Joseph Schumpeter in an attempt to free economics from its straitjacket and create room for a broader, interdisciplinary approach. Though their contributions are 'breakthroughs' in economics, they still suffer from two limitations. In the first place, explanations of the rise of new technologies still concentrate on too strict an economic framework and tend to take too little account of political, institutional and social factors; although they pay lip-service to these important factors, they do not analyse them. The study of these contextual factors is still in its embryonic stages and it takes a more interdisciplinary approach to enrich the pro-theories put forward by the innovation economists. Secondly, the rise of a new techno-economic paradigm is almost exclusively approached in relation to information technology. Though information technology, spawned by microelectronics, is presently central to the transformation of economic activity, it is not the only core technology on which the new techno-economic paradigm is being built.

This chapter assumes the postulation that the new techno-economic paradigm is a tripod, consisting of a clustering of innovations around information technology, biotechnology and the new materials. Technological revolutions take place between and around these new core technologies which not only dramatically change the current techno-industrial base but also put pressure on the existing socio-institutional structure to adjust to a new order of manufacturing, acting, consuming and communicating. In this chapter we shall argue that microelectronics, as the basic technology for a broad spectrum of applications, is only one of the three pillars on which the new paradigm is being built. For technological development is not restricted to microelectronics or information technology alone. Since the mid-1980s technological breakthroughs have occurred particularly in the fields of biotechnology and the new materials. These breakthroughs have already generated dynamics of change on a limited scale that will gather momentum in the foreseeable future. Together these three new core technologies provide the technological preconditions for a new political and economic organisation.

The New Core Technologies in a Conceptual Framework

Companies and countries have fallen under the spell of the new core technologies. Microelectronics and information technology were revolutionising the 1980s, and, if the forecasts are right, biotechnology will prove to be the alchemy of the late 1990s, while the new materials will enable every material to be tailored to size and weight. Information technology, as well as flexible automation, could make a considerable contribution to resolving the bottlenecks in the Fordist production method, which was based on mass production within a fragmented, partly automated production process. Biotechnology and the new materials might help to alleviate environmental and energy problems while also offering substitutes for scarce raw materials. The potential of these new core technologies is enormous and their applications seem endless.

The distinctive characteristic of the new technologies is that they enhance the unprecedented possibilities for management to increase control of labour and production costs, of quality and efficiency, of the use of energy and raw materials, of pollution and waste, as well as of many other aspects like better control of the international dispersal of production, increased control on logistics and subcontractors. These aspects can be seen in every production process, whether industrial, services-related, science-based or agricultural. The analysis of the substance and scope of the new technologies requires a new conceptual framework. The new core technologies are regarded as generic technologies, embracing far more than one single sector or field of application. Every branch in economic life, varying from health care to banking and from the automobile industry to agriculture is confronted with far-reaching changes due to the pervasiveness of the new core technologies, namely information technology, biotechnology and new materials. We are dealing with intermittent technological revolutions, which Schumpeter calls 'creative gales of destruction'. This type of technological change is characterised by its pervasive effects on all economic activities, creating what can be fairly regarded as a new techno-economic paradigm.

The paradigm consists of two parts. First, the technological part, and second, the economic part. The latter relates to the political–economic order that suits to the technological paradigm. Although it is obvious that the two parts are intertwined, most authors have concentrated on the role of technology in the economy rather than on the question of what the principal contents of a new political-economic order might be in the light of the new technologies. The most important contribution to the discussion on a new techno-economic paradigm is from the earlier quoted authors Freeman, Perez, and Soete. They have made a taxonomy of innovations. By paradigm change they mean a radical transformation of the prevailing engineering and managerial common sense for best productivity and most profitable practice, which is applicable in almost any industry. According to them, the organising principle of each successive paradigm is to be found in the dynamics of the relative cost structure of all possible inputs to production. In this context they speak about the 'key factors' of the paradigm or the specific set of inputs, that fulfils the following conditions:[3]

- clearly perceived low and rapidly falling relative cost;
- apparently almost unlimited availability of supply over long periods;
- clear potential for the use and incorporation of the new key factor or factors in many products and processes throughout the economic system.

The co-occurrence of these key factors triggers pervasive changes in the overall economic system, resulting in a new techno-economic paradigm. These criteria clearly show why information technology based on microelectronics changes the paradigm while nuclear technology does not. They confine their paradigm to information technology and call it the 'information technology paradigm'.

The Interlinkage Between the New Core Technologies

One can argue that the restriction to information technology is too cramped an approach, taking insufficient account of breakthroughs in the other two legs of the tripod, namely in the fields of biotechnology and new materials. Though information technology enjoys a far wider scope than biotechnology or the new materials for the moment, it should be borne in mind that the latter two are still in their embryonic stages and show great potential for future applications. Microelectronics creates the framework for the new technological paradigm, in which biotechnology and the new materials can be developed further. The pervasiveness of new materials and new-materials technology is substantial. New materials will be used from the engineering sector to the sports wear industry. Products will get complete new qualities, because of the use of specific 'tailor made' materials. New materials will lead to new standards, making traditional materials obsolete. The pervasiveness of biotechnology is often underestimated as well. If one takes into consideration the economic importance of agriculture, the food and beverages industry, the petrochemical and agrochemical industry, the pharmaceutical industry and the fine-chemical industry, as well as the waste-treatment sector, and the extent to which biotechnology is changing already corporate structures and international trade relations, one cannot leave out biotechnology as one of the pillars of the new paradigm. Moreover, biotechnology more than information technology puts pressure on traditional norms and values towards the manipulation or 'modification' of all kinds of living organisms. In this respect we only need to mention the economic and ethical debate on the admittance of bovine growth hormone (BST) to increase the production of milk per cow by about 20–30 per cent, or the debate on gene mapping and the consequences of the exploration of the human genome for health care, insurance companies, public-financed facilities for the disabled, social welfare, etc.

Another argument against the restriction of the new paradigm to the information technology solely, is that increasingly the further development of microelectronics as the base of information technology has become more and more dependent on developments in other technological fields, namely of biotechnology (e.g. the development of biochips and biosensors, but also to develop artificial intelligence and expert systems one needs to combine knowledge from neuroscience together with computer science) and new materials (e.g. gallium arsenide, superconductive materials, technical ceramics). Telecommunications, too, is eager to use developments in new materials such as optical fibre and heat-resistant materials used for building satellites. Biotechnology would be a lame duck without computers and electronic measurement and control engineering ensuring the correct composition in reactors. Biotechnology is also dependent upon developments in new materials, particularly in the field of membranes. Biotechnology stands to profit greatly from the ultrafiltration and hyperfiltration techniques as its processes involve vast amounts of process fluid (water). Finally, the field of new materials depends on CAD-CAM techniques as well as biotechnological knowledge on molecules and cell structures to design composite alloys for new (bio)materials. Because of their high degree of integration and interdependence, the three core technologies together constitute a new technological paradigm. For reasons related to the interdependence between the new core technologies one may expect that in the 1990s and beyond the most promising inventions and innovations will occur at the cross roads between these core technologies. The diagram in figure 5.1 illustrates the complementary nature of the new core technologies.[4]

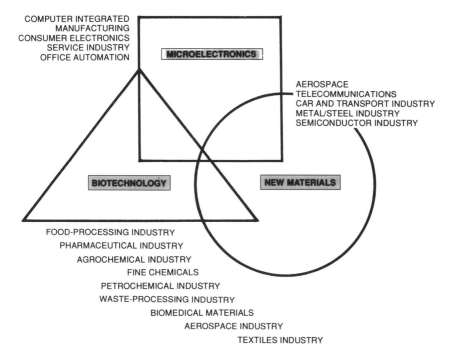

Figure 5.1   The integration and interrelation of new core technologies

The Technology Web around Biotechnology

For analytic use the technological contents of the new techno-economic paradigm can be split up into three different technology webs. We can locate a technology web around microelectronics/information technology, biotechnology and the new materials. Technology webs, a concept which has been worked out elsewhere,[5] are a helpful tool to indicate the pervasiveness of a core technology. It also helps to show how existing economic branches and new fields of economic activity are grouped around a core technology. It may be imagined that these branches and new fields are on the one hand centres of invention and innovation, but at the other hand important areas of applications of the new core technology. This explains the upsurge of new technological systems as a result of the internal dynamics of a branch or a new area of applications. At the same time the vivid interaction between the economic activities grouped around the core technology gives rise to the blurring of boundaries between separate activities. Whole filières or production chains, belonging to each of the activity fields, can be changed or can become completely intertwined, with sweeping consequences for the distinct parts of the filières.

The concept of technology webs is also useful in exploring the interrelations between different parts of a technology web, e.g. the close relation between the computer industry, telecommunications, and the software industry. Or, in the case of biotechnology, the close relation between the pharmaceutical industry and the human health-care sector, and the veterinary health-care sector. In short, besides the dynamics within each segment of the technology web, there is also a strong relation between the dynamics of the separate segments. It is this process of endless interaction of different dynamics that makes a new core technology so powerful and persuasive to its potential users.

The structure of each technology web is as follows. At the centre of the technology is a core technology. An extensive web of applications is, so to speak, spun around each technology (or perhaps better: a set of new basic techniques, upon which numerous applications are based). Figure 5.2 shows the technology web around biotechnology.[6]

As stated above, various areas of innovative activity can be found within a technology web. We can call these areas technology systems. These consist of several basic techniques spawned by the central technology together with techniques from a specific field of application. Examples of basic techniques include DNA-techniques, bioprocessing techniques and cell-culture techniques. A technology system brings together parts of a given web whose applications (i.e. products and processes) complement each other, e.g. DNA-techniques and veterinary health-care products, or DNA-technology and modifications of agricultural seeds. A technology system develops in accordance

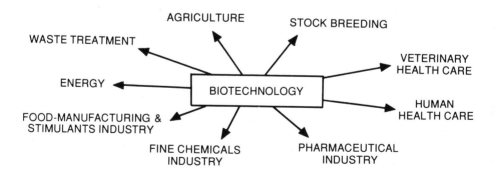

Figure 5.2 Technology web around biotechnology

with its own dynamics. This stems partly from the tendency for new applications to generate further innovations and adjustments. The dynamics of the process is also fed by developments in the central technology. Basically, there is a continuous interaction between advancing technical potential and expanding applications. The technology-push and technology-pull are inextricably united forces propelling the development of the trajectory, though in a given stage of development demand (pull) may exceed the supply of new technology (push). This will most commonly occur at the end of growth and during maturity. The reverse will happen in the initial stages when new applications have still to make their mark in the market.

Examples of technology systems developing around the biotechnology web include biotechnological applications in the field of agriculture in combination with the chemical industry and the veterinary industry: the 'farm of the future'. Further technology systems can be identified arising around the food-manufacturing and stimulants industry, the energy sector, the (fine) chemical industry, the pharmaceutical industry and the chemical equipment-manufacturing industry.

It is to be expected that the lines of demarcation between the various technology systems will become increasingly blurred owing to the coherence and closer ties between sectors and companies. As a result, the central or basic technologies will continually broaden their spectrum of applications, fanning out to a wide range of sectors. In the course of this expansion, technology systems will start to overlap each other and merge. The blurring of sectoral boundaries will radically change the complexion of the industrial structure. The main reason for this is that almost all fields of economic activity around a technology web draw from the same basic technology. Many techniques used in robotics are similar to office automation techniques while biotechnological techniques used for agricultural applications are often also employed in the medical sector. Consequently, companies might be tempted to introduce biotechnological techniques or applications into sectors originally based on a completely different central technology. This communal technological base could profoundly affect relationships between competitors in the national and international markets. At company level, this foreshadows an imminent fusion between the needs and interests of the pharmaceutical industry, chemical industry, seed industry and food-manufacturing industry. It is therefore hard to draw clear lines of demarcation between the various sectors. A fundamental difference with earlier diversification waves, such as in the 1960s, is that diversification was formerly based purely on commercial grounds, whereas now different industries are set up to tap the same basic technologies. As a result, technological development is providing the basis for the interpenetration of sectors, implying that many more companies than before will gain access to each other's markets through technology. In other words, it is no longer the product–market combination but the product–market–technology combination that is becoming increasingly dominant.

The integration of diverse technology systems will therefore not remain confined within the various technology webs, but will also take place between the webs. The main reason for this is that the new technologies can only fully develop if the applications of the different techniques complement each other, so creating a new composite whole. This brings us back to the interrelation between the core technologies, as sketched in figure 5.1.

It would appear therefore that the true technological breakthroughs will only be achieved if the different webs start to merge further and if companies start to combine different techniques in order to create fundamentally new products and processes. Until now, the integration of the new core technologies and the application of the combined scientific and technological potential has been developed on a very small scale. Many of the new products in biotechnology are mere substitutes for already existing products. This indicates that current developments are still in their infancy. One may expect, however, that the coming decades will witness radical breakthroughs as a result of the integration of sciences, technologies and application areas. This will have a tremendous impact on the existing industrial structures in countries and regions, in the competitive relationships between countries and companies, in the raw-materials trade, in the supply of energy and food and in the services sector. These will be complemented by socio-institutional adaptations at the level of the firm, at the level of national regulation and at international level.

Why Biotechnology is a Core Technology

The essence of biotechnology is that it gives control over the living microcosmos. New drugs for humans and animals, new crops, new fertilisers and new pesticides have already been launched. Many more products are awaiting the regulator's permittance. Experiments are going on with genetically enhanced livestock and human gene therapies to repair human genetic defects. Genetic engineering is cracking life's code. Never before in history has man been able to intervene in living processes with such precision.

Biotechnology is often defined as the integrated use of biochemistry, molecular genetics, microbiology and processing technology which aims to find applications of microorganisms, cell cultures or parts of microorganisms or cells for practical (industrial) purposes.[7] Biotechnology, therefore, is not one single technology but a group of various techniques culled from various disciplines around biology and chemical technology. Some claim that biotechnology is a time-honoured science, having been used for centuries to produce such staple foods as bread, cheese, wine, beer and yoghurt. They argue that the scientific developments and improvements in the bioprocessing industry which enabled large-scale production of penicillin and enzymes in the 1940s and 1950s stemmed naturally from these traditional uses. This view also sees the upgrading of plants and the new artificial cattle-breeding techniques as gradual developments. However, it fails to recognise the fundamental breakthroughs which have marked the history of biotechnology. What's more, it ignores the 'new biology's' ground-breaking capabilities in a wide range of areas. The most important developments in the basic sciences include the progress in genetics, which spawned the recombinant DNA technique and cell-fusion technique; the modern cell-culture techniques and the developments in enzymology, immunology and processing technology. Biotechnology should be seen as developing in phases or generations rather than as an evolutionary process.[8] That way, it is made clear that though the reservoir of existing knowledge and techniques is being continuously tapped – and remains important – periods can be discerned which saw accelerated development. The first phase of biotechnological development embraces the traditional applications. The second phase is characterised by the scientific and technical developments since the 1940s, leading mainly to new processing techniques and research into immobilised enzymes which can be used in various processes. This finally culminated in enhanced (occasionally even continuous) fermentation processes. Important breakthroughs since the 1970s, such as in the field of genetic engineering and biological processing techniques, have ushered us into a new, the third, phase of biotechnological development. Theoretical understanding has advanced to the point where it has become possible to control bio (techno)logical processes. As a result, biotechnology has changed from an empirically based technology into a science-based technology. For that reason, recent developments bear no comparison with former developments. Breakthroughs in genetics allowed the protein-producing part in the DNA chain (the genes in a cell) to be isolated and transferred to a vector, where the protein multiplies further. This gene-cloning technique is called the recombinant DNA technique. In principle, it can generate hereditary characteristics to precise specifications, being a much more reliable method than the one formerly used. Although cloning remains time-consuming, the procedure has been considerably shortened by the introduction of 'gene machines' which are capable of pre-processing DNA. Another way to manipulate genes is through cell fusion. This method is mainly used when a large proportion of the genes has to be transferred to another cell. The cells fuse, forming a hybrid which has the characteristics and properties of both original cells. This technique can be applied to produce mono-clonic antibodies which react with a specific antigen. Tissue-culture techniques can be used to mass-produce pure antibodies. It is a technique that is applied in various fields. In health care, for instance, it speeds up diagnosis. It also helps upgrade plants for a variety of applications. Seeds can be engineered to make crops resistant to certain insects, diseases or external conditions and so multiply harvests, improve quality and reduce the use of pesticides. Finally, great progress has been made since the 1970s in bioprocessing technology, focusing on the extraction, treatment, purification and conversion of usable materials. It is more common these days for parts of chemical processes to be turned into biological processes. Advances in the field of (computerised) measurement and control engineering have contributed significantly to maintaining the desired process in bioreactors.

Recent advances in recombinant DNA procedures, the chemical synthesis of genes and gene fragments, protein-structure determination, and computerised molecular modelling have brought about a new era of protein engineering. Protein engineering can be achieved either through direct modification of the amino-acid molecules that comprise proteins, or by altering the DNA molecules of the genes that produce the proteins. The newer methods for modifying proteins at DNA level are collectively referred to as in vitro mutagenesis. Protein engineering can be seen as a logical successor to genetic engineering. Instead of simply mass-producing a natural product, researchers tailor biological products to their own requirements. In protein engineering the characteristics of the new technological paradigm – miniaturisation, flexibility, dematerialisation, minimisation of waste, and a tailor-made design – are evident. The ability to make proteins function more efficiently, to operate under stressful conditions within the human body, or to create totally new proteins that do not exist in nature, are all possibilities of importance to the commercial development of novel human therapeutics, e.g. TPA (tissue plasminogen activator; a blood protein), interleukin-2 and EGF (epidermal growth factor, important for wound healing, including burns, and cataract surgery).[9]

What distinguishes the current third-generation biotechnology from the previous biotechnological developments is the scale and scope of the possible changes, as well as the ability to steer and programme genes in living organisms. As a result, man has come much closer to controlling nature in a broad sense. But, what is more important is that the new generation of biotechnological techniques has changed the whole set of basic groundrules, it has opened up new approaches to old problems, and new fields of development. Many traditional methods and techniques will retain their value, though being continually perfected by modern technology. The power derived from the ability to control and manipulate nature as well as the benefit of low energy intensity and the possibility of using replaceable materials as an industrial base material, put it beyond doubt that many sectors will either use or be influenced by the new biotechnological techniques. Already 40 per cent of the goods produced in the industrial countries is biologically based.[10] The new applications are sure to raise this figure even further.

The pervasiveness of biotechnology seems to justify it as a core technology. Biotechnology can be placed in a technology web, with genetic engineering, protein engineering, cell fusion and new biological-processing technology as its main set of techniques. Like microelectronics, biotechnology can be expected to spawn new technology systems and new applications in very different industries. This explains the pervasive effects of the group of techniques. In the next few years biotechnology will perhaps not so much create completely new products, but will provide primary new production techniques that lead to products with higher quality and higher value added. However, protein engineering will certainly result in breakthroughs in pharmaceuticals and health care. But it is expected that because of the uncertainties about regulation of biotechnological processes and applications, as well as uncertainities in social acceptance, the diffusion of biotechnology will follow an evolutionary path. By replacing existing products biotechnology is gaining ground in a large number of well established industries. Besides being implemented in traditional sectors, the new biotechnological applications will also influence the products and processes of sectors where formerly biotechnology played no role whatsoever, like for example in the waste-treatment sector.

Dynamics behind the Rise of Biotechnology as Core Technology

The question of why biotechnology is regarded as a core technology leads to an analysis of the problems and limitations of the current products and processes. Our thesis is that the rise of the three new core technologies, of which biotechnology is one, is not an accident, but the dynamics behind the rise are to be found in the limitations of the preceding ruling techno-economic concepts. The dominant preceding concept was Fordism. Fordism is the description of a concept that is based on the idea that in capitalism there is a strong relation between the economic structure and the political and socio-institutional structure. Starting already in the late twenties and thirties, but introduced on a large scale after the Second World War, Fordism became the dominant concept of establishing a balance between different structures on which society is based. The outstanding characteristics of Fordism are the integration of mass-production, the development of masspurchasing power and mass-consumption. To maintain this balance between economic and socio-institutional development (government) regulation is necessary. In the late 1960s Fordism came under fire as a result of altered international economic and political relations. In the years since, we have witnessed a growing mismatch between the economic development and the socio-political order and its institutions.

Just as a cluster of core technologies helped to instigate the transition to Fordism, so specific core technologies are again acting as an engine of transition from the existing to a new economic and socio-institutional order. For once again new core technologies, namely information technology, biotechnology and new materials, will trigger dramatic changes in the structure of production, productivity, industrial relations, labour organisation, the international and regional division of labour, as well as in the cost structure.

If we maintain that biotechnology is a core technology and an engine of transition we should first give an answer to the question to what extent biotechnology is capable of solving problems and limitations of the current products and processes. Various reasons can be given to justify the central role of biotechnology. Though all derive from apparently disconnected developments, they all in one way or another helped stimulate biotechnological research and applications in new products and processes.

In the first place, the discovery of genetic engineering techniques meant a significant breakthrough. These made it possible to exchange genes between cells and consequently to create completely new, upgraded combinations within a much shorter time. Though plants could formerly be cross-fertilised to achieve the required combinations of characteristics, the method used suffered from the following drawbacks:

- cross-fertilising techniques are extremely time-consuming;
- only genetic material of closely related varieties could be exchanged;
- it was impossible to attain the exact combination as the cross-fertilised cell was always tainted by undesired remnant hereditary characteristics.

As a result, the procedure not only entailed repeated cross-fertilisa-tions but also a string of follow-up treatments. These disadvantages were removed by genetic engineering which involves controlled transfer, though this is still mainly performed on a laboratory scale. Tissue culture, by contrast, has made large-scale production possible, thanks to the fact that full-blown plants can grow from one single cell.

In the second place, unstable prices of raw and other materials have made industry look to biotechnology as a potential source of cheap alternative and reusable materials. In the chemical-processing indus-try, raw materials account for 50 to 70 per cent of the production costs. Possible savings achieved through biological processes could tempt industry to switch to a new technology which thanks to its greater flexibility could also put an end to overcapacity. In the past years, the American chemical industry has been operating at only 70 per cent of its capacity, adding to the costs of this form of production. Biotechnol-ogy can provide reusable materials. These substitutes, mainly for petroleum derivatives, can be extracted from agricultural products or waste (agricultural, wood, cotton or even domestic waste and manure). An additional advantage is that the various substitutes are highly interchangeable, thanks to the fact that various crops (maize, sugar) and energy sources (coal, natural gas, petroleum) yield the same crucial intermediary chemical substances such as ethanol and methanol. This makes production less dependent on specific raw materials (especially petroleum). The production of single cell protein (SCP) in the former Soviet Union is a case in point. SCP can be produced from diverse materials, ranging from petroleum to agricul-tural crops and waste. SCP itself serves as a substitute for fodder crops, which will probably be imported less in the future.[11] The new potential of agricultural crops to be used as a substitute for other raw materials will lead to far-reaching industrialisation in the agricultural sector.

A third reason is that biotechnological methods can help mine mine-rals which otherwise would have remained unreachable. Petroleum can be extracted using microbes. Copper too can be extracted by microbes through a process known as 'mineral leaching'. The vast quantities of this and other minerals such as cobalt, nickel and platina which are located in sites that are normally difficult to access makes this method particularly attractive, especially as it requires little en-ergy. It could lead to a more efficient use of local minerals. In addition a fourth incentive can be mentioned. Biotechnology can be applied in various ways to clean up polluted areas and contaminated water or to counteract dangerous substances, which is vitally important in con-nection with soil pollution. Anaerobic microbial waste processing, water purification and biological decomposition could provide answers to the environmental problems of the mass-production industries.

A fifth consideration is that the chemical industry uses energy on a mammoth scale, clearly a liability in view of unstable oil prices. As most biological processes take place at relatively low temperatures (between 20 and 80 degrees Celsius), they could lead to tremendous energy savings.

A sixth incentive spurring the chemical industry to look for new avenues is its negative public image. In the past few years, the public protest against chemical environmental pollution has gathered momentum. Founded on mass-production, the Fordist method produced belched mass waste. But the consumer pattern also underwent a change during Fordism, symbolised in plastic packaging and disposable products. Mass-production not only dominated industry, but also encroached on arable farming and stockbreeding. Crops and soil were overexposed to chemical treatment which not only caused soil and water pollution but also meant that industry had to get rid of its toxic waste, dumping it in the sea or at illegal tips. Recent years saw the introduction of stringent legislation aimed at safeguarding and preserving the environment. These regulations mainly put an additional financial burden on industry, reducing profit margins. Biotechnology could supply an environmentally more compatible process through closed fermentation systems. Furthermore, biotechnological decomposition techniques could help to revitalise the soil. These techniques are preferable as they are cheaper than burying and incineration and because they do not kill off all the organisms in the soil.

A seventh motivation behind biotechnological research is the wish to grow more agricultural and market-garden products closer to the consumer markets, irrespective of climatological and soil conditions. When the United States faced severe erosion following decades of unbridled use of artificial fertiliser and agricultural chemicals, they had to adjust the crops in order to save their domestic agriculture. This led to research into plants which would be able to resist heat, drought, saline soil, dampness, cold and other climatological and soil conditions that impede growth. The resulting applications of this research, still very much in the laboratory stage, will put the industrial countries far ahead of the developing countries. It would be theoretically possible for the Netherlands to cultivate cocoa beans. In that case, developing countries could well find themselves no longer able to sell their produce to the West, already catered for by, for example, the Dutch.

Another aspect of biotechnological research into agricultural applications has focused on making plants resistant to insects. This could be realised by treating plant cells with monoclonal antibodies. There are also biological ways of fighting insects. Though slightly less effective, they do much less damage to the environment. Biological herbicides are also being studied, which are intended to counteract the current erosion brought on by previous herbicides. At the moment, many farmers cannot cultivate different successive crops on a particular patch of land as the herbicides they use render the soil unfit for other crops. Biological herbicides could permit a more intensive use of the land. Biotechnology can also provide a kind of built-in fertiliser by ensuring that little balls of rhizobia start growing out of the plant's roots, supplying it with sufficient nitrogen. Reducing the need for artificial fertiliser, chemical pesticides and herbicides by genetically manipulating crops would cut the overall use of energy in the chemical industry as well as in agriculture and market gardening, while benefiting the quality of products and the environment. This tendency to keep enhancing results with less means can be described as dematerialisation. Dematerialisation stems from a better understanding of the functioning of an organism, which leads to optimal use of scarce or expensive inputs, greater efficiency and less waste. As was mentioned above in the description of protein engineering, biotechnology offers tools and techniques to stimulate developments of 'lean production' at microcosmos level. This tendency is also evident in other fields of application. The sweetener industry, for instance, has in the past few years developed new sweeteners, such as aspartame, monelline and thaumatine, which are much sweeter than traditional sugar. As this tendency is also noticeable in microelectronics and the new materials, it will eventually culminate in a more compact production chain. The realisation of all these possibilities depends largely on the ability of the agrochemical industry to keep developing biological substitutes or other products. The seed industry, especially, which is at the basis of genetically engineered agricultural applications, will be a decisive factor in the agrochemical industry's bid to control the entire manufacturing process from crop-growing through to food processing. An eighth reason springs from the fact that the life cycle of several important pharmaceutical and agrochemical products is drawing to an end. Before long, a number of key patents of important pharmaceutical and agrochemical industry products will lapse. For that reason, the pharmaceutical and agrochemical industries are eager to experiment with new technologies which could improve or even replace the existing techniques. Its high profit margins give the pharmaceutical industry plenty of scope to experiment at an early stage.

Another motivation, number nine, which can be mentioned in this context, is the prospect of developing new products with an even higher added value by means of biotechnology. This development is obvious in the pharmaceutical industry, but also in agriculture. Biotechnica International, of Cambridge, Massachusetts, has produced genetically engineered alfalfa seed. Alfalfa, mostly used as cattle fodder, is America's fourth largest crop. By using genetic engineering to boost the protein content of alfalfa seed and build in resistance to pests, the company can ask a far higher price for this genetically altered seed than for ordinary seed. The company reckons it can sell its alfalfa seed to the farmer at $ 12.30 per pound, a premium of $ 10 over ordinary seed. According to Biotechnica the farmer will still make savings through a combination of increased nutritional value, reduced pesticide and fertiliser use and increased crop yields. The biotechnology firm will make an estimated 88% gross margin on the sale of its genetically engineered seeds, similar to the sorts of margins reported in the drugs industry and nearly double the 46% gross margins on the scale of ordinary alfalfa seeds. If this development should become the trend in other crops it will lead to an increase in input costs from an estimated 7% to up to 20%. However, much will depend upon the environmental regulation of biotechnologically engineered seeds and the degree to which farmers can be persuaded to buy much higher priced seeds.[12]

To end our list with motivations behind the research activities in the area of biotechnology, a tenth consideration can be brought up. In most industrial countries the national health-care systems have come under pressure from the increasing costs for medical treatment of mass-diseases, like cancer, diabetes, heart and vascular disease, as well as high costs for the mentally and physically handicapped. Many of these diseases are related to the dominant work and living circumstances in the Fordist era after the Second World War. Biotechnology could perhaps provide some answers to the treatment of these diseases. In the United States, large funds were pumped into cancer research which naturally stimulated research in this direction. These private and government funds were made available because cancer, like heart and vascular disease and diabetes, has become one of the major killers of our time, consequently costing the economy and society large sums of money. The same 'cold calculation' can be made in relation to the disabled, especially for those born with a more or less serious handicap. Prenatal diagnostic research can establish as early as the tenth week of pregnancy which, if any, genetic defects the foetus is suffering from. Should this test not be performed and the baby turn out to be seriously handicapped, the care of the child will cost the state (not to mention the National Health services and insurances) an enormous amount of money over the first ten to twenty years. It follows that there are clear economic incentives to carry out a certain type of genetic research. The pharmaceutical industry achieved quick results through focusing experiments mainly on well-known, simple enzymes. Products such as insulin, interferon, mono-clonal antibodies, growth hormones and new diagnostics are already on the market and are accounting for the largest marketshare of the biotechnology market.

Although this list of ten motivations behind the stimulation of bio-technological research is certainly not exhaustive, a combination of some of these reasons has prompted a wide variety of companies to get involved in biotechnology. The above mentioned dynamics and the economic impact of the limitations of current products and pro-cesses explain governments' interests in biotechnology and explain why governments have actively stimulated biotechnological research.

Biotechnology and Market Outlook

If we take a look at the markets for biotechnology we see that since th mid-1980s growth in biotechnology has been steady rather than spec-tacular, although sales revenues are expected to increase by a factor of more than seven in the mid 1990s, and more than 20 times over the next ten years. According to Steven Burill and Kenneth Lee of Ernst & Young, in 'Biotech 91: A Changing Environment', the indus-try revenues from product sales have increased to $ 2.9 billion in

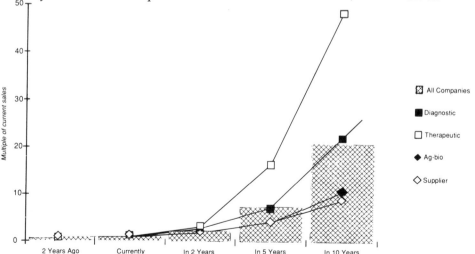

Figure 5.3 Projected sales growth by market segment (US firms)
Source: Steven Burill and Kenneth B. Lee, Jr, Biotech 91: A Changing Environment. Ernst & Young, San Francisco, California, 1990, p. 30.

the United States.[13] Particularly in the field of therapeutic drugs sales have increased. Genentech's tissue plasminogen activator (t-PA) (blood proteins) achieved sales of $ 196 in 1990. Another successful blood proteins product is Amgen's erythropoeitin (EOP), which reached nearly $ 200 million in the first year revenues. Human growth hormone, also developed by Genentech, is a $ 123 million product. Therapeutic companies project sales to increase by a multiple of 17 in the period 1990–1995 and even to more than 50 times the level of current sales by the end of the decade. Although one may have one's doubts about these forecasts, according to the above quoted Ernst and Young survey these estimates are quite logical. The therapeutic companies are relatively young and most of them, currently, have minimal sales.

Growth is less spectacular in other market segments of the biotechnology sector in the US. While the sales-growth projections of diagnostic companies mirror the industry averages, suppliers and biotechnological firms in agriculture lag (ag-bio; focus on plant genetics and the development of microbial pesticides and herbicides; companies applying diagnostic kits, enzymes in the processing of agriculturally-based food products). Suppliers were the earliest segment of the industry to commercialise manufacturing raw materials and equipment for the other segments of the industry. Ag-bio companies by and large are building their sales and distribution networks with traditional products – seeds, foods, and speciality chemicals. Their genetically modified products are still being tested and evaluated by regulatory authorities. These companies are just at the brink of their real commercial lives.

It should be noticed that biotechnology in the United States is very much 'human health' dominated. However, the developments and trends in the United States are an indicator of worldwide trends. The highest concentration of biotechnology firms can be found in the United States. Moreover, large European companies involved in biotechnology have set up laboratories in the US. European companies in this field are closely tied to American companies by the many joint-ventures that were concluded in the 1980s between American and European companies, as well as by take-overs of American specialist biotechnology firms by European companies.

In Europe the pharmaceutical and chemical industries also dominate biotechnological research, but the research activities are, more than in the US, spread out over other application areas, like agricultural biotechnology, particularly in plant sciences, but also the food-manufacturing industry, as well as research in the field of biological waste treatment. The European Commission calculates that the growth in sales of products derived from biotechnology (not includ-ing food and drinks) would increase from ECU 7.5 billion in 1985 to ECU 26–41 billion in the year 2000.[14] The global market for biotechnol-ogy-based products of ECU 5 billion ($ 5.5 billion) in 1991 could grow to ECU 80 billion by the year 2000, according to the Senior ...dvisory Group for Biotechnology (SAGB), part of the European chemical in-dustry federation (CEFIC) and one of the most influencial biotechnol-ogy lobby organisations in the European Communities.

Biotechnology: A Revolutionary Technology Hoping for an Evolutionary Take-off

The versatility of the new core technologies, and biotechnology in particular, described in the preceding paragraphs may have created the impression that the new technologies are capable of solving the restrictions inherent in the Fordist production mode. This is only partly true. Each of the three new core technologies can help effect the transformation from a rigid volume-oriented production concept to a concept characterised by flexibility, adjusted production scale, quality, built-in intelligence, miniaturisation and dematerialisation, compatibility and personalisation. Biotechnology will certainly play an important role in the transformation of important economic sec-tors, particularly agriculture, food manufacturing, the pharmaceutical industry, the chemical industry and the health-care sector.

But although new core technologies can provide an answer to many of the technical limitations of Fordism, technology itself can never be a cure for a concept of control in decline, for the problems of Fordism are not purely technical, but are related to changes in the political–economical framework, as well as the changing norms and values in the broader social context and the environment. Therefore, in order to find an answer to the control problems of Fordism we need an integrated innovation concept which centres on cohesion between the industrial–economic sphere and the socio-institutional context. Voices advocating increased attention for these kinds of innovations have been gaining strength. In an OECD report, *New Technologies in the 1990s. A Socio-economic Strategy*, one of the recommendations of the Group of Experts was:[15]

We stress the need for a long-term socio-economic strategy for new technologies. By this we mean a set of interrelated policies which take into account that neither the technical, nor the economic potential of major new technologies can be fully realised without concomitant, even anticipatory, social and institutional changes at all levels.

In this respect it is interesting to mention a recent study which has been carried out in line with the recommendations of the OECD study with regard to the importance of socio-economic aspects in relation to biotechnology.

In a recent study to the barriers to the diffusion of biotechnology in the food industry we made an international comparison between the Netherlands, Denmark and Germany. [16] Aim of the study was to find out which factors affected the diffusion of biotechnology in the different countries.

From the numerous national reports and the over 250 interviews with experts held for this study it appeared that important steps have been taken over the past few years to develop biotechnology in the food industry. However, the application of the research results had almost not yet taken place. It turned out that many factors may influence the diffusion process, but the main factors or facets are: commercial factors and market trends, public perception, regulation, and technological options. The most important barrier in all countries of our research turned out to be the public perception, followed by commercial and technological factors. Regulation was least import-ant, although we had expected to find this as a very important barrier. The facets are surrounded by all kind of stakeholders, public and private institutions. All exert influence on the facets. We have called this approach 'multifaceted'. The multifaceted approach has success-fully been applied to four basic products in the food industry: beer, cheese, potatoes and snacks, and veal/beef.

Biotechnology, like the other core technologies, is a tool, and some-times a very powerful tool for change, but the core technologies can never be the ultimate answer to complex problems of economic, pol-itical, social and institutional change. With R&D continuing to spawn fresh innovations and applications being introduced on an ever-larger scale, the need will increase for a new social consensus capable of providing the basis for a new type of regulation or new control con-cept, like Post-Fordism. The social embedmentáof technology seems a prerequisite for a balanced development of the two sides of the new techno-economic paradigm. Without the social embedment of bio-technology and the acceptance by the public at large the biotechnol-ogy revolution will be halted before it has begun.

NOTES

[1] Annemieke J. M. Roobeek, *Beyond the Technology Race. An Analysis of Tech-nology Policy in Seven Industrial Countries*, Elsevier Science Publishers, Am-sterdam/Oxford/Ne York/Tokyo, 1990.

[2] Carlota Perez, Structural Change and the Assimilation of New Technologies in the Economic and Social System, in: *FUTURES*, October 1983, pp. 357–375; Christopher Freeman and Luc Soete (eds), *Technological Change and Full Employment*, Basil Blackwell, 1987, pp. 49–70; Christopher Freeman and Carlota Perez, *Structural Crisis of Adjustment*, in: Giovanni Dosi, Christopher Freeman, Richard Nelson, Gerald Silverberg, and Luc Soete (eds), *Technical Change and Economic Theory*, Frances Pinter, 1988, pp. 38–66.

[3] Christopher Freeman and Carlota Perez, Structural Crisis of Adjustment, in: Giovanni Dosi et al. (eds), *Technical Change and Economic Theory*. London/New York: Frances Pinter, 1988, p. 48.

[4] This figure was first published in: Annemieke J. M. Roobeek, De rol van technologie in de ekonomische theorievorming, Scheltema/Holkema/Vermeulen Publishers, Amsterdam, 1987, p. 149. See also: Annemieke J. M. Roobeek, *Beyond the technology race*. Amsterdam/New York: Elsevier Science Publishers, 1990, p. 35.

[5] Roobeek, 1987, pp. 131-51; Roobeek, 1990, p. 36–38

[6] Annemieke J. M. Roobeek, *Beyond the Technology Race*, 1990, p. 63; earlier published by the same author in a research paper for the Dutch Ministry of Economic Affairs in 1985.

[7] This definition is used by the European Federation for Biotechnology (EFB) and by the Commission of the European Communities, as well as in several national biotechnology programmes.

[8] Margaret Sharp, The New Biotechnology. European Government in search of a Strategy, Sussex European Papers, no. 15, Industrial Adjustment and Policy: VI, SPRU, University of Sussex, 1985, pp. 14–17.

[9] U.S. Congress, Office of Technology Assessment (OTA), New Developments in Biotechnology: U.S. Investments in Biotechnology – Special Report, OTA-BA-360, U.S. Government Printing Office, Washington DC, 1988, p. 166.

[10] European Commission, FAST Programme I, Brussels, 1982, p. 11.

[11] Annemieke J. M. Roobeek, Biotechnology in the Soviet Union, Working Document, European Parliament, Committee on Energy, Research and Technology, Directorate-General for Committees and Interparliamentary Delegations, Brussels, 23 January 1985, PE 95.434; Annemieke J. M. Roobeek, Biotechnology and grain production in the Soviety Union: their effects on East–West relations, in: *International Spectator*, Vol. 39, No. 4, April 1985, pp. 207–215; Anthony Rimmington, Single-cell protein: the Soviet revolution?, in: *New Scientist*, 27 June 1985, pp. 12–15.

[12] A Survey of Biotechnology, *The Economist*, 30 April 1988, p. 13 of the Survey.

[13] Steven Burill and Kenneth B. Lee Jr of Ernst & Young, *Biotech 91: A Changing Environment. Fifth annual survey of business and financial issues in America's most promising industry*, Ernst & Young, San Francisco, CA, 1990, p. 29. The data on product development and sales quoted in this paragraph are all derived from this survey, pp. 29–32.

[14] *European Chemical News*, 29 April 1991.

[15] OECD, *New Technologies in the 1990s. A Socio-economic Strategy*, Paris, 1988, p. 13.

[16] Annemieke J. M. Roobeek, Mariska de Bruijne and Jeroen Broeders, Barriers to the Diffusion of Biotechnology in the Food Industry. A study commissioned by the Dutch Ministry of Economic Affairs, University of Amsterdam, Amsterdam, 1993.

Part II

The Technologies

Introduction to Part II

In this section four of the main biotechnologies are examined: recombinant DNA, monoclonal antibodies (or hybridoma), cell fusion for use in agriculture-related applications, and bioprocessing. In this introduction a number of more general comments are made about the significance of the technologies.

The first three of these technologies are good examples of 'science-led technologies' in the sense that it was scientific activities, undertaken mainly in universities and government health-related institutions, that provided the initial input for the development of the technologies. However, it would be a mistake to see the causation as operating linearly from 'science' to 'technology' to 'industrial application' to 'economic and social impact'. Causal influences also operate simultaneously in the reverse direction. For example, the industrial application of biotechnology and its economic payoff have influenced the scientific agenda by throwing up puzzles and problems and by defining financial incentives indicating where further scientific research might be financially profitable. In this way research has been influenced in industry, government research institutions, and universities.

The above implies that the technology is never static but is constantly in a process of evolution, influencing the economy and society which are simultaneously influencing it. One of the important trends that has been evident in the evolution of the technologies has been the tendency to automation and the reduction of the tacit and experience-based elements in the technologies. As Walter Gilbert, for example, has put it: 'Biology has been transformed by the ability to make genes and then the gene products to order . . . all of these experimental processes – cloning, amplifying and sequencing DNA – have become cook-book techniques. One looks up a recipe . . . or sometimes simply buys a kit and follows the instructions in the inserted instructional leaflet' (Gilbert, 1991). In turn these technological trends further influence the scientific process.

While the main aim of this section is to introduce some of the major biotechnologies in an understandable way, the thrust of this introduction is to stress that the science and technologies being discussed here are not separable from the applications and impacts, which are discussed in later sections in this book, but interact very closely with them.

REFERENCE

Gilbert, W. Towards a paradigm shift in biology, *Nature*, 10 January 1991, p. 58.

6

Fermentation and Fermenters

Rod N. Greenshields and Harry Rothman

Introduction

Fermentation is the most mature area of biotechnology, and in many respects it might be said to be the most fundamental. Fermentation has been used as a productive craft technique for thousands of years to process foods, such as oriental fermentations, and beverages, and to produce a myriad other useful substances.

Fermentation technology has gone through several periods of major change: the development of the brewing industry in the nineteenth century, especially under the influence of scientific knowledge; in the first three decades of the twentieth century when fermentation-based chemical production, for example Weizmann's famous acetone process, and the production of bakers' yeast had a period of commercial success; and after the Second World War in response to the needs of the new antibiotics industry. There have also been major changes in the marketability of fermentation-produced products; the post-war growth of the petrochemical industry made synthetic methods of production more economic than most of the industrial fermentation processes for chemicals and almost caused the latter industry to disappear, although the reverse of this is now happening. The purpose of this very brief introduction is to remind the reader that fermentation technology over the last hundred years has been subject to several phases of major expansion and contraction caused by both scientific and economic influences. It is now entering a further expansionist phase in association with biotechnology, the influence of the Green approach, and the diminishing availability and increasing cost of petrochemicals. This on the one hand poses new challenges for fermentation technology, but on the other will in large measure determine the economic success of biotechnology. Current uses of fermentation products include a broad range of products in: bulk organic chemicals, fine organic chemicals, pharmaceuticals, energy production, food, and agriculture (see table 6.1).

Table 6.1 Current uses of fermentation

Production of cell matter, biomass, e.g. bakers' yeast, single cell protein

Production of cell components, e.g. enzymes, nucleic acids

Production of metabolites, i.e. chemical products of metabolic activity: including primary metabolites, e.g. ethanol, lactic acid; and secondary metabolites such as antibodies

Catalysis of specific, single substrate conversions, e.g. glucose to fructose; penicillin to 6-aminopenicillamic acid

Catalysis of multiple substrate conversions, e.g. biological waste treatment.

Terminology

One of the surprising features about public discussion of biotechnology is the tendency to ignore this fundamental role of fermentation. For this reason we need to discuss some terminological definitions. Of course, not everybody agrees what biotechnology is; nevertheless the following definitions may supply a starting point.

'Biotechnology – is the study of the commercial exploitation of biological materials, living organisms and their activities'. This being the case, it follows that biotechnological studies require means of studying living organisms that allow them to be grown under controlled conditions. Thus, we would argue that 'fermentation – is the biochemical activity of a microorganism in its growth, physiological development and reproduction, possibly even senescence and death'. Then *per se* a 'fermenter – is the container, whether conceptual or physical, which contains the fermentation'. The origin of the term is from the Latin verb *fermentare*, to boil; as in brewing where the yeast *Saccharomyces cerevisiae* causes a 'boiling' ferment of sugary solutions. A 'fermentor – is the operator of the fermenter or the controller f the fermentation'.

One way of determining whether or not someone is a biotechnologist is to see whether they use an 'organism growth box' – a fermenter

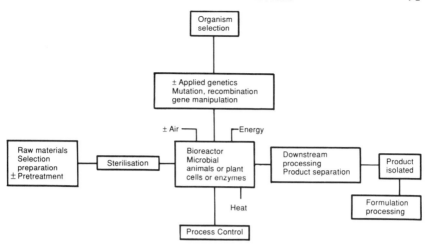

Figure 6.1 Schematic overview of a biotechnological process
Source: Smith (1984).

or a bioreactor! Until relatively recently biotechnology dealt with microorganisms such as yeasts, other fungi, and bacteria. Today that limitation is disappearing and the fermenter may contain all kinds of cells: plant, animal, human, insect, protozoa, algae, viruses. It may even contain parts of cells such as organelles or enzyme complexes. In addition, further variety to this vast array has been provided by the novel techniques of genetic manipulation which can increase the functional possibilities of cells and organisms. An understanding of what the fermenter does, its operation and the influence that it has on the living cells, or cell parts, that it contains is central to the development of biotechnology. Figure 6.1 shows this central role that the bioreactor or fermenter plays in a biological process. Each of the input and output, upstream and downstream processes, are currently being extensively studied to determine their optimal effect on the fermentation. Our understanding of fermenter design and control is now entering an exciting phase of development since biotechnology has at last combined such disciplines as genetics, microbiology, biochemistry and chemical engineering. Further, these are profoundly influenced, like many other disciplines, by advances in microelectronics and computerisation.

The Fermenter: Basic Concepts, Development, and Types

First, let us consider some quite basic concepts that have led to the development of the fermenter. On the definition that we have

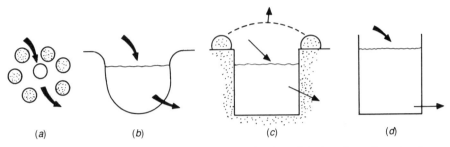

Figure 6.2 Development of fermenter design: (*a*) droplet of water in soil (wastes in soil); (*b*) ditch or pond (human waste disposal); (*c*) dug hole or ditch with bunds, sometimes covered (linen retting, human waste disposal, biogas production); (*d*) simple box made with plant material – leather, wood, metal, plastic (wine, cider, beer, alcohol, vinegar etc.)

provided it can be seen that the simplest form of fermenter is a drop of water in the soil or on vegetation (figure 6.2*a*). It has an integral shape and volume due to the surface tension of water; it is able to absorb nutrients and thus provide a haven for microorganisms, which are found everywhere. The resulting activities within the droplet fermenter would be a fermentation. Such a fermentation could be of many possible types; it could be aerobic or anaerobic, batch or semi-continuous or even continuous under some circumstances.

If we progress beyond the droplet we can imagine a range of natural environmental fermenters such as water-filled holes, ditches, or ponds (figure 6.2*b*). These are the oldest fermenters in the service of mankind. People have used them, and are still using them, in most parts of the world for many important tasks. For instance removal of liquid and solid wastes, and in the relatively sophisticated crafts such as the 'retting' of flax or cotton. When shallow, such fermenters are aerobic but they can be made to provide anaerobic and aerobic conditions (facultive aerobic) by deepening. If soil is dug out of the ditch to form higher sides or 'bunds' (figure 6.2*c*), then the hole may be closed over and a fully anaerobic fermentation can take place, this can be continuously operated to produce biogases. With the right proportions of methane (marsh gas) to carbon dioxide, this is a useful form of energy which can be conserved and distributed. In China there are many thousands of these biogas fermenters providing the local energy for cooking for families and even villages. In the Western world, refuse tips if sited in a sufficiently deep valley or hole, and properly covered over, can be drilled for the same biogas to supply local industry (up to four kilometres away) with a source of energy over a period of some 20 to 25 years. If we are being strictly accurate we can argue that this covered hole type, rather than the natural environmental fermenter, is the first true fermenter since some deliberation and control over the system is exercised to produce a product.

It is but a short step from this to the box fermenter (figure 6.2*d*). Such a fermenter is made deliberately and out of a wide range of materials whose historical lineage is probably plant leaves and seed containers (for example, the 'bottle' tree or gourd), wood, leather, later ceramic or glass, then metal and now from a whole variety of man-made plastics. In the Far East the ubiquitous plastic bag, which is used to hold almost everything including liquids, has been used for a variety of their fermentations. Indeed this is not restricted to the East for in Britain we have similar fermenters for home-brewing beer and wine-making. On a much larger scale, ICI have lined mine shafts with plastic to form 'deep-shaft' fermenters for treatment of urban wastes.

Box fermenters have been used for the classic fermentations of wine, cider, beer, alcohol, bread making, vinegar, yogurts, fermented foods, and silage. Such a fermenter allowed controlled fermentations and gave the opportunity for that wide range of skills and arts constituting a traditional or craft biotechnology to be developed into a science. Pasteur in the nineteenth century pioneered fermentation science and became known as the 'father of microbiology'. As a consultant to the French wine and beer industry he overcame the problem of fermentation sickness by quite simple, but logical and elegant, experiments using simple fermenters to demonstrate the cause of 'ferment' and how microorganisms were responsible for specific activities.

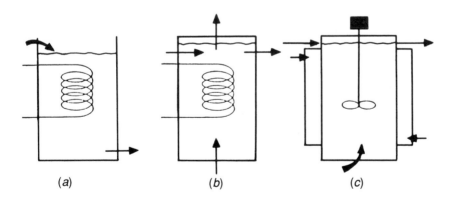

Figure 6.3 Development of fermenter design: (*a*) open fermenter, as in brewing, with temperature control (beer, wine, cider, alcohol); (*b*) sterile closed box with controlled fermentation (yeast, special biochemicals); (*c*) stirred tank fermenter (STF or STR fermenter) (antibiotics)

Control over temperature, aeration and, to some extent, the micro-organisms was possible; this allowed the sophisticated craft skills of such complex processes as the manufacture of koji and soy sauce and vinegar to be developed (figure 6.3*a*).

This type of fermenter remained in use for centuries until the latter half of the nineteenth century. Since then, under the interacting influences of commercial necessity and scientific advances, a continuing, though by no means even, flow of modifications have occurred in fermenter design and operation. The development of the science of microbiology, through Pasteur's work, brought the concepts of asepsis, pure culture and pure uncontaminated products. The organic chemical fermentations, bakers' yeast production, and war-time and post-war antibiotic production all made growing demands on fermenter design and operation (figure 6.3*b*). Increasingly precise control over the organisms, over pH, temperature, substrate and product sterility, concentration and composition were demanded and thus the box fermenter was closed. Economic incentives for scale advantages also forced changes and for certain tasks fermenters began to be operated commercially in semi-continuous and even continuous modes. Such changes have also necessitated advances in upstream and downstream engineering.

Fermenter Designs

As the fermenter came to be controlled with the confidence of scientific knowledge a fuller realisation of the potential of microorganisms and their biochemistry became possible. This led to a better consideration of the engineering aspects of fermenter design to ensure required levels of operation as one moved from laboratory-scale fermenters to large-scale production units.

The addition of stirring apparatus to ensure mass-balance and mass-transfer efficiency was an important engineering advance. The result was what is now the standard fermenter system, a stirred tank fermenter (STF) (figure 6.3*c*). Whilst there have been many modifications for ease of construction and maintenance, as well as improvements in peripheral equipment, the basic design remains almost the same today. Most microbiological studies have been made with this equipment and many commercial fermentations developed from it.

Despite its flexibility the STF does have certain disadvantages. It has to be carefully designed to give proper mixing and aeration; it has problems over dead spots; it tends to affect microbial morphology; it is difficult to scale up; and heat transfer can become a problem. Thus, we find that although there have been, over the years, few major

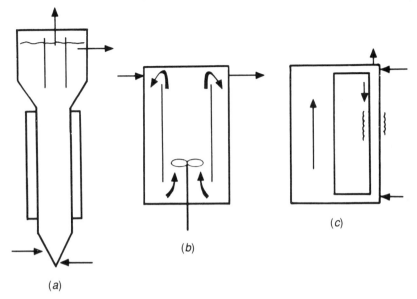

Figure 6.4 Development of fermenter design: (*a*) tubular tower fermenter, APV design (beer, wine, cider, vinegar); (*b*) internal recycle airlift fermenter (yeast from oil); (*c*) external recycle airlift fermenter (bacteria from methanol)

innovations in the STF, there have been many variations to accommodate particular fermentation requirements. For example, sequential STFs, vortex aeration, internal and external recycle, with and without baffles, aeration from above, aeration from below, stirring from above or below, pressure systems, etc.

The STF was principally conceived by microbiologists, but it was translated into hardware by chemical engineers. By the 1960s it was clear to biochemical engineers (chemical engineers who had specialised in fermentation) that the full range of engineering possibilities open to them was not being utilised. It was becoming clear to the bioengineers that the STF ignored all the potential of shaped reactors, fluidised-bed systems, differential recycling (figure 6.4*b*), air-lift systems and pressurised containment, all of which had been in common use for chemical reactions in the chemical engineering field for some decades.

Continuous Fermenters

Apart from some STFs which had been elongated in the antibiotics industry just after World War II and one or two simple blown aeration fermenters (as opposed to mechanically stirred and aerated units), perhaps the first commercial major change in STF design came with the APV system for the continuous production of beer (figure 6.4*a*). This design, envisaged by biochemical engineers, brought a number of new and interesting features to the concept of fermenter design and fermentation activity. Apart from the high aspect ratio of the fermenter (6:1 to 10:1), the time dimension of fermentation was converted to a space dimension. High microbial concentrations were possible with the concept of flocculent morphology and gave rise to fast continuous fermentations (reducing fermentation time from days to hours, even minutes). Careful microbiology was required to ensure a stable culture free of infection and achieve a self-maintaining population. Continuous fermentation in the brewing industry was a technical success, but it did not prove an economic success (except in New Zealand) since it was not applicable to the overall economics of the beer business. A major factor in this situation was that beer sales,

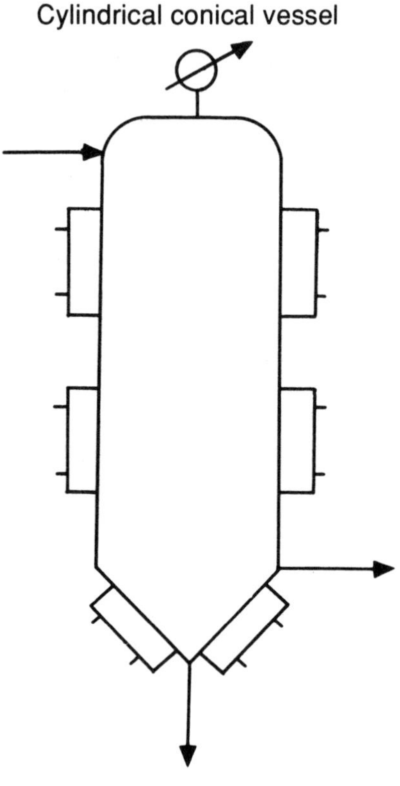

Cylindrical conical vessel

Figure 6.5 Nathan vessel: a semi-continuous fermenter now frequently used in brewing

and therefore production, are subject to seasonal variation, multiple types and units. This gives batch processing a distinct advantage since it is more flexible. A number of continuous fermentation processes in brewing have since been abandoned in favour of traditional and more flexible semi-continuous and high-speed batch systems. Nevertheless, the technical influence of continuous fermentation was dramatic and it will doubtless be applied in other fields, such as the production of fuel alcohol and biochemicals on a large scale.

It is now more common to find semi-continuous systems in brewing operations. These are equally as demanding as continuous systems, but suit the economic and product requirements more flexibly. One of these, in particular, the cylindrical conical vessel (CCV) or Nathan vessel (figure 6.5), is in widespread use and vessels with volumes up to 5, 000 bbl (180, 000 gallons or 800, 000 litres) have been successfully operated. The original CCV system was introduced in the 1930s but only came into its own in the 1960s; mainly because of the lack of communication between microbiologists and biochemists and their counterparts in engineering. There was, of course, no reason to; the common call of biotechnology had not emerged. Similar systems had been used in the aerobic fermentations for vinegar manufacture at an earlier date, but since this industry has only a small market it has had little commercial impact.

Tubular Fermenters

Tubular reactor concepts, in fermentation, have become more common since 1970 and have led to the stretching of many STFs used in a variety of fermentation processes (figure 6.6). An understanding of the fluidised-bed reactor kinetics has now given further opportunities in fermentation for recycle systems, airlift and pressurised airlift fermenters.

The main disadvantage of the STF was the problem of the introduction into the fermenter of the mechanical stirrer, this required complex equipment to overcome leakages and prevent contamination. More serious were the disadvantages involved in the dispersion of the mechanical energy introduced in the form of heat, and the problem of ensuring good mixing. Yet another consideration is the effect of the sheer forces of the mixing blade on the morphology of the organism. This was not significant with bacteria but became a crucial factor when fungal species were concerned. Since New Biotechnology involves the growth of plant and animal cells this factor has become pertinent to fermenter design. Moreover, since precise physiological control related to morphology is often needed to ensure an organism's

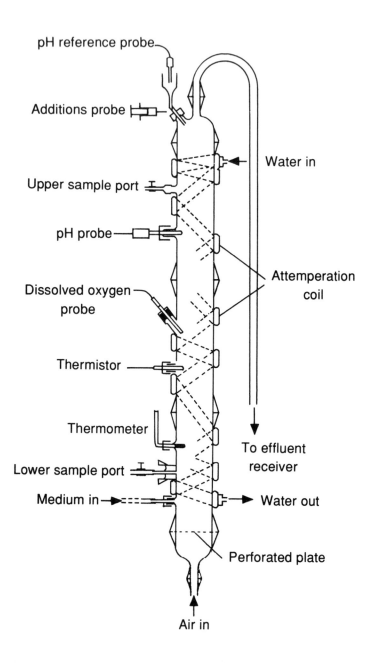

Figure 6.6 Continuous tower fermenter: Greenshields' design for continuous production of filamentous fungi. Malima process: agricultural or food process waste to fungal single cell protein

or cell's biochemical situation then this freedom for its environment will be required.

Perhaps the best example of this latter type of fermenter is the 1, 000 cubic metre continuous airlift fermenter developed by ICI for the manufacture of Pruteen (high protein bacterial biomass from methanol) (figure 6.7). A very high level of technical achievement was necessary to develop such a large and sophisticated fully continuous sterile-media fermenter able to operate with such a precise biological control. Amongst their technical achievements was the design of new sterile valves, special instrumentation and computer control using complicated software, and anti-foaming agents. Foaming, that occurs in most industrial fermentation, proved a major technical problem

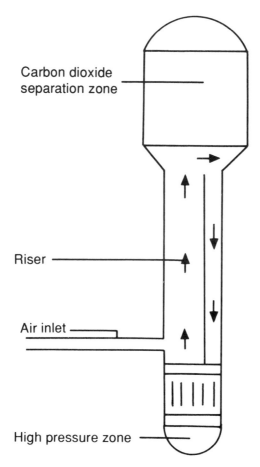

Figure 6.7 Continuous airlift internal recycle defermenter. ICI Pruteen process: methanol to bacterial single cell protein

because of the scale and because conventional anti-foaming agents were unsatisfactory. Many of them, for example, were unable to withstand sterilisation. ICI discovered and developed a novel anti-foaming agent, whose structure is a proprietary secret. ICI paid a high price to enter biotechnology on this scale: taking into account the research and development, the nutritional studies and plant construction, costs are believed to have been in the region of £150 million. Unfortunately for ICI, Pruteen has not been able to compete successfully in price with soya.

Some New Trends

The design experience of the Pruteen fermenter has served to underline the important fact that biotechnology now involves a multidisciplinary approach drawing upon the skills of both engineers and biologists, who have to learn a common language. This is emphasised by the considerable design possibilities now available for fermentation equipment. Despite the many scientific advances, fermenter design and development contains still a high degree of empiricism and 'hands on' skills. The new trends emerging in fermenter design are not confined to hardware. There are also important developments in terms of the design and control of the organisms' environment.

Apart from fermenter scale, the main emphasis of bioreactor design and development will be the use of unconventional shapes to combine simpler construction with efficient mass transfer. Nevertheless, it is unrealistic to expect that any single configuration of fermenter will satisfy the demands of different processes and organisms. Developments in genetic manipulation could create novel forms of organism with perhaps unusual degrees of environmental adaptation that would accommodate these situations, but conversely they could create sensitivities that would call for even more precise fermenter control.

It is feasible to forecast new fermentation units in a general manner by listing some of the situations which we believe they will have to be designed to deal with. Of particular interest will be fixed-bed or immobilised microorganisms which will enable better chemical-engineering design principles to be used in bioreactor operation, because under these conditions microorganisms will behave more like the catalyst spheres familiar to chemical engineers. Not only are microbes more robust in this condition and have longer 'shelf-lives' but they also have longer reaction life and shorter reaction times, coupled frequently with a greater resistance to extreme conditions. Moreover, the organisms can also be of almost any type, (plant or

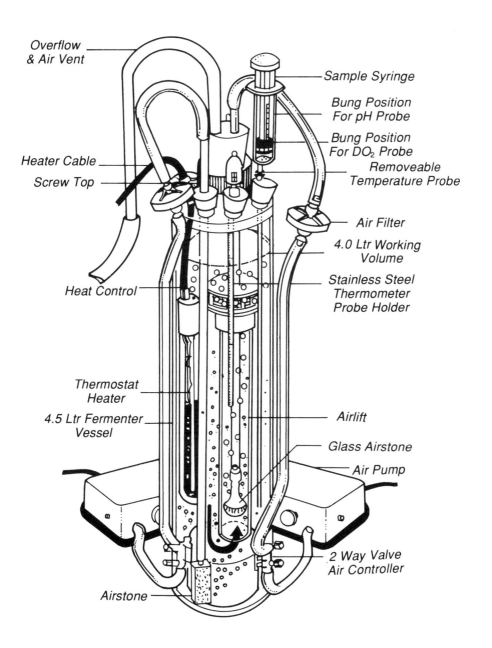

Figure 6.8 Industrial R&D laboratory fermenter (G. B. Biotechnology Ltd)

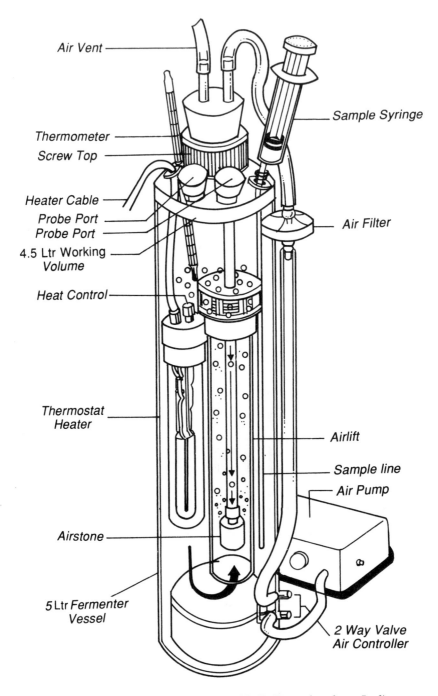

Air Vent

Sample Syringe

Thermometer

Screw Top

Heater Cable

Probe Port

Probe Port

Air Filter

4.5 Ltr Working
Volume

Heat Control

Thermostat
Heater

Airlift

Sample line

Air Pump

Airstone

5 Ltr Fermenter
Vessel

2 Way Valve
Air Controller

Figure 6.9 Schools airlift fermenter (G. B. Biotechnology Ltd)

animal cells) and not necessarily microorganisms (yeasts, bacteria or fungi). A final advantage could be that in a fixed bed, sequential reactions which are common in the living cell could be more readily controlled and accomplished, the classic example being that of steroid modification.

Perhaps one of the best areas which illustrates the value of modern fermenter designs is in research and education. The first attempts to build units for this requirement were merely miniaturisations of STF types, and whilst successful were complex and expensive. Using air-lift concepts and equipment borrowed from laboratory microbiology it has been possible to build simple but effective units which can illustrate the basis of the technology or be utilised in preliminary research work without the usual heavy investment in equipment. Figures 6.8 and 6.9 show one unit for industrial R&D and another for schools.

REFERENCES

J. H. F. van Appeldoon (ed.). *Biotechnology: a Dutch Perspective* (Delft: Delft University Press, 1981).
J. Bu'Lock. Process design and bioreactor choice, *Conference on Fermenters: their Impact in Biotechnology* (London, Nov. 1982).
R. N. Greenshields. Fermenters and biotechnology, *Conference on Fermenters: their Impact in Biotechnology* (London, Nov. 1982).
B. McNeil and L. M. Harvey. *Fermentation: A Practical Approach* (IRL Press at Oxford University Press, Oxford, 1990).
OTA. *Commercial Biotechnology: An International Analysis* (US Congress Office of Technology Assessment, OTA-BA281, Washington, DC, 1984).
J. E. Smith. *Biotechnology Principles* (London: Van Nostrand Reinhold, 1984).

7

Enzyme Technology[1]

Z. Towalski and H. Rothman

What are Enzymes?

Enzyme technology as a term is a problem to define.[2] Conventionally, when used generically it includes the production, purification and immobilisation of enzymes on the one hand, and on the other hand, when used specifically, enzyme technology describes the application of an enzyme within a particular process where it is used to catalyse a chemical reaction.

To understand the technology we first need to understand what an enzyme is and how it works. Today enzymes are described in terms of chemical catalysis, an ironic twist for in 1836 Berzelius crystallised his new collective concept of catalysis from scattered observations that included the decomposition of hydrogen peroxide by fibrin (or, as we now know, the enzymes within it). The quantitative verification of the catalytic concept was determined by Oswaldt in 1893 when he noted the difference between the transfer of heat or electrical energy and the transfer of chemical energy in chemically catalysed reactions.

The term enzyme was coined by Kühne in 1878, from the Greek for 'in yeast'. In so doing he sought to distinguish the enzymes' specific roles as catalysts from the general processes of fermentation that are carried out by the cell. For at that time it had been discovered that a cell-free juice could be obtained from yeast that would ferment sugar to alcohol and carbon dioxide, this is now called zymase (which is in fact a complex not a single catalyst). Previous to that it had been thought that fermentation was a vital mysterious property of the cell itself.

The development of the modern meaning of the term enzyme emerged from a number of independent threads of research concerned with unravelling the nature of the life processes. This development is best understood in terms of the emergence of an 'enzyme paradigm'. Thus investigations concerned with the nature of digestion, understanding the process of alcoholic fermentation, those directed towards developing the cellular theory and the chemical nature of life all became unified once the relevance of the enzyme paradigm gained general acceptance.

This enzyme paradigm itself emerged over a number of years during which the meaning of the term enzyme became refined.[3] Thus the concept that enzymes were highly specific emerged from the work of Fisher, who in 1894 demonstrated that only sugars with similar optical activity to glucose would be fermented. The requirement of enzymes for co-factors, that is thermostable dialysable non-protein molecules that act as carriers which facilitate the transfer of chemical groups between molecules and without which certain biochemical reactions could not proceed, was demonstrated by Bertrand, although he described the co-factor as the enzyme and the enzyme as the co-factor! It was left to Harden to develop the modern usage of the term correctly. Croft-Hill showed in 1898 that the enzyme maltase was capable of catalysing the synthesis of maltose; and the concept of the reversible nature of enzyme catalysis emerged. Despite these advances it was not until 1909 before Sorensen showed that pH influenced enzyme catalysis and was able to explain away the erratic results of previously reported catalytic reactions. The fact that enzymes were proteins had to await the work of Sumner, who purified the enzyme urease and obtained it in crystalline form, thus meeting chemical criteria for purity. It then became possible to determine an empirical formula for an enzyme and show it to be equivalent to that of other proteins. The resolution of the complex three-dimensional structure of enzymes and the nature of the catalytic site had to await the pioneering work on proteins by X-ray crystallographers like Bernal. The lock and key hypothesis of the nature of enzyme action as postulated by Fischer was later modified by Koshland in terms of an induced fit reaction between enzyme and substrate. The need for knowledge about enzyme kinetics developed in response to the needs of enzyme users such as brewers; the work of Brown at the British School of Malting and Brewing laid the foundations for these developments. He showed that the rate of hydrolysis of cane sugar by the enzyme saccharase was independent of the sugar concentration over a wide range of concentrations, but dependent on the concentration of the enzyme. He also suggested that the enzyme and the sugar formed an enzyme–sugar complex that broke down into the enzyme and products, thus recycling the enzyme. He concluded that enzyme-catalysed reactions differed from chemical reactions in that the rate increased to limiting value when the enzyme became saturated by the substrate. Henri, working in Paris, reached similar conclusions, but also produced a mathematical model of the process. In 1913 Michaelis took up these ideas and developed the enzyme–substrate kinetic model further. In 1925 Briggs and Haldane showed that the Michaelis kinetics were a special case of a more powerful model of enzyme kinetics that they had developed. These studies provided engineers and enzyme technologists with a powerful tool for the quantification of enzyme-catalysed reactions when applied to production processes.

Enzymes, therefore, are proteins, thermolabile and highly specific. They can usually be extracted from the cells in which they are produced. To achieve their task they do not need to be in a high state of purity, and small quantities are able to act on and change large quantities of reactants, or substrate. Enzymes operate under a fairly narrow range of physical conditions, within which they show high affinity for their substrate. Like other proteins they possess complex three-dimensional structures, and they are usually distinguished from other proteins by possessing a marked cleft wherein lies their catalytic or 'substrate binding site'. The substrate on which the enzyme acts is induced into this site in a specific steric fit, so precise as to be analogous to a key and lock. This degree of precision partially explains why enzymes can work selectively on one substrate within a cellular environment containing many. Once an enzyme–substrate complex forms it lasts only long enough to allow the reacting groups to interact, then it breaks down to release products and enzyme. The freed enzyme is then available to repeat the process.

The Practical Value of Enzymes

Enzymes are useful as industrial catalysts because they are non-polluting and biodegradable. They operate in a condition of mild pH (4–8) and at normal temperature and pressure. They may be produced at relatively low costs in virtually unlimited quantities, and hold out the potential for reuse and extension of their operating range through immobilisation. Enzyme users therefore are able to reap the benefits of energy savings and reduced process costs. Their adoption also makes possible savings in fixed capital costs since enzyme-catalysed processes operate under milder conditions of pH, temperature and pressure than their chemical counterparts.

Enzyme-utilising processes, however, have their own particular problems. Their complex protein structure, so vital for their function, is a source of vulnerability making them susceptible to denaturation and inhibition even by slight changes to their physical environment or mild forms of chemical change. Enzymes are also susceptible to a range of poisons, and therefore require for their use a pure substrate and handling conditions. Many enzymes are only active in the presence of specific metal ions; such requirements vary from enzyme to enzyme and need to be determined for each enzyme individually.

Other problems with enzymes relate to the difficulties of being able to produce them in a form suitable for assay. This is made more complex in animals by the presence of enzymes catalysing the same reaction with similar ranges of molecular weights but differing by their amino-acid compositions. These iso-enzymes, as they are called, vary considerably in their properties, such as pH and temperature optima, and their reaction kinetics. The polymorphous state of these enzymes enables similar reactions to proceed under the different conditions in different organs of the body.

Other complications arise as some enzymes may require co-factors; these are thermostable, dialysable, non-protein molecules that act as 'carriers' to facilitate the transfer of chemical groups, and without which the reactions will not proceed. Co-factors are very costly to produce and may need to be regenerated. Further difficulties include the fact that upon being 'purified' enzymes often have poor keeping qualities. Some metal ions can enhance enzyme stability; others bring about the degradation of the catalysts. Salts are often added to stabilise commercial enzyme preparations. Alternatively, immobilisation may sometimes be used to stabilise the enzyme; however, this is always accompanied by some loss of enzyme activity. Although there are broad generalisations as to the range of operating conditions there are no clear-cut rules. For use of specific enzyme catalysts each individual case requires careful analysis of the conditions and process trade-offs.

Classification of Enzymes

Enzymes are named and classified according to the system developed by the International Union of Pure and Applied Chemistry (IUPAC) and the International Union of Biochemistry (IUB).[4] This classification uses three principles to arrive at a unique identification code for each enzyme. First, that single enzymes should end in the suffix 'ase' and enzyme systems containing more than one enzyme should be clearly identified as such. Second, enzymes should be classified and named according to the reaction that they catalyse. Third, that enzymes should be divided into groups on the basis of the reaction type that they catalyse. This system coexists with the pre-existing 'trivial name' classification, which because it is often shorter is often preferred for general use (e.g. maltase for alpha-D-Glucosidase).

There are four code numbers allocated according to class, sub-class, sub-sub-class, and serial number in the sub-sub-class respectively. For full details readers are recommended to examine the latest IUPAC–IUB volume on enzyme nomenclature. There are six main enzyme classes:

    EC1 – oxido-reductases
    EC2 – transferases
    EC3 – hydrolases
    EC4 – lyases
    EC5 – isomerases
    EC6 – ligases (synthetases)

The Development of Enzyme Technology

Over 2000 enzymes have been identified, of which only about 20 are manufactured on a large scale. In addition there are a few hundreds that have some commercial applications, as for example those used for genetic manipulation, i.e. restriction endonucleases, ligases and editing enzymes. Many enzymes remain to be discovered; calculations based upon theoretically possible amino-acid configurations for proteins with molecular weights in the enzyme range suggest up to 10^{1300} different possible combinations. Clearly this figure will be much smaller if we consider the nature of the catalytic site alone, but it should keep those protein engineers concerned with modifying enzymes busy for the foreseeable future. Today Nature is no longer the only source of potential new enzyme creations, synthesis or semi-synthesis of enzyme analogues opens up new possibilities of designer catalysts.

What then of the technology? The demonstration of enzyme activity associated with successful isolation or partial purification of an enzyme was often followed by a specific application of that enzyme. Many examples of these early uses are still with us today. Most of these belonged to the hydrolases group. Their early application is probably attributable in part to the fact that they were the active components in processes where plant and animal extracts or microorganisms were already in use, and partly to the fact that many of these enzymes functioned well outside the cell, i.e. were already sufficiently 'stabilised' to work in an extra-cellular aqueous environment. This made them relatively easy to prepare and use, and enabled purified enzymes to displace the empirically derived crude 'active' extracts. These developments were pioneered by individual entrepreneurs who identified a market need and developed an enzyme to supply it.

For example, Christian Hansen in 1874 produced a standardised rennet, Otto Rohm in 1917 developed an improved leather bate, and Leo Walterstein in the 1930s developed an enzyme for chill-proofing beer.

Table 7.1 provides a summarised listing of such early technological substitution by enzymes.

Table 7.1 Some processes in which enzymes have replaced other industrial methods

Process	Method Substituted
1. Brewing	Enzymes used to supplement some of the enzymes in malt.
2. Cheese Making	(i) Animal rennets replace the use of plant extract and bacteria used for clotting milk.
	(ii) Microbial rennets substitute for animal rennets.
3. Leather Bating	Enzymes replace the action of dog and bird faecal extracts.
4. Meat Tenderising	Enzymes supplement the action of natural cathepsins.
5. Starch Conversion	Enzymes replace acid hydrolysis.
6. Textile Desizing	Enzymes replace acids as means of removing starch from fibres which in turn replaced the soaking of fabrics in stagnant ponds (the 'rotten steep').

The growing ability to prepare enzymes from microbial sources freed the enzyme user from problems of limited or seasonal supplies by allowing the availability of bulk supplies. In addition they expanded the available range of products. Amongst the early pioneers, in 1891, were Calmette and Boidin in France, who developed a method of cultivating the fungus Amylomyces to produce amylases used in beer production. In 1914 Takamine, in the USA, transferred and refined the solid 'koji' fermentation techniques from Japan to also produce crude fungal amaylases, called 'takadiastases', which eventually found a use in the desizing of woven cotton cloth.

New uses began to emerge in which the unique properties of enzymes could be exploited. Such developments occurred, for example, in the fruit-juice industry. In the 1930s German and American merchandisers began to clarify fruit juice using pectinases. From this beginning a series of industrial applications has emerged allowing treatment of apples, stone fruits, citrus fruits, grapes, etc. Developments in processing have led to an increasing specialisation of pectin-digesting enzymes and also the application of amylases and cellulases to the processes. These enzyme developments formed an important component of the techno-economic process which enabled an enormous growth in the world fruit-juice market.

The development of enzymic clarification of beer provides an example of an existing enzyme product, a papain, being put to a new use. When beer is chilled a protein–tannin complex is formed which makes the beer go cloudy; pretreatment of beer with papain prevents this, allowing a clear cold beer to be marketed.

The next major development in enzyme technology was the adoption of the deep fermentation techniques for microbial enzymes production. The techniques were developed during the Second World War for large-scale production of penicillin. Vat culture enabled large amounts of microbial cells to be produced, cells that could be used for the production of enzymes. The progress of science brought a new understanding of the relevance of genetics and of the culture environment to enzyme production, gradually transforming it from an empirical craft into a technology with a sound scientific and engineering basis.

In medicine the availability of animal digestive enzyme preparations opened the way for treating people suffering from enzyme deficiencies. Boudalt was an early pioneer, producing and marketing 'pepsin' as a digestive aid. Other applications followed; enzymes were used to clear wounds, to lyse blood clots, as anti-inflammatory agents, and in anti-cancer treatments. The practice of using enzymes in diagnosis, either to provide simple evidence of tissue change, or as constituents of reagents, began in the 1960s and has grown rapidly. By combining immobilised enzymes with monoclonal antibodies, and their use with spectrophotometers, fluorimeters and micro calorimeters, automation of basic biochemical diagnostic tests has been achieved. In the pharmaceutical field the production of semi-synthetic penicillins by the enzyme penicillin acylase produces a penicillin precursor from a penicillin produced by fermentation. The precursor is then modified chemically to produce a so-called semi-synthetic molecule. This technological process is also used to produce semi-synthetic cephalosporins and may be applied likewise to other economically important antibiotics.

The use of enzyme electrodes that are highly specific for biochemicals has been advancing steadily since their development in 1966 by Updike and Hicks. Utilising different enzymes, these electrodes can be linked up to a variety of metered displays that measure current, voltage or resistance, and are used for the quantitative determination of a large range of substrates, e.g. glucose, urea, amino acids and alcohols. A large number of such electrodes are available, some have found commercial applications, such as the electrode developed at Cranfield for monitoring blood sugar levels in diabetics and monitoring the freshness of meat. Other developments are also possible, ATP-utilising enzyme-catalised reactions can be monitored by chemiluminescent means, or some reactions can be monitored by thermometric devices which can measure the temperature output of certain enzyme-catalysed reactions.

Contemporary Applications

There are today five distinct areas of enzyme applications:[5]

1. As scientific research tools; generally production is in very small amounts and distribution is through specialist scientific suppliers.
2. For cosmetic uses.
3. For medical diagnostic purposes.
4. For therapeutic use.
5. For use in industry.

Table 7.2 summarises some of these applications.

Less than 10 % of known enzymes have been practically applied so far. We have moved during the twentieth century from a situation where scientific knowledge trailed empirical practice to one in which application might be said to lag behind scientific knowledge. However, application doesn't follow theory in a deterministic fashion. The application of new knowledge requires skill, ingenuity, intimate knowledge of an existing product sector and a will to try it out. That more use will be made of enzymes seems certain; whether they will be used to replace existing processes or to develop new ones is more problematic. In the past, major developments have occurred largely under the influence of strong market demand, this in turn is not an autonomous process and can be influenced by the availability of technical knowledge. The history of innovations in enzyme technology resembles technological innovation in general in as much as it is a reflection of the interaction of market-demand pull factors and science push. Enzymes have been developed to clean up industrial wastes; for example, ICI has a product, cyclear, based on extracted and freeze-dried fungal mycelia which use the enzyme cyanide hydratase to convert cyanide to formamide. The Dutch company DSM has developed an enzyme reactor using the enzyme urease immobilised in sand to treat waste waters from its urea plants.

Table 7.2 The types of applications to which enzymes are currently being put

Region of application	Sector	Specific areas of use with enzyme examples
i. Analytical Research & Genetic Engineering		Most enzymes available from suppliers at a relatively high cost, for unit activity.
ii. Cosmetic		a – dental hygiene (e.g. dextranase) b – skin preparations (proteases)

iii. Diagnostics		a – blood glucose (glucose oxidase)
		b – urea (urease)
		c – blood/urine alcohol
		d – cholesterol (cholesterol oxidase)
		e – blood triglycerides (lipase)
		f – blood CO_2 carbonic anhydrase
		g – urine steroids p-Glucoronidase
		h – EMIT & ELISA systems
		    EMIT – Enzyme Multiplied Immunoassay Technique
		    ELISA – Enzyme-Linked Immunosorbent Assay
		i – enzyme electrodes

iv. Therapy		a – anti-thrombosis agents (e.g. streptodornase) b – digestive aids (pepsin) c – anti-tumour treatments d – poison ivy treatments e – wound cleaning (trypsin) f – anti-inflammatory (super oxide dismutase) g – hypotension control (Kininogenase) i – anti-bacterial
v. Industrial	1. Food & Drink	a – brewing and wine-making b – baking c – dairy products d – fruit juice production (pectinase) e – extraction of other plant products (pectinase) f – production of protein hydrolysates (pepsin) g – modification of toxic or unwelcome food components (melibiase) h – starch modifications (amylases) i – antioxidants or glucose removal (glucose oxidase) j – flavourings k – leather (proteases) l – sugar and confectionary (invertase) m – production of modified fats (lipases)
	2. Chemicals	a – detergent formulation (subtlisin) b – paper making (amylases) c – fuel alcohol (amylases) d – lacquer production (phenolperoxidase) e – amino-acid synthesis (proteases) f – inhouse modification of pharmaceuticals (ergosteroloxidase) g – immobilised enzymes used to obtain inhibitors
	3. Textiles	a – desizing cotton (proteases) b – degumming silk
	4. Waste Treatments	a – reclaiming wastes b – improving waste-treatment management

Enzyme technology has already entered a new era. Pioneering work by Klibanov showed that enzymes exhibit totally unexpected stabilities and reactions when they are placed in organic solvents; alterations in the polarity of the medium in lipases, for example, extends their stability to temperature ranges as high as 100 degrees C., whilst their catalytic activities appear to increase five-fold. The properties of using enzymes in organic solvents at low water activities have been patented by companies such as Unilever, who have developed processes for the modification of triglycerides; 1–3 dipalmitoyl-2-mono olein (POP), the major constituent of the relatively cheap mid-fraction palm oil is modified into 1–3 palmitoyl-3l (stearoyl 1–2-mono olean (POSt) and 1–3 disteroroyl 2-mono olean (StOSt), the main triglycerides in expensive cocoa butter, by a lipase under low water conditions. The Danish company Novo has recently reported the production of sugar fatty esters that work well in detergent formulations as surfactants. These can be produced easily by using a crystalline sugar mix with fatty acids in conjunction with their enzyme Lipolase (TM) with which Novo hopes to capture part of the Green market.

Regio-selective reactions of enzymes in organic media have triggered a great deal of research into new ways of producing speciality chemicals. The ability to recycle co-factors has been solved and three routes for their recycling have been successfully demonstrated on a laboratory scale.

Research into designing peptides for food additives, artificial sweeteners, and drugs has prompted the search for new approaches to their synthesis. Likewise the synthesis of biocatalysts by protein engineering is making progress. The targets for this are the design of enzymes for particular substrates and the desire to remove industrially disadvantageous properties that limit their applications at present. For example, poor temperature stability, inability to tolerate extremes of pH, reactive chemical reagents, and inhibition of certain enzymes by their product or substrate.

The recent realisation that the high specificity and binding ability of anti-bodies can be used to lock onto substrates and, given the right conditions, trigger a catalytic reaction has provided the impetus for research into so-called Anti-body enzymes or Abzymes and this research promises much. Commercial success in enzyme technology has been at the expense of the chemical catalysts; they have outperformed them consistently in terms of specificity and precision of chemically catalysed reactions. However, giant strides have also been made by the catalytic chemists to a point where catalysis is emerging as a science in its own right. The emergence of chiral crown catalysts threatens to take back some of the ground lost to enzyme technology. At present, therefore, we consider that enzyme-based technologies are poised on the edge of a new stage in their development, a stage which shows much promise.

NOTES

[1] This text is largely based on T. Godfrey and J. Reichelt, *Industrial Enzymology: The Application of Enzymes in Industry* (London: Macmillan, 1983). Z. Towalski and H. Rothman, *Enzyme Technology*, in S. Jacobsson, A. Jamison and H. Rothman (eds), *The Biotechnological Challenge* (Cambridge: Cambridge University Press, 1986) pp. 37–76.

[2] A. L. Lehninger, *Principles of Biochemistry* (New York: Worth, 1982).

[3] Z. Towalski, *The Integration of Knowledge Within Science, Technology, and Industry: Enzymes a Case Study* (Ph.D. Thesis, University of Aston, Birmingham, 1985).

[4] IUPAC–IUB, *Enzyme Nomenclature* (New York: Academic Press, 1979).

[5] G. M. Frost, *Industrial Enzyme Applications*, in R. Greenshields, *Resources and Applications of Biotechnology: The New Wave* (London: Macmillan, 1989) pp. 150–184.

8

Application of *In Vitro* Techniques in Plant Breeding

Gisbert A. M. van Marrewijk

Introduction

Plant breeding can be defined as the sum total of activities directed at the improvement of cultivated crops to meet human needs. Defined this way plant breeding started when men gave up their nomadic existence of collector and hunter and settled in agricultural communities. For a very long time 'plant breeding' has been performed on a sheer intuitive basis. Farmers selected plants (or seeds from plants) which were bigger, stronger, better adapted, and less vulnerable to pathogens and other stress factors. By repeatedly doing so, cultivated forms evolved from the wild species which only could survive by the protection of man, because in natural conditions they would not be competitive against their wild relatives. This process of 'taming plants' is called domestication.

After the redetection of sexuality in plants in around 1600, botanists started crossing experiments, and new variants, mainly interspecific crosses, became available. It took another 300 years, till the rediscovery of Mendelian laws in 1900, before plant breeding had a scientific base. By application of carefully considered selection procedures and making use of genetic, statistical, cytological and physiological tools, production potencies and many other characters of crops were greatly improved.

In the last 90 years the production level of all food crops has increased fourfold; an estimated 40% of the increase being due to 'genetic improvement' (plant breeding) and the remaining 60% resulting from improved cultural practices, including crop protection measures.

When in 1978 Melcherts et al.[1] published their sensational article on the 'pomato', an interspecific hybrid between the potato and the tomato, obtained after fusion of naked cells (*protoplasts*), there were many speculations about a 'breakthrough' in plant breeding. In fact it was nothing more than a further step in a process that had started some 20 years earlier: the application of *in vitro* techniques in plant breeding research. Since then several novel cellular and molecular techniques have become available, which, after some hesitation, are now readily adapted by the plant breeder as additional tools for the attainment of his aforementioned goal.

This chapter does not restrict itself to techniques which are usually included in 'biotechnological approaches', 'genetic engineering' or 'genetic manipulation' but covers the broader scale of *in vitro* techniques, ranging from vegetative propagation to molecular applications.

Vegetative Propagation *In Vitro*

Principally there are two ways of propagating plants: (i) by sexually originated seeds – *generative propagation*, and (ii) by somatic plant parts – *vegetative propagation*. All methods of vegetative propagation have in common that they produce completely identical offspring, generally called *clones*. Until recently vegetative propagation was mainly restricted to species with specific propagation devices like corms, bulbs, tubers, bulbils etc. In agricultural practice artificial vegetative propagation occurred in a number of crops through leaf or stem cuttings, graft scions, buds and other plant parts.

The restriction of all *in vivo* propagation techniques is that the reproduction rate is low. Anoher shortcoming of the system is that viruses contained by the mother plant are generally distributed over all the progeny. Well-known examples of this situation are citrus crops (*tristeza* virus) and potato (various viruses). This means that a complex control system is needed in production of propagation material in these crops. Since the mid 1980s *in vitro* propagation procedures have been developed for a great number of crop species using various plant parts (Pierik, 1987). Apart from the general advantages, such as rapid and disease-free propagation, there are specific benefits for the plant breeder:

a. Inbred lines, which are often difficult to propagate sexually because of inbreeding degeneration or the incapacity for self-fertilization (*self-incompatibility*) can be maintained in a vegetative way: cabbage crops.
b. Genotypes which cannot be reproduced unchanged sexually can be propagated *in vitro*. Many applications: interspecific crosses,

haploid, triploid and aneuploid plants, cybrids, male sterile plants, etc.

c. Whole plantlets can be stored at low temperatures in gene banks.

d. *In vitro* propagation furnishes the basic material for genetic manipulation procedures.

Essential for use in plant breeding is 'genetic stability' of the *in vitro* material. This forms the major limitation of the method. Various vegetative plant parts and tissues are used for divergent objectives: meristems, axillary buds, single nodes, stem explants etc. The ultimate form of vegetative propagation is single cell regeneration. This has been obtained in several crops, e.g. tobacco, carrot, endive, asparagus, rape and citrus. As single cell culture always passes a callus interphase the danger of genetic instability is enormous with this method.

Somaclonal Variation

The occurrence of 'mutants' amongst plants regenerated *in vitro* was observed already some 30 years ago. The phenomenon is called *somatic variation.* Usually a big part of the variation is of epigenetic or physiological origin and disappears after some time, but part of the variation results from genetic changes (*mutations*) in the cells or from expression of pre-existing somatic mutations. Along with abnormal chromosome numbers (polyploids and aneuploids), minor mutations affecting morphological or physiological traits may also occur. The frequency of somaclonal mutants is highly affected by the method of vegetative propagation but also by the genotype. High mutation rates are observed in such different crops as freesia, begonia, tomato, potato and sugarcane.

Though somatic variation is usually an unwanted side effect of *in vitro* procedures, it sometimes can positively contribute to the genetic variation within a crop by producing mutants which so far have not been obtained in other ways (spontaneous, chemical treatment, radiation). Already in 1959 mutants for tolerance against extreme temperatures have been selected in snapdragon (*Antirrhinum majus*). In sugarcane plants resistant to two important diseases, eye-spot disease, caused by the fungus *Helminthosporium sacchari*, and the Fiji virus disease have been selected.

Many research groups are working on inducing somaclonal 'plus' mutants of important crops. The attractiveness of the approach is that one can add one positive mutation without disturbing the genetic structure of an existing variety. Recently mention was made of tomato plants with improved resistance to *Fusarium* wilt disease and tobacco mosaic virus (TMV).

A major problem is how to select random positive mutants out of a mass of unchanged and inferior mutant plants in an efficient way.

Mutant induction *in vitro*

The cellular level also offers opportunities to induce and select for specific mutants. Variation can occur spontaneously or be artificially induced. Essential for the approach is the availability of a selecive agent in the culture medium. Several pathogenic fungi produce specific pathotoxins. By adding the toxin to *in vitro* material, sensitive cells or tissues will die or suffer and the insensitive material is selected. Exemplary is the case of Southern corn leaf blight in corn, caused by the fungus *Drechslera maydis*. In 1970 the US maize crop was attacked by a new race of the fungus, which was virulent on all hybrid varieties containing a particular type of cytoplasm called T(exas)-cytoplasm. T-cytoplasm causes male sterility and was generally applied in the female inbred line of hybrid combinations. In the hybrid its expression is counteracted by the presence of specific fertility restoring genes provided by the pollinator line. The epidemic spread rapidly over all the US maize-growing area and caused losses of some hundred million US \$. Several research groups have tried to disconnect the cytoplasmic male-sterility trait and susceptibility to the fungus toxin by growing callus tissues of T- plants on media containing *Drechslera*-toxins. In this way, indeed, toxin-resistant strains have been obtained. So far, however, all toxin insensitive mutants obtained are also male fertile, suggesting that male sterility and toxin susceptibility are pleiotropic effects of the same genetic factor. Improved resistance to pathogens in plants regenerated from toxin-treated material have been obtained in various host–pathogen combinations (table 8.1).

In the tabulated cases resistance has been shown to be inherited. Several other combinations provided resistant regenerants but so far have not been tested in the next generation.

The above method cannot generally be appied, as most pathogens do not produce pathotoxins. Another serious drawback is that resistance on the cellular level is no guarantee of resistance on the mature-plant level and vice versa.

In vitro mutant induction is also very helpful for obtaining 'marker' traits, which are essential for selection of cells changed by protoplast fusion or transformation. Especially antibiotics-resistant mutants have won great popularity. Some antibiotics (e.g. streptomycin and kanamycin) are toxic for plant or plants organelles (plastids). By adding antibiotics to the culture medium all cells (or tissues) will be negatively affected except the mutant cells which have gained insensitiveness to the antibiotic. So, selection for streptomycin resistance occurs on a medium on which callus normally turns green (because of chloroplast formation). As a result of toxification by streptomycin all calluses remain white except the mutant cells, which can easily be selected and transferred to a regeneration medium. In this case 'streptomycin resistance' is extranuclear (coded by the chloroplast DNA). This is very useful for the selection in protoplast fusion experiments.

Table 8.1 Host–pathogen combinations with improved resistance after exposure of *in vitro* material to pathogen-generated toxins.

Crop	Pathogen
rape	*Phoma lingam*
maize	*Helminthosporium maydis*
rice	*Helminthosporium oryzae*
celery	*Fusarium oxysporum*
tomato	*Fusarium oxysporum*
tomato	TMV (virus)
tobacco	TMV
tobacco	*Pseudomonas tabaci*
tobacco	*Alternaria alternata*
tobacco	*Pseudomonas syringae*

Haploidization

A *haploid* is an organism which contains half the 'normal' chromosome number. Haploids constitute a valuable tool for the plant breeder for various reasons:

(i) Haploids from diploid plants have only one chromosome set. This enables the direct induction and selection of recessive mutants (A→a); in (homozygous) diploids recessive mutations (AA→Aa) are shielded by the dominant allele.

(ii) Haploids are also useful for the production of inbred lines required for hybrid production; by duplication of the chromosome number a pure line is generated (aBcDEf———→aaBBccDDEEff). In the classical way this would take 5–7 generations.

(iii) Haploidization of tetraploids reduces the chromosome number to two sets. This facilitates breeding work and enables crosses with diploid relatives.

(iv) Haploids can also be used in cell fusion experiments, so restoring the chromosome number to the generally preferred diploid level.

Haploid production sometimes occurs *in vivo*. Well-known examples, exploited by plant breeders, are the cross between particular genotypes of the tetraploid potato and its diploid wild relative *Solanum phureja*, resulting in diploid, unfertilized potato embryos, and the cross between oats and *Hordeum bulbosum* which produces a hybrid

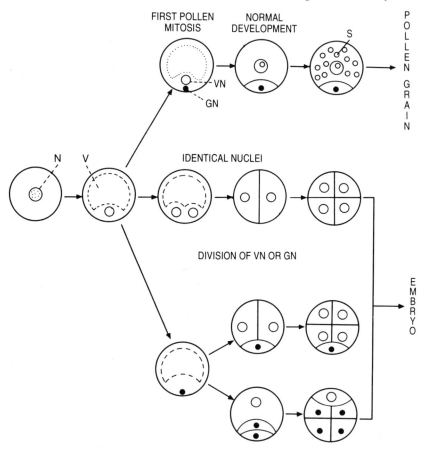

Figure 8.1 Different pathways of *in vitro* haploid formation from isolated microspore
N   = nucleus;   V = Vacuole;   S = Starch;
GN = generative nucleus; VN = vegetative nucleus.

zygote from which the *bulbosum* chromosomes are gradually eliminated, leaving only one set of oat-chromosomes.

Haploids can also be induced *in vitro*, the most common method being *anther culture*. The first successful regeneration of haploid plantlets from *Datura stramonium* anthers was reported in 1964. Since then anther culture has been successfully applied in about 250 species belonging to some 35 plant families, including the agronomically important and recalcitrant *Gramineae* (rice, wheat, barley, corn).

The effectiveness of anther culture is largely dependent on the genotype and the physiological condition of the donor plant, but especially on the stage of pollen development. This and the occurrence of diploid embryos originating from the diploid anther tissues is a real constraint on the applicability of anther culture. To avoid the latter the developing pollen grains are sometimes removed from the anther and grown as *pollen* or *microspore culture* (fig. 8.1). However, pollen culture requires a much more complicated growth medium and so far has been very difficult to incite to embryo formation. Successes have been achieved in a number of nightshade species, wheat, barley and cabbage.

A specific application of anther culture is found in asparagus. In this dioecious crop male plants are greatly preferred because of higher shoot production. Femaleness depends on the presence of two X sex-chromosomes; males are X–Y. So haploid plants derived from anther culture are either X or Y. By subsequent duplication we get homozygous XX and YY. Intercrossing of well-combining XX (♀) and YY (super ♂) lines produces completely male (XY) hybrids.

An alternative way to get *in vitro* haploids is *ovary* or *seed bud culture*. In this the unfertilized embryosac nuclei are stimulated to develop embryos by placing dissected ovules or ovaries on a culture medium. Though less effective than anther culture the procedure has been successfully applied in such different crops as tobacco, onion, wheat and rice. One further step is *embryo culture*. This is necessary when embryo abortion occurs in the ovule, caused by failure of endosperm development.

Embryo Culture

Embryo culture is the oldst application of tissue culture in plant breeding. This implies excision of the embryo from the seed bud, to be grown on an artificial medium. Embryo culture may be used to circumvent seed dormancy, to shorten the generation cycle or to prevent embryo abortion. The last application has proved to be successful in several interspecific cross-combinations with poorly developed or absent endosperm, e.g. in common bean, lily, flax, tomato and rice. Embryo culture is also used after crossing tetraploid × diploid parents, resulting in triploid plants with poor endosperm. The success of embryo culture is highly dependent on the stage of development of the embryo. Well-developed and differentiated embryos can easily be stimulated to germinate, but immature embryos must complete their embryonic development *in vitro* before they are able to form seedlings.

Sometimes embryo culture is used in an indirect way. In most gramineous crops, regeneration of plants from somatic cells is (nearly) impossible. As embryonic tissue has an increased regeneration capacity it can serve as protoplast donor. In this way regeneration of plants from embryo-deduced protoplasts has been realized in *Pennisetum americanum* (pearl millet) and in wheat.

Artificial Seed

Natural seeds consist mainly of an embryo and a nutritive tissue, the endosperm, encapsulated by one or more seed coats. In dry conditions seeds can be stored for one or more years. By water uptake the seed becomes metabolically active and starts germinating. Earlier it has been shown that in various plant species somatic embryos can be developed on a large scale from cell cultures. There is much speculation about the possibility of using these embryos for the production of 'artificial seeds' by encasing embryos in a nutritive and protective coating which is permeable and biologically degradable.

'Artificial seed' might offer a solution for large-scale propagation and storage of sterile offspring of wide crosses, and plants which by nature produce no seeds (bananas), or are propagated vegetatively (potato, yam, cassava). A further application is the direct reproduction of hybrid varieties, thus circumventing the need for maintenance of the parental lines. However, there still are a lot of snags connected with commercial application of 'artificial seed'.

To get uniform seedling emergence all embryos must be in the same developmental stage. This requires highly conditioned production procedures. Furthermore the embryos have to be resistant to transport, storage and sowing damage, contain the required nutrients to develop into a seedling, be permeable for moisture uptake, and protected against soil pathogens. Recently, microspore-derived 'artificial seeds' of barley have been produced consisting of embryos encapsulated in sodium alginate and maintaining their germinating capacity for over half a year.

Somatic Hybridization

As demonstrated with the example of the 'pomato' presented in the introduction, it is well possible to fuse cells of different plant species and to regenerate the hybrid. This process of fusing unrelated somatic cells is called *somatic hybridization*. It essentially differs from sexual hybridization in the following points:

(i)   With sexual hybridization between diploid parents both parents contribute a haploid gamete, resulting in a diploid descendant. Somatic cells are diploid, so somatic hybrids are tetraploid.

(ii)   At gamete formation exchange of genetic information *(recombination)* between homologous chromosomes occurs by crossing over between chromatids. Consequently gametes and embryos resulting from gamete fusion are usually genetically non-identical. Somatic hybrids are principally identical as long as no mutations occur.

(iii) With sexual hybridization the pollination in most cases only contributes a naked nucleus without organelles, whereas the maternal parent supplies a complete cell apparatus (the egg cell). Consequently extranuclear genetic information is maternally inherited. With somatic hybrids the cytoplasm of the two fusion partners is blended, inclusive of the mitochondria and chloroplasts.

The production of somatic hybrids requires four essential steps.

(i)   Release of cells from differentiated tissues and degradation of the cell-wall structures: *isolation*.
(ii)   Protoplast *fusion*.
(iii) Detection and separation of fusion products: *selection*.
(iv)   *Regeneration* of plants.

'Naked' cells or *protoplasts* are isolated by enzymatic degradation of cell walls in a routine way. Protoplast fusion is stimulated by neutralization of the medium with Ca^{2+} and low acidity (pH 9–11) and agglutination of the cells with polyethyleneglycol (PEG). This method has in recent years largely been replaced by *electrofusion*. With this, protoplasts arrange in long chains in an electric field generated by an alternating current. Via a direct current surge of high voltage the membranes 'melt' at the points facilitating their fusion.

Selection of protoplasts is a crucial step. After fusion treatment a blend of cell types, parental types, fusions between equal cells and the required heterofusions, results. Mechanical isolation using a micromanipulator is amenable when the heterofusions are visually discernible from parental cells and homofusions. The simplest way is combining green choroplast-containing leaf mesohyll protoplasts and

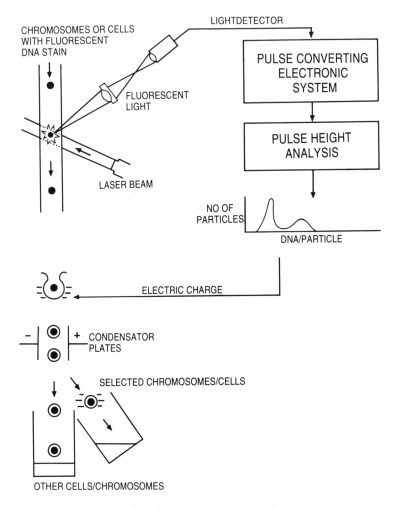

Figure 8.2 Principle of a flow cytometer with cell or chromosome sorter

The DNA of the particles (cells, chromosomes) is stained with a non-toxic fluorescent stain. Individual particles pass a laser beam and produce fluorescence signals of differing strength depending on the DNA content. The signals are electronically analysed. When a signal has a value belonging to the required particle the flow is electrically charged for a short while, resulting in deflection of the particle.

white protoplasts from cell suspensions, or labelling one (or both) protoplast types with a non-toxic fluorescent stain. The attractiveness of this approach is the possibility of segregating the fused cells with a cell sorter connected to a flow cytometer (fig. 8.2).

Selection is greatly simplified on a medium on which non-fused protoplasts die or are restrained in their growth. This can be realized by introducing selectable *marker genes* (e.g. antibiotic or herbicide resistances; amino-acid requiring or enzyme-defective mutants). By combining protoplasts of petunia sensitive to the antibiotic actinomycin D and those of the wild relative *Petunia parodii* tolerant to the antibiotic but deficient for nitrate reductase, only heterofusions will grow out on a medium containing actinomycin and nitrates.

The final step in the production of a somatic hybrid is regeneration of a plant from the heterofusion. For a long time this has been the most serious drawback as regeneration was restricted to only a few species. However, in the most recent years a great number of successful regenerations have been recorded. Till 1989 some 210 higher plant species, representing 96 genera and 31 families, have regenerated into embryo-like structures or even into plantlets. It has been found that plant families largely differ with respect to regenerating potency. With the exception of two *Pinaceae*-tree crops all regenerable species are Angiospermae and within that category *Solanaceae* (67 regenerating species) predominate. The agronomically important *Gramineae* and *Papillionaceae* appear to be very recalcitrant. Positive results have nevertheless been obtained in for example rice, sugarcane, soyabean, pea and alfalfa.

Regeneration of parental cell types is no guarantee of heterofusion regeneration. For some time researchers expected to overcome the crossing barriers by somatic hybridization unlimitedly. This is obviously not possible. Most regenerants are obtained from closely related fusion partners. So somatic hybrids are obtained from fusion products of *Brassica oleracea* (cabbage) and some other *Brassica* species (rape, turnip). Also fusions with the related genus *Arabidopsis* resulted in flowering regenerants. The earlier mentioned potato × tomato hybrids ('pomato'), however, regenerated only completely sterile plants. Interfamiliar hybrids reported so far have appeared genetically instable, i.e. they gradually lost one of the genomes and generally also sorted out their organelles. Genetic instability is also observed in many intergeneric and to a lesser extent in interspecific crosses. It has become more and more clear that somatic hybridization is no panacea to overcome intertaxon incongruity. However, in the majority of somatic combinations complete hybrids are not the final objective of a breeding programme, but fusion is used as a mechanism to realize genetic recombination either on the nuclear or on the organelle level. This situation, for example, is prevalent when one wants to introduce one characteristic of a wild species into a cultivated crop. These so-called *asymmetric hybrids* can be obtained by irradiation of the wild 'donor' genome prior to fusion. The donor chromosomes are rapidly eliminated but one chromosome may remain, giving rise to an *addition line*, or a DNA-fragment of the donor can be integrated in the recipient genome. A classical example is the introduction of some carrot traits into tobacco after irradiation of the carrot genome. In a comparable way it has been possible to restore nitrate reductase deficient tobacco by introgression of the NR-gene of barley.

Another goal of somatic hybridization can be to replace or hybridize only the organelle DNA contained by chloroplasts and mitochondria (*cybridization*). A popular case is 'cytoplasmic male sterility' (CMS). This character is of great importance for the production of hybrid varieties. Presence of the male-sterility trait inhibits the seed line from self-pollinating and so guarantees the pure hybrid nature of seeds after crossing with a pollinator line. CMS in most reported cases is located on the mitochondrial DNA. Several research groups are trying to introduce the trait into cultivated crops by somatic hybridization with related wild species or distant-relatives, e.g. in sugarbeet, cabbage and tomato. In these cases elimination of the (un-wanted) donor genome is activated by irradiation or centrifugation of the protoplasts (fig. 8.3). A major constraint in cybridization is the unpre-dictable and random sorting out of parental organelles, especially plastids. In some cases this has resulted in welcome novel organel–genome complexes, e.g. *Petunia* nucleus with tobacco chloroplasts. The most suc-cessful cybrid combinations have been realized in *Brassica*. Pelletier and co-workers of the INRA, Versailles, succeeded in restoring chlorophyll-deficiency in a cytoplasmic male-sterile (sexual) hybrid of radish and rape by replacing the radish chloroplasts by rape chloroplasts (in combination with radish mitochondria). In another fusion product resistance to the herbicide atrazine from turnip was incorporated into CMS rape.

Genetic Transformation

Genetic transformation, the incorporation of defined DNA- sequences (genes) into a receptor genotype is the most direct method of crop improvement because it is supposed to leave intact the structure of the genome, adding only one or a few new traits to existing commer-cial varieties. Whether this assumption is completely true has still to be proved. It is quite possible that the site of integration of alien DNA and the number of integrated copie will also influence the perfor-mance of the transgenic plants for other traits than the ones controlled by the new gene. Nevertheless, transformation offers a number of

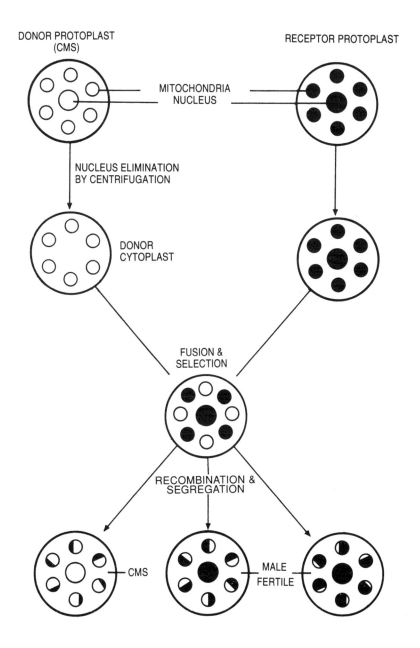

Figure 8.3 Transfer of cytoplasmic male sterility (CMS) by mitochondrial DNA recombination after protoplast fusion

attractive possibilities. In conventional breeding, required genes are often closely linked to undesirable traits and it takes much effort (several backcross generations) to get rid of this 'linkage drag'. Transformation with pure genes is much more efficient. Transformation is also helpful in exchanging genetic material between very distantly related organisms such as plant species and bacteria or fungi.

To realize successful transformation the following conditions have to be met:

(i)   The desired gene has to be isolated. This condition is the main bottleneck in transformation. Though some progress has been made in recent years the number of available genes is restricted.
(ii) A mechanism is needed to transfer the alien DNA into the recipient cell or tissue.
(iii) Transgenic products must be selectable at an early stage.
(iv) The novel information has to be expressed in the regenerated plant and in further generations.

Various systems for DNA transfer have been developed. The most simple approach is direct exposure of protoplasts to isolated DNA. The chance of hitting the target of course is very low as the DNA may degrade in the medium or somewhere on its passage through the cell into the nucleus. Using high dosages of DNA, workers of the Swiss Friedrich Miescher Institute succeeded in introducing resistance to the antibiotic kanamycin (a bacterial gene) into tobacco protoplasts. Though the method is not very efficient it is a good substitute for species which cannot be transformed with *Agrobacterium*, e.g. cereals and grasses. Recently, transgenic plants have been obtained by direct gene transfer in rice and sugarcane.

To prevent the alien DNA being degraded by nuclease-enzymes in the medium it may be presented as *liposomes*, i.e. in a coating of fatty substances. In this way successful transformation of tobacco with the kanamycin-resistance gene Km^r has been realized. An even simpler means of protection is co-cultivation of the protoplasts with the bacteria in which the alien DNA is amplified. After cell–bacterium fusion the alien DNA can enter the receptor cell and integrate. Transformation percentages up to 3% have been obtained with this approach in tobacco.

Microinjection

Microinjection of DNA directly into the cell might be an alternative in cases where direct DNA transfer is not amenable or isolation of protoplasts is not yet realized, e.g. in several gramineous crops. The method requires much labour and skill and the number of treated cells is low compared with other systems. On the other hand, by injecting many DNA copies into the cell or directly into the nucleus the probability of integration increases. Especially with nucleus-injection transformation, percentages over 10% have been realized.

Microinjection may develop into a promising tool for DNA transfer into microspores, germinating pollen grains and organelles. Especially the last application is challenging, because so far no techniques are available to incorporate alien DNA in extranuclear DNA.

Some spectacular successes with microinjection have been recorded already. In rape, transformants were obtained with an efficiency of about 50% after microinjection of embryoids. Even more appealing to the imagination is the production of Km^r-containing rye plants after injection of young inflorescences. To increase the number of treated cells new ballistic techniques have been developed known as 'gunshot' procedure or 'particle bombardment'.

Vector-guided Transformation

With direct gene transfer the alien DNA has to penetrate the host nucleus and to integrate in the genome autonomously. The major part of the DNA gets lost and the number of transformed cells is rather low. Gene transfer can become more efficient by connecting the alien DNA to a vehicle which transmits DNA to the host cells. Plant viruses look well-suited for transfer of genetic information, as they rapidly replicate. Besides, many viruses are very infective and easily spread through a whole plant. Consequently viral vectors would allow transformation on the plant level, so circumventing the often critical step of cell or tissue culture.

The main limitation for viral vectors is that most viruses contain RNA instead of DNA and therefore are not amenable as a DNA vector. Another drawback is that virus-transmitted DNA usually is not integrated in the plant genome but replicates autonomously with the virus. This implies that the novel genetic information is not expressed in the next generation as germ cells and consequently seeds are generally virus-free. This restriction is not applicable to vegetatively propagated crops like potato, sugarcane and bulb flowers. A positive point of autonomous virus replication is that the novel DNA is available in many copies and may come to expression massively.

Two plant DNA-viruses have been more extensively studied: *cauliflower masaic virus* (CaMV) and *gemini viruses*. CaMV and other caulimoviruses have only a limited host range; gemini viruses exhibit a wider host range and may be more suitable for general application. It is claimed that Km^r and a few other bacterial genes have been transmitted to cultured wheat and maize cells by way of the wheat dwarf virus, a member of the gemini family.

A CaMV vector has been successfully applied to introduce resistance to the toxin methotrexate into rape.

The Ti-plasmid

Many higher plants can be infected by the soil bacterium *Agrobacterium tumefaciens*. The organism causes cancerous outgrowth of wounded stem parts called 'crown gall'. Tumour cells are able to grow on a medium without plant growth hormones (auxins and cytokinins) and they produce characteristic compounds, the so-called opines, which normally are not produced in plant cells. Apparently the bacterium provides the plant cells with genetic information responsible for tissue proliferation and opine-production.

It has been shown that virulent strains of *Agrobacterium* contain a large plasmid along with the chromosomal DNA. This plasmid, an autonomously replicating and circular DNA fragment, is responsible for the tissue proliferation and is therefore referred to as tumour-inducing or *Ti-plasmid*. Tumour formation results from the insertion of a small part of the Ti-plasmid, called *T-region* (= transfer region) into the host genome.

In the integrated T-DNA fragment the so-called *onc*-genes which turn on tissue proliferation are localized, as well as the opine genes. The virulence genes, necessary to 'attack' and 'invade' the plant cells, are not located in the T-DNA but elsewhere in the Ti-plasmid. The Ti-plasmid also contains specific genes foropine digestion.

Research groups in different laboratories in Europe and the US have constructed various vector systems based on *Agrobacterium tumefaciens* and its relative *A. rhizogenes*, which causes excessive root formation ('hairy root' disease). Essential in all systems is the replacement of the proliferation genes by desired gene(s) provided with adequate regulatory sequences (promotor and terminator) which allow gene expression in the host plant (*chimaeric genes*). Though the opine genes might be maintained as marker genes, they are often substituted by new markers, e.g. antibiotic resistances.

The *Agrobacterium* vector systems have been made appropriate for co-cultivation with plant protoplasts, resulting in transformation percentages up to 10% and more, e.g. in tobacco. A much simpler method is incubation of leaf discs in an *Agrobacterium* suspension followed by a culture procedure for adventitious shoot induction. This procedure by-passes the often problematic protoplast culture but is only applicable with species that are able to regenerate into plants from leaf parts, e.g. tomato, potato, tobacco and other nightshades and some leguminous crops. The leaf-disc method has been used for the transfer of tolerance to glyphosate-containing herbicides into petunia, potato and tomato. This case and other resistances or tolerances to herbicides have found wide publicity because of the possible ecological risks (e.g. incorporation of the resistance in wild relatives), diverting attention from commercially even more promising results.

A clear example is found in changing starch composition in potato. Starch of potato consists of two main elements, *amylose* and *amylopectin*. Sometimes at starch extraction the components have to be separated, thus making potato-starch production laborious and expensive. Researchers of the Groningen University in the Netherlands succeeded in obtaining an amylose-free potato mutant (*amf*) and in isolating the controlling gene (fig. 8.4). Subsequently this gene has been retransmitted to *amf*-mutants, resulting in 'repair' of amylose production (*complementation*).

Figure 8.4 Tuber starch granules of amylose-free mutant in potato

In both tomato and tobacco, plants have regenerated with resistance to the TMV-virus on incorporation of the gene that codes for the viral coat protein. The phenomenon resembles the naturally occurring resistance to aggressive viruses after infection with mild strains (*cross protection*). The procedure is now being tried out by several research groups in the Netherlands and elsewhere to improve leading commercial potato varieties by adding resistances to various virus diseases.

A completely different approach has been chosen to transmit insect resistance. Larvae of several insect groups (butterflies, flies, mosquitoes) are killed after digestion of spores of the bacterium *Bacillus thuringiensis*. The crystalline spores contain proteins which are broken down into smaller, toxic particles during their passage through the intestinal tract. The company Plant Genetics Systems, in Ghent, Belgium, was the first to clone a *B thuringiensis* crystal gene and to transfer it into tobacco cells. Some of the regenerated plants appeared to be resistant to several caterpillars. Elsewhere insect-resistant tomato plants have been produced in a similar way.

RFLPs

Efficient selection is greatly hampered by the lack of knowledge of the number and location of agronomically important genes (AIGS). This is especially the case with polygenic traits such as growth rate, yielding capacity, protein composition etc. If the breeder could recognize the individual genes, directly or indirectly, selection would become much more easy.

Many genes are expressed only in a late stage of development, e.g. resistance to fruit-rot-causing pathogens manifests itself not until fruits have developed. This constitutes a further handicap for selection efficiency. The problem can be overcome by making plants with genes closely linked to AIGS which show themselves at an early stage.

In the past, *morphological* and *pigmentation markers* have usually been used. Disadvantages of this approach were the limited number of this kind of genes, and the late expression of marker traits. A step forward was the use of *isozyme markers*, based on the occurrence of different forms of enzymes. These diferences can often be observed already on the cellular level. A revolutionary step however was the development of a *molecular marker*, the so-called Restriction Fragment Length Polymorphisms or *RFLPs*.

The principle of RFLP-detection is shown in figure 8.5. Any restriction enzyme cleaves DNA on specific sites by recognition of specific base sequences. Mutations in the DNA (base substitution, inversions,

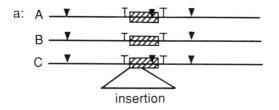

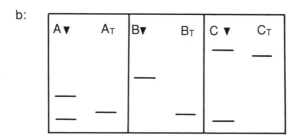

Figure 8.5   Principle of an RFLP

a. A, B and C are DNA fragments. A is the 'wild' type; B contains a base substitution and C an inserted DNA segment.
   ▼ and T represent the cleavage sites for two restriction enzymes.
b. Hybridization patterns after electroforesis and hybridization with a labelled probe ▨. Both enzymes show the insertion-mutation (C), but only enzyme ▼ displays the base substitution (B).
   (After Wagenvoort and Den Nijs 1988; Prophyta 42: 275.)

deletions) result in changes in cleavage sites compared with the original situation. These differences can be made visible by electrophoresis and hybridization with a (radioactively or otherwise) labelled DNA probe. In figure 8.5 the cleavage sites of two restriction enzymes are presented for the original ('wild') type (A) and two mutants (B and C). Mutant B lacks a cleavage site for enzyme ▼, resulting in a bigger DNA fragment and a higher position in the electroferogram (fig. 8.5b). Mutant C contains an insertion influencing the fragment size for both endonucleases ▼ and T.

RFLPs are commonly used in human genetics, especially in prenatal diagnostics, since as early as 1974, but their potential application in plant breeding was only put forward in 1980. RFLPs have a great number of advantages compared with classical genetic markers. Their number is principally unlimited – though large differences in number between plant species occur – they have a complete Mendelian inheritance, show no pleiotropic effects and are completely co-dominant and insensitive to environmental factors. Finally they allow early detection of characteristics which become manifest only at a late stage of development.

RFLPs can be used in plant breeding in various ways (table 8.2), which will not be discussed in detail. The most attractive aspect is the potential use in selection for quantitative traits. In various laboratories all over the world RFLP research is executed in, for example, potato, tomato, pea, cabbage, barley, maize, lettuce and onion. Some preliminary results of these efforts are linkae of RFLPs with resistance genes for the TMV-virus and the fungus *Verticillium* in tomato and with the *Cercosporella* resistance locus in wheat.

The main limitation of the RFLP procedure is the high cost connected with finding appropriate probe-restriction fragment combinations.

Table 8.2 Application of RFLPs in plant breeding

I *Identification*
– characterization of cultivars
– characterization of genotypes in a breeding programme
– determination of genetic stability of *in vitro* products
II *Indirect selection*
– localization of useful genes
– isolation of useful genes
III *Relationship analysis*
– analysis of relationship between plant taxa
– determination of genetic diversity in a breeding stock.

Anti sense RNA

A few years ago a new approach became available which is based on suppression of gene activity, coined *anti sense RNA* technique. The essence of the procedure is presented in figure 8.6.

The technique requires the availability of copy DNA of at least a part of the target gene in reversed sequence (*as*-gene). The *as*-gene serves as a template for the production of *as*-RNA. The *as*-RNA (which is complementary to the mRNA produced by the target gene) associates with the mRNA strand, resulting in a (partially) double-stranded RNA, and inactivation of the mRNA.

An advantage of anti sense constructs in comparison with mutants is the dominant expression of gene suppression; mutants are usually recessive and only become manifest in the absence of the dominant allele. Anti sense is also to be preferred to transformation as insertion of alien genes so far is completely random and may negatively affect other useful genes. Anti

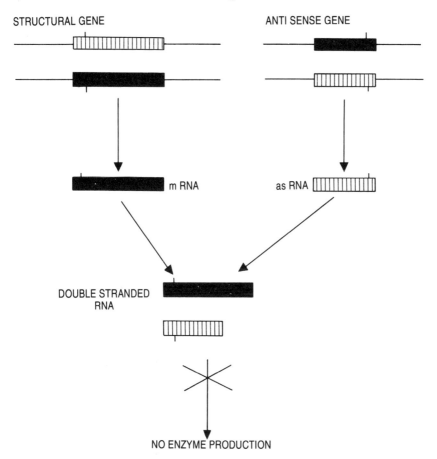

Figure 8.6 Selective suppression of gene expression by anti sense RNA

sense does not influence other genes whatsoever. It still has to be proved, however, that anti sense suppression is stable and complete.

The technique is applied in the aforementioned research project on modification of starch composition in potato. In this, amylose-free anti sense-modified plants can be compared with *amf*-mutants with respect to intensity and stability of expression in different plant tissues.

NOTE

[1]   Melcherts, G., M. D. Sacristan and A. A. Holder. 1978. Somatic hybrid plants of potato and tomato regenerated from fused protoplasts. *Carlsberg. Res. Comm.* **43**: 203–218.

136   G. A. M. VAN MARREWIJK

REFERENCES

Cocking, E. C. 1986. The tissue culture revolution. In: L. A. Withers and P. G.
   Alderson (eds), *Plant tissue culture and its agricultural applications*. London:
   Butterworths, pp. 3–20.
Eucarpia. 1989. Science for plant breeding. Procs XIIth Congress of Eucarpia,
   27 Feb.–4 March 1989. Göttingen. Berlin etc.: Paul Paney Scient. Publ.,
   477 pp.
Mantell, S. H., J. A. Matthews and R. A. McKee. 1985. *Principles of Plant
   Biotechnology. An introduction to genetic engineering in plants*. Oxford, etc.:
   Blackwell Scientific, 269 pp.
Negrutiu, I., S. Hinnisdaels, A. Mouras, B. S. Gill, G. B. Gharti-Chhetri, M. R.
   Davey, Y. Y. Gleba, V. Sidorov and M. Jacobs. 1989. Somatic versus sexual
   hybridization: features, facts and future. *Acta Bot. Neerl.* **38**: 253–272.
Pierik, R. L. M. 1987. *In vitro culture of higher plants*. Dordrecht etc.: Martinus
   Nijhoff, 344 pp.
Schell, J., B. Gronenborn and R. T. Fraley. 1989. Improving crop plants by the
   introduction of isolated genes. In: J. L. Marx (ed.), *A Revolution in Biotech-
   nology*. Cambridge etc.: Cambridge Univ. Press, 130–144.

9

Protein Engineering

Lindsay Sawyer

Introduction

Since the mid-1980s, there can be little doubt that the techniques of genetic engineering have revolutionised molecular biology in the broadest sense of the term. Further, the consequences of this revolution have barely begun to be felt, remarkable though those already achieved appear to us today. One example is the ability to produce a protein from one species in the milk of another, which has opened up novel approaches to the commercial preparation of clinically important molecules like Factor IX (Clark et al., 1987). This chapter will address three questions which should put into perspective the more detailed descriptions of some of the applications of the new biotechnology as they appear in later chapters. First, what is protein engineering? Second, how can it be achieved? Third, why engineer proteins? In answering these questions, definitions of some of the terms or 'buzz-words' will emerge, together with the current problems and unanswered questions which make the field such an exciting one in which to be working.

What is Protein Engineering?

Proteins are fundamental to all life processes. They fulfil roles as enzymes, the catalysts which facilitate essentially all biochemical reactions within the cell. They provide much of the structural scaffolding of an organism, whether this be a single-cell bacterium or a human being with a large number of different cell types, all of which have important, often complementary, functions. They provide the receptors on the cell surface to allow cells to communicate with each other;

protein hormones, like insulin, which control cell processes, interact with these receptors. Antibodies, which afford protection to the animal, are proteins, as are the molecules like haemoglobin and albumin which act as intercellular transporters. Tens of thousands of different proteins have already been identified and many of these have been isolated and their specific properties examined.

All proteins are polymers made form a pool of the same 20 amino acids, although in the mature molecule several of these basic building blocks may have been modified. The cell constructs a protein with a precise linear sequence of amino acids specified by (translated from) an RNA message which itself has been previously transcribed from the DNA in the gene. This DNA contains the genetic blueprint for the particular protein. Thus, the information flow is from DNA to RNA to protein. A particular amino acid sequence contains the information necessary for the protein to fold into its unique functional form with little or no external assistance, and as the folding process is spontaneous, the active protein molecule is a relatively fragile entity. The code by which the protein folds is redundant in that, within limits, similar structures can be derived from different amino acid sequnces and considerable, though so far unsuccessful, efforts are being made to decipher it fully (Anfinsen, 1973; Chothia and Finkelstein, 1990).

That the three-dimensional arrangements of the atoms in molecules are responsibl for the highly specific nature of biochemical reactions was realised nearly 100 years ago by the German chemist Fischer, who likeneáthe molecular interactions in enzyme catalysis to those of a lock and a key, with the enzyme being the lock and the substrate (or reactant) molecules being the key. In fact, since molecules, especially large ones like proteins, are not rigid and inflexible, a better analogy is a hand and glove. However, it is only relatively recently that the necessary techniques for protein structure determinatin have become sufficiently routine to permit the details of many specific molecular 'locks' to be revealed. To continue the analogy, a locksmith who knows the structure of a lock can describe accurately how it works, how to make a key for it and, within limits, how to modify it to work with a different key. At the molecular level, the protein engineer is trying to copy th locksmith.

Formally then, protein engineering refers to the modification of a protein to achieve desired changes in its structure and hence in its behaviour (Oxender and Fox, 1987; Rees et al., 1992). This can sometimes be achieved by chemical or enzymic modification of the purified protein but it is by genetic manipulation that both subtle and radical changes can be conveniently introduced. Since the 1980s, the techniques for isolating, sequencing, characterising and manipulating

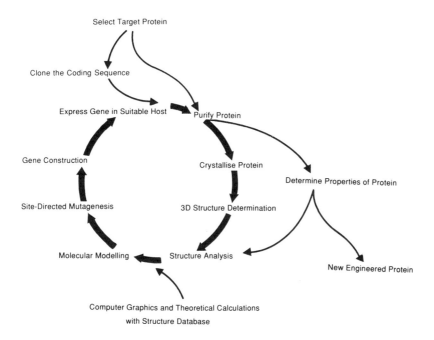

Figure 9.1 Protein engineering cycle

genes have become standard but are still being improved and/or automated. Thus, it is now possible to modify the amino acid sequence of a protein more or less at will. In order to engineer a protein or even to design *de novo* one with specified properties, the folding code must be understood and the detailed relationship between structure and function must be known so that any change or changes introduced will give the predicted result. Figure 9.1 shows the general strategy of the protein engineering cycle.

Having decided upon the target protein and characterised it as fully as possible, the stages in a protein engineering project are:
- to define the change in function to be made;
- to use the available information, which, as we have seen, must include a model of the 3-D structure, to determine how to achieve the required changes. This last process usually involves some form of computer modelling involving sophisticated interactive graphics facilities;

- to clone and express the gene or cDNA for the protein as a prelude to mutagenesis;
- to carry out the predicted changes to the cloned DNA, to express the mutant protein and to purify it; and
- to characterise the mutant protein to see that it conforms with the expected functional modifications.

How is a Protein Engineered?

Many texts now describe the details of what is involved (see for example, Brown, 1990) and excellent 'recipe books' are also available (e.g. Sambrook et al., 1989). Another chapter covers the details of the techniques involved. In broad outline, the DNA coding for the protein of interest is introduced into a piece of DNA, either a plasmid (an extra, non-chromosomal piece of DNA which is passed on from generation to generation, but which is also passed between bacteria – the spread of antibiotic resistance involves plasmid borne genes) or a bacteriophage (a small, bacterium-specific virus), which is then transported into a host cell where it is multiplied as the host divides, producing a clone of identical cells each with the extra piece of DNA. It is then necessary to grow up these cells to produce large quantities of the required protein, which is then purified. The presence of the gene in the cell is no guarantee that the protein will be synthesised (expressed) at all since even in the simplest bacterium not all of the available coding sequences are translated at any one time. It is advantageous if the cell secretes the protein into the surrounding medium, but in any event, it is often this expression stage which leads to a hold up in obtaining the protein in significant amounts for study and use.

The piece of DNA whose coding sequence is that ultimately expressed, is usually obtained initially from a library known (or at least, thought) to contain it. A library in this context is a random series of short lengths of DNA prepared from a given cell type, and selection of the particular gene from the library involves synthesising chemically a short, radiolabelled section of the coding sequence which can be used as a probe.

DNA is a double-stranded molecule each strand of which is made up of a linear polymer of four nucleotide bases, A, C, G and T, whose sequence contains some of the genetic blueprint for the organism. (Organisms like bacteria have one chromosomal piece of DNA, others like fruit flies, galinules or wombats may have several.) The property that A and T, and C and G, are able to 'recognise' each other by means of weak yet specific interactions is the basis of heredity – given one strand of DNA, the complementary one can be synthesised. Given both strands as in a chromosome, the cell separates and copies both halves, generating two daughter molecules each containing one old and one new strand. These processes of separation and regeneration can be carried out in the test-tube, where the separated strands can be recombined. Also, a small piece of the complementary DNA sequence (a synthetic oligonucleotide) will recombine with the complete single strand to form a hybrid molecule which can then be 'finished off' *in vitro*, thereby generating a complete, double-stranded, semi-synthetic daughter molecule. If the DNA fragment used to form the hybrid molecule is synthesised with radioactive bases complementary to those coding for part of the sought protein, only the radiolabelled daughter molecules will (or at least should) contain that coding sequence, allowing them to be identified and separated. Chemical synthesis of the radiolabelled fragment is now routine and automated but there is still a problem of choosing which sequence to make. This is because the genetic code is degenerate and, in most cases, an amino acid can be specified in more than one way. Thus, back-translating from part of a protein sequence is going to produce a family of possible DNA sequences. It is normal to try to use sequences containing the less commonly occurring amino acids, which have correspondingly fewer redundancies in the genetic code, to limit the size of this family. However, in this way it is possible to obtain a DNA clone with the coding sequence for the protein.

Modifications to this basic sequence are made by hybridising a suitable synthetic oligonucleotide sequence to a single strand of the DNA into which the coding sequence has been introduced. Although the pairing of A with T and C with G is highly favoured, if a single change is made in an oligonucleotide of around 20 bases, a mismatch can be accommodated and passed on to the next generation of molecules. It is also possible to rearrange pieces of DNA coding for parts of different proteins to make chimaeric molecules.

However, there are still problems associated with the technique. It is too easy to assume that a scheme devised for one protein will work when adapted for another: although the manipulations involved are all fairly standard, there is much still to be discovered. Thus, cloning the full coding sequence of the protein may require several attempts including synthesising different oligonucleotides chemically. Expression, as mentioned above, is by no means always clearcut and, even when the protein is over-produced by the cell, it may not be properly folded and may therefore be inactive. Purification may also pose problems, particularly if the protein is labile and only produced in small quantities. Thus, whilst it is straightforward in theory to clone a particular piece of DNA, in practice it may prove time-consuming.

To make rational changes to a protein, it is essential to have a 3-dimensional model of it – one which shows the relative positions of all of the atoms, inclding if possible the positions of substrate or ligand atoms. The main technique for determining such a molecular structure is X-ray crystallography, in this case usually referred to simply as protein crystallography (Blundell and Johnson, 1976; Wyckoff et al., 1985; Stout and Jensen, 1989; Rees et al., 1992). Recently, nuclear magnetic resonance (NMR) spectroscopy has been successfully used to solve the structures of small proteins (Wuthrich, 1989; Wright, 1989; Rees et al., 1992) but, whilst its use has the advantage of looking at the protein in solution, there is a theoretical limit to the size of protein which can be tackled. On the other hand, crystallography is producing accurate models up to the size of viruses and macromolecular complexes but, in every case, the sample must be a single crystal of the protein (Blow, 1988). Crystallisation is usually achieved by increasing the salt concentration of a concentrated protein solution whilst keeping the pH and the temperature constant – tens of milligrams of pure protein are likely to be needed for the full structure determination (McPherson, 1990). Without good quality crystals, perhaps 0.2–0.4mm in each dimension, the technique cannot be applied. To try to overcome this bottleneck, automatic methods are being applied to what is still essentially trial-and-error (Ducruix and Giege, 1992). Purification of the necessary quantities of material (Harris and Angal, 1990) can also pose problems if the preparation uses the natural source and that is why over-expression in a convenient host is important at this stage too.

It is the nature of physics that in order to resolve or 'see' individual atoms, radiation of the same wavelength as the interatomic separation must be used. This radiation is in the X-ray region of the spectrum but, unlike visible light, focusing of the X-rays scattered by the object to obtain a magnified image in the manner of a conventional microscope cannot be achieved by any lens. The refocused, and magnified, image must therefore be obtained by calculation. In order to detect the X-rays scattered by a protein, an ordered array of molecules as found in a crystal, is required. The crystal imposes restrictions upon where the scattered X-rays can be observed, and the crystallographer needs to move the crystal in the X-ray beam in such a way as to collect as many as possible of these scattered rays, called reflections. Each reflection consists of 3 indices which are derived from the position of the scattered peak on the detector and the orientation of the crystal, and an intensity which is related to the arrangement of the atoms within the crystal. The X-ray data for a protein crystal thus consist of several thousand (in the case of viruses, several hundred thousand) reflections each of which must be accurately measured. Computer-controlled, position-sensitive, proportional counters (area detectors) or photographic film, subsequently processed in a special scanner, are used routinely for this process and high-intensity X-ray sources called synchrotrons extend the amount of data which can be observed (Helliwell, 1992). In order to calculate the magnified image of the scattering matter in the crystal, a contribution from each and every reflection must be combined in the correct register, that is with the correct phase. Unfortunately, the phase information is lost during data collection, and it is this which gives rise to the phase problem in crystallography.

It is ironic that if the positions of the atoms are known then the phase can be readily calculated! However, techniques exist by which it is possible to 'bootstrap' a way to the final phase. One such, the most widely applicable, is multiple isomorphous replacement. When a crystal of a protein is soaked in a solution containing a heavy-metal complex, the complex percolates into the crystal and, with luck, binds to only a few, specific sites on each molecule in the crystal. X-ray data for this and probably several other 'heavy-atom derivatives' are collected and from these it should be possible to derive an image of the protein molecule. The other technique which is widely used is called molecular replacement and relies upon knowing the molecular structure of a related protein. For example, the proteolytic enzyme trypsin has been used to determine the structure of thrombin (Bode et al., 1989), an enzyme of closely related structure. The known model is carefully positioned with respect to the unknown structure, whereupon the calculated phases will be sufficiently close to the true values so that when they are combined with the X-ray observations, the unknown structure can be determined. Considerable advances in the technology of X-ray diffraction have taken place in the past few years: synchrotrons provide a high-intensity, tunable X-ray beam (Helliwell, 1992), various types of detector have been introduced, perhaps the most versatile of which is the image plate, and novel computational techniques are beginning to be applied to the phase problem for proteins (Dodson, 1988).

This last point highlights the basic need in all crystallography for high-speed computers. Since phases must be calculated for several thousand reflections, which are measured with a computer-controlled instrument, and the model once obtained must be refined (or optimised with respect to the experimental observations), it should be clear that without computers the crystallographer would be lost. Further, when it comes to examining the model, interactive computer graphics are essential and it is this last application which is fundamental to protein engineering, since, without careful planning, considerable resources will be wasted. The planning of how to tailor the protein to its new function is the core of protein engineering and th computer and graphics facilities are used to try out the mutations and observe their likely effects. Many of the available computer programs have ways of calculating and minimising the energy of the structure based upon semi-empirical energy terms derived from the host of known molecular structures. Thus, only those mutations likely to achieve the desired result need to be synthesised.

Since a model of the protein structure is seen as an essential prerequisite to any protein-engineering project, what can be done when no X-ray structure is available? Two approaches are possible depending on what is known about the structure. First, protein structures seem to be grouped into families, and if the structure of another member of the family is known, this can be used as the basis for model-building by superimposing the new sequence onto the structure which is already known (Blundell et al., 1988; Greer, 1990; Sali et al., 1990). The computer programs for energy minimisation and so on, referred to above, provide the means. Such modelling can be improved by using short regions of 'consensus structure' derived by comparisons of several proteins which have known structures for similar amino acid sequences (Jones and Thirup, 1986; Wodak, 1987; Claessens et al., 1989). Thus a 'spare-part' approach to modelling also helps both the interpretation and improvement of the initial X-ray model. Together with secondary-structure prediction methods (Bishop and Rawlings, 1987; Fasman, 1989), it also provides a means for generating structures for which no X-ray data are available (Greer, 1990; Schiffer et al., 1990). This is the second approach and is still highly speculative. However, as more is learnt about the folding rules for proteins, such predictive techniques can but improve (Akrigg et al., 1988; Blundell et al., 1988; Sali et al., 1990).

Characterisation of the protein, either in its natural form or after mutation, depends rather upon the type of protein that it is. For enzymes, measurements of the rates of the catalysed reactions as functions of temperature and pH are most valuable and are normally supplemented by the technique of circular dichroism. Measurement of the stability of the enzyme to denaturing agents like urea is also a valuable way of assessing the effects of mutagenesis. These same techniques are also applicable to other proteins where the kinetic measurements are replaced by binding studies which are often less easy to perform. NMR techniques are also finding applications here.

Why Engineer Proteins?

We have seen how, with recombinant DNA technology, protein crystallography and molecular modelling, it is possible to engineer a protein to obtain a predicted hange in function. The applications which have been made to date have shown the potential of the technique but have also shown that there is a great deal still to be learned. On the purely scientific side, answers to the effects of changing given amino acids in given environments are beginning to emerge – it is still not always as expected (Wells and Estell, 1988; Carter and Wells, 1990). However, as these types of experiment continue, the database of the relationship which exists between amino acid sequence and 3D-structure will also increase, thereby improving and refining the protein folding rules, still one of the most sought-after goals in biochemistry. Coupled to this, but in the longer term, is the relationship between protein structure and function (Knowles, 1987; Carter and Wells, 1990). Protein engineering has also benefited the earlier stages of X-ray structure determination in providing novel ways of preparing heavy-atom derivatives (Forest and Schutt, 1992).

But what of the applied side? Many of the industrial applications of enzymes rely on hydrolytic enzymes, those which break down carbohydrate (or starch), fats and even proteins themselves. Typical problems are associated with the lack of heat stability or the limited specificity of the enzymes in question, and several enzymes have already been engineered to make them more thermostable (Nosoh and Sekiguchi, 1990) or to modify their specificity (Wilks et al., 1988). The problem of a significant change, say 3 pH units, in the pH optimum of an enzyme, is still unsolved. The ability to make chaemeric proteins has important applications in the design of new antibodies, where the part of the protein responsible for recognising part of a foreign cell can be grafted onto a toxin which, rather like a letter bomb, is then appropriately targeted (Chaudhury et al., 1988). Or, an antibody to some highly toxic substance can be raised in mice, the recognition region determined and a human antibody engineered to have the same exquisite specificity (Reichmann et al., 1988). Pharmaceutical proteins are an obvious case for engineering and Brange et al. (1988) have already shown that it is possible to improve the absorption of insulin by making a monomeric form of the protein. Finally, rational approaches to drug design are dependent on many of the techniques described for protein engineering (Hol, 1987; Blundell et al., 1988). Since a drug either interacts with a receptor to produce an effect or interacts to abolish an effect, studying this interaction at the molecular level, ideally by producing a 3-dimensional model of the drug–receptor complex, will provide a basis producing new and better drugs.

At the present time, only relatively small changes are being introduced, which do little more than tinker with the function of the protein (Knowles, 1988). Experiments are under way to design novel proteins (Richardson and Richardson, 1989) and even to use antibodies as the basis for designing new enzymic activity (Shokat and Schultz, 1990; Presta, 1992). A much more dramatic prospect, indeed the goal of the protein engineer, is to design and produce *de novo* a protein with a specified activity which also has the correct solubility and stability properties. When this is achieved, protein engineering will truly have come of age.

REFERENCES

Akrigg, D. et al. (1988). *Nature*, **335**, 745–746.@REF = Anfinsen, C. B. (1973). *Science*, **181**, 223–230.
Bishop, M. J. and Rawlings, C. J. (1987). *Nucleic Acid and Protein Sequence Analysis*. IRL Press, Oxford and Washington, DC.
Blow, D. M. (1988). *Bull. Inst. Pasteur*, **86**, 47–54.
Blundell, T. L. and Johnson, L. N. (1976). *Protein Crystallography*. Academic Press, London and New York.
Blundell, T. L., Sibanda, B. L., Sternberg, M. J. E. and Thornton, J. M. (1988). *Nature*, **326**, 347–352.
Bode, W. et al. (1989). *EMBO J*, **8**, 3467–3475.
Brange, J. et al. (1988). *Nature*, **333**, 679–682.
Brown, T. A. (1990). *Gene Cloning: an Introduction*. Chapman & Hall, London.
Carter, P. and Wells, J. A. (1990). *Proteins: Str. Funct. Genet.*, **7**, 334–342.
Chaudhury, V. K. et al. (1988). *Nature*, **335**, 369–372.
Chothia, C. and Finkelstein, A. V. *Ann. Rev. Biochem.*, **59**, 1007–1039.
Claessens, M., van Cutsem, E., Lasters, I. and Wodak, S. (1989). *Prot. Eng.*, **2**, 335–345.
Clark, A. J., Simons, P., Wilmut, I. and Lothe, R. (1987). *TIB Tech.*, **5**, 20–24.
Dodson, G. (1986). *TIBS*, **11**, 309–310.
Ducruix, A. and Giege, R. (1992). *Crystallisation of Proteins and Nucleic Acids*. IRL Press, Oxford and Washington, DC.
Fasman, G. D. (1989). *TIBS*, **14**, 295–299.
Forest, K. and Schutt, C. (1992). *Curr. Opin. Str. Biol.*, **2**, 576–581.
Greer, J. (1990). *Proteins: Str. Funct. Genet.*, **7**, 317–334.
Harris, E. L. V. and Angal, S. (1990). *Protein Purification: Methods*. IRL Press, Oxford and Washington, DC.
Helliwell, J. R. (1992). *Macromolecular Crystallography with Synchrotron Radiation*. Cambridge University Press, Cambridge.
Hol, W. G. J. (1987). *TIB Tech.*, **5**, 137–143.

Jones, T. A. and Thirup, T. (1986). *EMBO J.*, **5**, 819–822.

Knowles, J. R. (1987). *Science*, **236**, 1252–1258.

McPherson, A. (1990). *Eur. J. Biochem.*, **189**, 1–23.

Nosoh, Y. and Sekiguchi, T. (1990). *TIB Tech.*, **8**, 16–20.

Oxender, D. L. and Fox, C. F. (1987). *Protein Engineering*. Alan R. Liss Inc., New York.

Presta, L. G. (1992). *Curr. Opin. Str. Biol.*, **2**, 593–596.

Rees, A. R., Sternberg, M. J. E. and Wetzel, R. (1992). *Protein Engineering: a Practical Approach*. IRL Press, Oxford and Washington, DC.

Reichmann., L., Clark, M., Waldemann, H. and Winter, G. (1988). *Nature*, **332**, 323–327.

Richardson, J. S. and Richardson, D. C. (1989). *TIBS*, **14**, 304–309.

Sali, A., Overington, J. P., Johnson, M. S. and Blundell, T. L. (1990). *TIBS*, **15**, 235–240.

Sambrook, J., Fritsch, E. F. and Maniatis, T. (1989). *Molecular Cloning – a Laboratory Manual*. 2nd edition. Cold Spring Harbour Press, New York.

Schiffer, C. A., Caldwell, J. W., Kollman, P. A. and Stroud, R. M. (1990). *Proteins: Str. Funct. Genet.*, **8**, 30–43.

Shokat, K. M. and Schultz, P. G. (1990). *Ann. Rev. Immunol.*, **8**, 335–363.

Stout, G. H. and Jensen, L. H. (1989). *X-ray Structure Determination*, 2nd edition. John Wiley, New York.

Wells, J. A. and Estell, D. A. (1988). *TIBS*, **13**, 291–297.

Wilks, H. M. et al. (1988). *Science*, **242**, 1541–1544.

Wodak, S. J. (1987). *Ann. N. Y. Acad. Sci.*, **501**, 1–13.

Wright, P. E. (1989). *TIBS*, **14**, 255–260.

Wuthrich, K. (1989). *Science*, **43**, 45–50.

Wyckoff, H. W., Hirs, C. H. W. and Timasheff, S. N. (1985). *Meth. Enzymol.*, **114**, 115.

10

Bioprocessing

K. Schügerl

Introduction

Biotechnological processes play an important role in different parts of our life: in agriculture, food, and the chemical industry, in the energy sector as well as in environmental protection. Also, if one does not consider the food industry (baker's yeast, bread, wine, beer, etc.) and environmental protection (biological waste and waste-water treatment), the world market volume for biotechnological products still amounts to about 20 billion $ per year.

The pharmaceutical industry has the largest share in it (60%); of that, the antibiotics have the highest market volume of around 10 billion $ per year. The annual increase of biotechnological production is an estimated 8%.

Biotechnological processes have their largest impact in the pharmaceutical industry, because pharma products are usually complex organic compounds which often exist in several stereoisomers (with chemically identical composition, but different spacial structure), of which only one has the desired biological effect. The chemical synthesis of the desired stereoisomers of a high yield and purity is uneconomical or sometimes even impossible. Reactions catalysed by enzymes are stereoselective, therefore, the microorganisms usually form or transform only one of the stereoisomers. On account of this, the biotechnological processes in the pharmaceutical industry are without competition. The same holds true for the formation of natural polymers, e.g., biologically active proteins (enzymes, lymphokines, growth factors, etc.) and polysaccharides (xanthane). Biotechnological production processes are not competitive in the manufacturing of bulk products (simple organic compounds, such as ethanol, butanol, acetic acid for industrial use), which are usually produced by chemical synthesis from inexpensive petrochemical products.

Biological Prerequisites

For the manufacture of products by microorganisms, cells or parts of them, suitable microorganisms or animal cells are needed which are stable enough and form the desired products (primary or secondary metabolites or proteins) in high concentration.

The first thing for developing such a production process is the search for suitable microorganism strains or cell lines. The improvement of these isolated cultures is performed by mutation (with radioactive rays or mutanogeneous chemicals) by the crossing or hybridization of different strains, by protoplasm fusion and nowadays also by inserting foreign genes into the organisms by means of so-called vectors (e.g., plasmids). In the latter case, the foreign gene can either originate from a foreign organism or can be chemically synthesized. This strain improvement is called genetic engineering or recombinant DNA technique. These improved high-producing and genetically stable strains are maintained in the industry in primary stock cultures. The economically less interesting strains are maintained and stored in public culture collections available to everybody.

Advantages of biotechnological production processes are that they proceed at low temperatures and pressures with high selectivity and optical purity (only one particular stereoisomer is formed, which is characterized by its optical rotation value, i.e., how they change the angle of polarized light). Their disadvantages are the low product concentration in the aqueous solution, since organisms are not viable at high substrate or product concentrations. The substrate is the source of energy and carbon (nitrogen, phosphate) compounds, which the organisms need for their maintenance, growth, and product formation. Several accompanying compounds can be present in the solution, the separation of which is often rather difficult. These accompanying compounds originate either from the nutrient (molasses, pharma medium, yeast extract, meat extract, cornsteep liquor, etc.) added to the liquor, or they are secreted by the organisms. On account of the high dilution and the complex composition of biological systems, the recovery of the products from the liquor and their purification are more difficult and expensive than the product recovery from chemical synthesis. In many cases, the economy of the process depends heavily on efficient product recovery or purification.

Monocultures

If the product is well-defined, e.g., an antibiotic or an enzyme, the next step of the process development is the evaluation of the optimal conditions for the cultivation of microorganisms or cells that are to be used for the product formation.

The isolated strains which have been naturally selected over many millions of years, have lost their ability to compete with other organisms in the environment through the artificial modification of their genetic pool. The artificial mutants, which are forced to produce compounds which are unnecessary or harmful to their survival, would quickly be overgrown by their competitors. The foreign organisms are kept off by carrying out the bioprocess in a closed vessel, called a reactor, in which the foreign organisms are killed by chemicals or heat (chemical or heat/steam sterilization) before the organisms are injected into the reactor (inoculation). During the production process, all manipulations are avoided that could bring foreign germs into the reactor, i.e., would cause insterility of the reactor/process. Therefore, these production processes work with monocultures, i.e., only with a single strain in the absence of other organisms.

The first step in production is the propagation of the organisms. In order to attain the maximum rate of cell growth, the environment of the organisms (pH-value, i.e., acidity, as well as temperature, composition of the nutrient solution, dissolved oxygen concentration, etc., in the liquor) is maintained at optimal levels, which were evaluated beforehand by separate measurements. After the desired concentration of the organisms has been attained, the optimal conditions for product formation are established and maintained, which are evaluated by separate measurements also. Finally, the product has to be recovered from the liquor and purified.

Culture Media

The organisms consist of proteins, nucleic acids, and organic compounds which play a role in the energy reservoir (polysaccharides, lipids) and inorganic components. For their growth, they need sources of carbon, hydrogen, oxygen, nitrogen, phosphor, sulfur, potassium, sodium, magnesium and calcium, as well as trace elements.

The carbon source is at the same time a hydrogen and oxygen source (carbohydrates) and eventually also a nitrogen source (proteins). The carbon compounds serve as energy source. Most popular carbon sources are glucose in the laboratory and sucrose (molasses) in industry. Also oils/lard/fats are often used, and in special cases ethanol, methanol, and methane.

As nitrogen source, inexpensive plant proteins (soy flour, peanut flour, pharma medium, potato protein liquor, etc.), or inexpensive protein hydrolysates (cornsteep liquor, yeast extract, peptone, etc.), meat extract, urea, or inorganic nitrogen compounds are used. As phosphr and sulfur sources, usually inorganic phosphates and sulfates are employed. Last, but not least, trace elements and vitamins, and in animal cell cultures, growth factors are needed as well.

Cell growth and product formation strongly depend on the chemical composition of the culture medium (aqueous solution in which the cells grow). Therefore, an evaluation of the optimal composition of the culture medium is necessary. This is usually carried out at constant temperatures in small flasks; several of them are put in shaking machines to ensure good mixing, and in the case of aerobic microorganisms or cells (which need molecular oxygen for living) they are also supplied with oxygen. The reproducibility of these shaken cultures is not very good and their quantitative evaluation is difficult, because the cells go through different growth phases during cultivation, and the composition, dissolved-oxygen concentration, and the pH-value of the nonbuffered medium varies while being measured.

When using continuous cultures, the organisms are in constant environmental conditions and a well-defined growth phase in the steady state. The influence of the medium composition on cell growth and product formation can easily be evaluated, if the concentrations of the organisms, products and, most important, key components are measured. Their on-line measurement makes a computer-aided automatic medium optimization possible.

The use of culture media consisting of well-defined components (synthetic media) allows the quantitative determination of the ingredients on the physiology of the organisms. However, it is often not possible to attain the same high growth rate and productivity with synthetic media that can be achieved with chemically complex ingredients (yeast extract, cornsteep liquor, pharma media, etc.). On account of the latter, and also because of their lower costs, complex media are used in industrial production.

In culture media of animal cell cultures, the number of ingredients is very large (60 to 80). Only rarely is it possible to use synthetic or serum-free media, because the growth factors in (foetal calf) serum are essential for sufficient cell growth and product formation in most cases.

Bioreactors and their Operation

Bioreactors are used for maintaining the optimal, usually aseptic, environmental conditions for the growth of organisms and product formation. Therefore, they are often provided with instruments for measurement and control of temperature, pH-value, dissolved oxygen (for aerobic organisms), and sometimes for some of the key medium components (substrates, products, precurser, inductor, etc.). In principle, every closed vessel that can be made aseptic is suitable for the cultivation of microorganisms and cells. However, high-performance, large-size bioreactors are usually carefully constructed in *in situ* steam-sterilizable aerated stirred tanks, in which the dispersion of the air for a sufficient oxygen supply, the mixing of the liquid phase for adequate substrate mixing, and heat transfer for keeping the temperature in the reactor at the optimal level, are maintained by fairly high specific power input (2–3 kW/m^3) and controlled by the stirrer speed. Especially in cell cultures, high stirrer speed and direct aeration (with dispersed gas phase) can cause cell damage. Therefore, for these cultivations, special reactors have been designed.

Reactors for the cultivation of microorganisms can be operated in batch, fed-batch and continuous mode.

The culture medium is inoculated, the organisms are propagated, and the product is formed successively in batch reactors.

In fed-batch reactors, after a batch phase, the key substrate concentration is kept at a low level by feeding the substrate solution gradually to the culture medium in order to avoid growth inhibition caused by high substrate concentrations. At the end of the cultivation, the feeding is stopped, and the substrate is consumed in batch operation.

In continuous cultures, after a batch phase, the culture medium is continuously fed into the reactor and removed from it. With the variation of the throughput of the medium, the so-called dilution rate D, which is the ratio of the medium throughput to the medium volume in the reactor, and the specific growth rate of the microorganisms can be varied.

The specific growth rate μ is the change of the cell concentration X with the time t with regard to X.

As long as the dilution rate D is lower than the maximum specific growth rate μ_{max}, steady-state condition exists. If D is increased above μ_{max}, the organisms are washed out of the reactor, no steady-state culture is possible.

In general, the cell concentration in continuous cultures is lower than in batch or fed-batch cultures. In order to increase the cell concentration and the productivity (product formed per culture medium volume and time) of continuous cultures, the cell concentration is raised in the reactor by retention or recycling of the microorganisms or cells by means of cross-flow microfiltration membranes or by their flocculation. By the sedimentation of the flocs in a calming sector of the reactors, the concentration of flocculated microorganisms can be increased considerably.

Another possibility for increasing the productivity in case of product inhibition (high product concentration reduces growth rate and product formation) is the *in situ* removal of the product from the reactor during product formation.

This is an integrated product-formation–product-recovery system, which is discussed in the interated processing section.

Downstream Processing

As already mentioned, the cells and the products at the end of the cultivation process are in low concentration in the culture media, accompanied by several metabolites, substrate residues, and secreted proteins or proteins from lysed cells.

The cells can be the product, e.g., in the case of baker's yeast. There the cells are recovered from the culture medium by a centrifugal separator.

In the case of low molecular products, they are usually secreted from the microorganisms into the culture medium.

In the case of proteins, they accumulate in the organisms or are secreted into the culture medium.

The recovery of extracellular products differs from that of intracellular products.

Extracellular products

The first step of downstream processing usually is the separation of the cells from the liquid medium. This can be carried out after suitable conditioning by centrifugation (e.g., bacteria or yeast), dead-end or cross-flow filtration (e.g., fungi), or sedimentation (e.g., flocculated microorganisms).

The next step is the recovery of the products from the filtrate. This can be performed by ultrafiltration (e.g., enzymes), flocculation (e.g., enzymes), precipitation (e.g., amino acids), adsorption (e.g., cephalosporin C, streptomycin), extraction (e.g., penicillin), distillation (e.g., ethanol, acetone/butanol), or crystallization (e.g., amino acids).

Intracellular products

In the case of intracellular products, the microorganisms have to be disintegrated. This can be performed by mechanical stresses (wet milling, high-pressure treatment, homogenization), chemical manipulation, or biological (enzymatical) treatment. The cell debris and the product in solution are separated by centrifugation, extraction, or filtration.

The product enrichment from the solution can be performed similarly to that of the extracellular products by ultrafiltration, precipitation, adsorption, or extraction.

Product purification

The most important product-purification process of low-molecular products is crystallization.

Purification of high-molecular products is performed by different types of chromatography: adsorption-, ion-exchange-, hydrophobic-, affinity- and gel-permeation-chromatography.

The product (solute) is separated from the solution by binding it on the column. The recovery of the solute from the column is performed by an eluent. During elution, the bound solute moves along the column and can be recovered at the column end.

The adsorption chromatography has a relatively low specificity, because it is based on the physical surface forces which are involved in retaining the solute on the surface. It is the simplest, but very effective technique for the separation of mixtures of nonpolar substances and compounds of low volatility. Silica, alumina, activated carbon, and macroporous nonionic polymers are the most frequently used adsorbents, and low-molecular-weight organic solvents are the most common eluents.

In ion-exchange chromatography, true heteropolar chemical bonds are formed reversibly between ionic components in the mobile and stationary phase. The ion-exchange beads may be either hydrophobic or hydrophylic matrices which have been functionalized with ionizable groups. The eluent is usually water. Large numbers of proteins have been purified by high-performance ion-exchange chromatography.

In reverse-phase liquid chromatography, polar-charged fermentation products, such as peptides and glycoproteins, can be separated by the hydrophobic stationary phase consisting of silica gel to which a C_8 to C_{18} hydrocarbon has been covalently attached, and the polar aqueous mobile phase.

Gel-permeation chromatography is also used to separate water-soluble compounds, e.g., polypeptides and enzymes. The advantage of this technique is that the enzyme activity is fully maintained during recovery.

Affinity chromatography is based on a selective biospecific recognition between the solute molecule and a ligand (a specific compound) immobilized onto a support material. The solute is selectively bound to the column. After the unadsorbed contaminants have been washed out of the column, the solute is eluted and recovered. The main problem with affinity chromatography is the strong binding between the solute and the ligand, which is partly irreversible. This reduces the frequent reuse of the affinity matrix and the solute yield of the affinity purification.

As this review shows, several techniques are at our disposal for the purification of fermentation products. The decision as to which of them should be applied depends on the properties of the product and contaminants as well as on the purity requirements.

Integrated Bioprocessing

In some instances, the products impair the growth and product formation (product inhibition) considerably. In order to improve product recovery, the product concentration is increased; this again, however, reduces the productivity. High productivity can only be attained at low product concentration. In order to eliminate this product inhibition, *in situ* product recovery is desirable. During product formation, the product should be removed from the culture medium.

Such integrated systems are difficult to develop, since the separation process is not allowed to impair the cultivation process, and the cultivation medium is not allowed to impair the separation process. Thus, it only works in some instances.

Extraction is impaired by the low partition coefficients of the product between the solvent and aqueous culture medium if the solvent is biocompatible. Solvents with high partition coefficients caus damage of the organisms. Therefore, a direct extraction in solvent contact with the cells is not practical; for this type of extraction, a preceding separation of the medium and the organisms, and a low solubility of the solvent in the aqueous medium are necessary.

This problem can be avoided if one uses two aqueous phases. For example, during the microbial transformation of hydrocortisone into prednisolone, the product can be extracted from the medium by biocompatible polyethyleneglycol (PEG) solutions.

Evaporation of the solvent at reduced pressure (vacuum fermentation) has the disadvantage that only the highly volatile components are removed. Components with higher boiling points (e.g., higher alcohols) continue to accumulate. They are sometimes more toxic for the microorganisms than the more volatile products.

Membranes can be used alone if the molecular weight of the product is much lower than that of the educt (original compound which is to be converted into the product). This is the case when polymers are decomposed enzymatically, and the low-molecular-weight products, which often act as enzyme inhibitors, are removed by having them permeate through an ultrafiltration membrane.

If educts and products have similar molecular weights, the use of membranes must be combined with another separation technique.

Pervaporation is a combination of evaporation and membrane permeation. This technique has the disadvantage that it works only with higher ethanol concentrations. Membranes that ensure selective ethanol permeation at low product concentrations are not yet on the market.

Perstraction is a combination of extraction and membrane permeation. Solvents with high partition coefficients can be kept below the critical level that is toxic to the cells. Silicon membranes have been used for alcohol separations of this type.

Persorption is a combination of adsorption and membrane permeation. The solubility of the product in the membrane can be increased considerably if the sorbent is included into the membrane (e.g., silicalite into silicon membrane). When using this membrane for pervaporation, increased separation rates can be achieved.

Electrodialysis uses cation- and anion-exchange membranes and the potential difference between the two sides of the membrane. It is suitable for the removal of salts, acids, and bases from cultivation media by means of the electric charges of the particles. It is usually necessary to remove the organisms from the broth before the electrodialysis (e.g., by means of microfiltration modules). This technique has been applied to the recovery of lactic acid and various amino acids from cultivation media.

Outlook

Modern biotechnology deals with extremely complex systems. Therefore, it requires an interdisciplinary cooperation of biologists (molecular biologists, microbiologists, botanists), chemists (biochemists, technical chemists), physicists (biophysicists) and engineers (chemical, electrical/control, agricultural engineers).

Bioprocessing is the application of biotechnology in order to manufacture useful products. Its success depends not only on the advancement of fundamental research in biology, biochemistry and biophysics (e.g., protein design) and its transfer into industry by scientists and engineers in their field of application, but also on the economy of these processes in competition with chemical processes as well as on the legislation of the countries where it takes place, which is strongly influenced by public opinion. In some countries in Europe, the acceptance of genetic engineering and its application in biotechnology is hampered by people confusing biotechnology with animal-breeding technology and medical application of genetic engineering, e.g., gene surgery.

Bioprocess development takes much time and requires sufficient capital. Several small biotechnological companies have gone bankrupt, because they did not have enough capital to develop products that were successful on the market. On a long-term basis, onlylarge companies will survive. The small ones which do not manufacture products, but sell their know how, lose their right to exist, because the large companies quickly adapt the new techniques by employing young scientists from the universities, who are ahead in basic research.

In recent years, considerable advancements have been made in bioprocessing, especially in the field of the production of biologically active proteins. However, on account of the increased requirements with regard to product quality, more and improved techniques are necessary. Protein engineering is the biological/biochemical tool of the future. In addition, improved cultivation and purification techniques must be developed. Several new separation techniques are used in analytical laboratories. They have a good chance of becoming efficient techniques for product recovery and purification.

Part III

Applications

Introduction to Part III

The main aim of this part is to give an overview of the potentially broad range of products and processes based upon the new biotechnology. The rate of biotechnology commercialisation and the factors affecting that rate vary among industrial sectors.

Applications of biotechnology can be found in many industries from pharmaceuticals to food manufacturing and from agriculture to waste treatment. The pervasiveness of the new biotechnology is due to its basic techniques which can be applied in numerous industries in various processes and in many different products. The economy-wide effects of biotechnology depend upon the progress in research, development and design over the next 10–20 years. Cost reduction will be a critical prerequisite for widespread diffusion of biotechnological techniques in the process industry, such as the chemical industry, oil recovery, the energy industries and the food industry. Until solutions have been found to reduce cost in the scaling up process of bulk commodities, biotechnological applications will probably remain concentrated in two main areas, pharmaceuticals and agriculture.

Many new products are in the pipeline, particularly in the pharmaceutical and health-care sectors. Health applications of biotechnology are likely to diffuse much faster than agricultural applications. The rate of introduction of new therapeutic drugs has increased since the second half of the 1980s. The main products are insulin, human growth hormone, interleukins, and growth factors. In terms of market shares the market for diagnostics, prevention and immunology is more important. Developments in new vaccines are important, because biotechnology has an enormous potential to prevent infectious diseases and reduce disabilities and deaths around the world. Vaccines could provide a relatively cheap and simple solution to mass diseases in developing countries. Another field that gets a lot of attention is the human genome mapping. The ultimate goal of the Human Genome Project (HGP) is the determination of the nucleotide sequence of the human genome, to disentangle the complex causes of human genetic disease. Although some testing has taken place already in the field of gene therapy it is not expected to lead to large-scale applications before the end of the century. New reproductive techniques have been developed. Antenatal diagnosis is used to detect genetic defects. At the same time the techniques have created a set of moral and psychological dilemmas.

In agriculture, horticulture, and floriculture there will be a need for faster and more selective breeding methods, improved varieties, new varieties tailored to meet demands set by the customer, by the agro-processing industry and by the need for environmentally acceptable culturing techniques. Biotechnological techniques, like, for example, in vitro plant-tissue culture techniques, are finding their application in tomorrow's agricultural practice, not only in horticulture and floriculture, but increasingly also in the major crop cereals: barley, maize, wheat, rye and rice. However, there is still a long way to go in biotechnological applications in agriculture. Often it is forgotten that plant biotechnology is still in its infancy. Most of the agronomically important genes have neither been identified at the molecular level nor cloned. But the trends are clear, and they show that plant biotechnology will bring us from the green revolution to the gene revolution, but in an evolutionary way.

Biotechnology plays an important role in agricultural inputs, like herbicides and pesticides. A critical examination of company strategies teaches that many of these applications are troubling for a sustainable agriculture. A better way of using the new biotechnological research would be to figure out how to eliminate pesticide use rather than conjuring up new and exotic ways to extend and expand pesticide use through genetic engineering.

Biotechnology will be applied to livestock production in many different ways. The main application areas are reproduction and breeding, animal health, animal nutrition and physiology of growth and lactation.

Fermentation-based fuel ethanol industries are heavily using biotechnological techniques. Particularly in the USA and in Brazil, ethanol industries have become well established. It will be necessary to develop new techniques that reduce the cost of production. In this respect there are significant technical opportunities, like, for example, starch liquefaction by acid hydrolysis, cascade fermentation and the use of molecular sieves or dried corn grits for dehydrating. At the development stage are systems that use immobilized yeast.

Many sections of the food and beverage industry use enzymes in their production. From starch in the beer-brewing industry to pectinase enzymes in the fruit-juice industry. But also in meat-tenderisation processes and in the cheese-making industry we find enzymes.

A totally different field of biotechnological applications is mining. The recent emergence of new biotechnology-based alternatives to traditional extraction methods offers dramatic possibilities of pollution and cost reduction, productivity improvements, flexibility of scale and improved value-added of final metal products as well as possibilities for maintaining or expanding employment. Biotechnology might even offer developing countries some windows of opportunity. Since the 1980s, efforts by Chilean and Peruvian mining companies to introduce biotechnology provide evidence of the success that seizing such opportunities early can bring.

Major environmental applications of biotechnology include confined-use systems (fermenters) for degrading xenobiotics, semiconfined systems such as waste-water treatment plants, and the deliberate release of appropriate organisms.

In a sense biotechnology is also a dual-use technology. Besides many civilian applications, biotechnology can also be applied in weapons. In this respect toxins are important, because they are considered as working faster and in a better controllable way than weapons based on bacteria or viruses. Furthermore, the design and production of cell-specific cytotoxic agents ('immunotoxins'), which have been explored for civil purposes, may become of military relevance. Although quite some research has been done in this field, these military applications of biotechnology are still in their infancy.

11

Applications of Biotechnology: An Overview

Margaret Sharp

Introduction

The applications of biotechnology are potentially very broad, but at present concentrated in two main areas, pharmaceuticals and agriculture. There are those who see biotechnology as a potential successor to information technology in the breadth and depth of its impact on both production methods and lifestyles. This chapter concurs with the conclusions of the OECD study, Biotechnology, Economic and wider impacts (OECD 1989, chapter III) that for biotechnology to be as pervasive as information technology, requires more of a 'revolution' than currently looks like taking place. In the short term the main impact has been on the medical/pharmaceutical sector where, in spite of the publicity given to therapeutic proteins, developments should be seen in the context of the total revolution in approach taking place in the whole biomedical sector. In the longer run, the major impact will be on the agro-food sector, where developments in plant biotechnology seem likely to underpin a new Green Revolution. Whether it will go further, and take us to a world where most food stuffs are cultured in factories rather than grown in fields, has yet to be seen. Other applications in the chemical and energy sectors are, as yet, limited. Sadly, an area of great potential gain – the biodegradation of much waste matter – waits on resource commitments and a tough regulatory stand from governments.

The Main Applications of Biotechnology

Table 11.1, taken from the Office of Technology Assessment's report on US Investment in Biotechnology (OTA, 1988), gives a good indication of sectors currently seen by US firms as the most important areas of application. Since the US has the most lively group of specialist biotechnology firms as well as hosting a large number of major multinationals in the chemical/pharmaceutical sectors, this gives a fair representation of world-wide trends. It will be noted that, for both large and small companies, pharmaceuticals are the most active sector, followed by diagnostics and reagents for the small-firm sector, while chemicals and plant agriculture top the bill for the larger firms.

Pharmaceuticals – Drugs and Human Health Care

Early diffusion of any new technology tends to concentrate on high value-added products where the possibility of very high returns offsets the costs and risks of early developments. The first products of the 'new biotechnology' which emerged from the breakthroughs in recombinant DNA (rDNA) techniques were what are termed the *therapeutic proteins* – insulin, the interferons, the interleukins, various growth factors. Table 11.2 lists the seven protein products which had received FDA approval by mid-1988, and Table 11.3 lists other protein products known to be under commercial development, of which two, Tissue Plasminogen Activator (TPA) and Erythropoletin (EPO), had received FDA approval by 1989.

Table 11.1 Areas of primary R&D focus by biotechnology companies

Research area	Dedicated biotech companies No.(%)		Large, established companies No.(%)	
Human therapeutics	63	(21%)	14	(26%)
Dlagnostics	52	(18%)	6	(11%)
Chemicals	20	(7%)	11	(21%)
Plant agriculture	24	(8%)	7	(13%)
Animal agriculture	19	(6%)	4	(8%)
Reagents	34	(12%)	2	(4%)
Waste disposal/treatment	3	(1%)	1	(2%)
Equipment	12	(4%)	1	(2%)
Cell culture	5	(2%)	1	(2%)
Diversified	13	(4%)	6	(11%)
Other	31	(18%)	0	(0%)
Total	296	(100%)	53	(100%)

Source: Office of Technology Assessment, 1988.

Table 11.2 Biotechnology-based human therapeutics with FDA market approval

Trade Name/Generic Name	Use	Company Receiving Market Approval
Humulin/Human Insulin	Treatment of diabetes	Eli Lilly and Company
Protropin/Human Growth Hormone	Treatment of children with inadequate secretion of growth hormone	Genentech, Inc.
Humatrope/Human Growth Hormone	Treatment of children with inadequate secretion of growth hormone	Eli Lilly and Company
Intron A/Alpha Interferon	Treatment of hairy-cell leukemia	Schering–Plough Corporation
Roferon-A/Alpha Interferon	Treatment of hairy-cell leukemia	Hoffman–La Roche, Inc.
Orthoclone OKT 3/ Monoclonal antibody against T-cells	Treatment for reversal of acute kidney transplant rejection	Ortho Pharmaceutical Corporation
Activase/Tissue Plasminogen Activator	Treatment of cardiac arrhythmia	Genentech, Inc.

Source: Office of Technology Assessment, 1988.

Table 11.3 Known or expected therapeutic applications of some human gene products under commercial development

Atrial Natiuretic Factor (ANF). One of the peptide hormones secreted by the heart; acts to regulate blood pressure, blood volume, and water and salt excretion; possible applications in treatment of hypertension and other blood pressure diseases and for some kidney diseases affecting excretion of salts and water.

Epidermal Growth Factor (EGF). A protein growth factor that causes replication of epidermal cells (those cells on the outermost layer of tissues); expected to have applications in wound healing (including burns) and cataract surgery.

Erythropoietin (EPO). A protein hormone growth factor normally produced by the kidney; causes the production of red blood cells; anticipated treatment for anaemia resulting from chronic kidney disease; some potential for curing anaemias associated with AIDS and other chronic diseases.

Factor VIII: C. A protein involved in blood clot formation; major application in prevention of bleeding in haemophiliacs (deficient in factor VIII) after injury.

Fibroblast Growth Factor (FGF). A protein that stimulates growth of blood vessels; may be useful in wound healing and treating burns.

Granulocyte Colony Stimulating Factor (G-CSF). One of a larger class of colony stimulating factors that stimulates production of the lass of white blood cells called granulocytes; could be useful in treating leukemia and AIDS, possibly in concert with other chemotherapeutics.

Human Growth Hormone (hGH). A peptide hormone naturally occurring in the pituitary gland; used as a treatment for childhood dwarfism; expected to have broader therapeutics potential in wound healing or treatment of Turner's syndrome and small stature.

alpha-Interferon (a-INF). A lymphokine protein used as a treatment for hairy-cell leukemia; possible broader applications in treatment of venereal warts, Kaposi's sarcoma (associated with AIDS), lymphoma, bladder cancer, and malignant melanoma.

gamma-Interferon (g-INF). A lymphokine protein that activates macrophage cells and interferes with viral replication; potential treatments for various cancers, AIDS.**Interleukin-2 (IL-2).** A lymphokine protein hormone that causes immune-system responses; potential treatment for various cancers.

Interleukin-3 (IL-3). A blood protein colony stimulating factor that promotes both red and white blood-cell production at the earliest stages of cell development; potential applications in treatment of white blood-cell deficiency in AIDS patients or that induced by radiation and chemotherapy exposures in other cancer patients.

Macrophage Colony Stimulating Factor (M-CSF). A colony stimulating factor that acts only on white blood cells of the monocyte/macrophage type; potential applications are expected for treatment of infectious diseases, primarily parasitic, but some bacterial and viral diseases; possible cancer therapy.

Superoxide Dismutase (SOD). An enzyme that seeks out superoxide free radicals in the blood and prevents damage when oxygen-rich blood enters oxygen-deprived tissues; applications in cardiac treatment and organ transplants.

Tumour Necrosis Factor (TNF). A protein growth factor with possible broad applications in antitumour and antiviral therapy.

Tissue Plasminogen Activator (TPA). A blood protein that activates plasminogen, a naturally occurring blood protein that breaks down fibrin blood clots; used for dissolving the coronary artery blood clots associated with myocardial infarctions, or heart attacks, with other possible blood-clot dissolving applications.

Source: Office of Technology Assessment, 1988.

Other important new products are the 200-plus diagnostic tests, many based on *monoclonal antibodies* (MABs).[1] MABs and enzyme-based tests have now substantially replaced older testing procedures based on the more expensive (and potentially more dangerous to handle) radio-immunoassay tests. MABs have until recently largely derived from murine (mouse) cell lines, but new techniques of mammalian cell culture have enabled chimeric, humanised MABs to be developed. Some of these will be used in drug delivery and drug therapy (see below).

Table 11.3, which lists the 15 leading applications of protein therapy, is put into perspective by the fact that there are some 50,000 proteins known to exist and to have some function within the human body. For the majority of these we know neither their function (let alone *how* they function) nor their gene sequence. Once the gene sequence that encodes for the protein is known, recombinant DNA technology now means that the protein can be 'manufactured' in pure form in a relatively large quantity, which in turn enables the scientist to study its structure (a protein is a string of amino acids, but its function is dictated not only by its sequence but by its three-dimensional folded structure) and hence how it functions within the cell. This helps to explain why so much emphasis is currently being put on two scientific areas – *the mapping (or sequencing) of the human genome and protein engineering*. The former will open up knowledge about the many proteins that exist but are presently unknown within the human body and will undoubtedly reveal many that have potential therapeutic value. The latter will help unravel precisely how these proteins function and lead eventually to novel man-made proteins, designed to fulfil a narrow function within human biochemistry.

Indeed, the advent of *protein engineering* has clarified what is sometimes called the 'technological trajectory' for biotechnology within the pharmaceutical industry. The first phase, which is now nearing its end, brought us the first generation of protein drugs – human insulin, the growth hormones, the interferons and interleukins. We are now in the middle of the second phase of development, which is bringing the early applications of protein engineering – altering the characteristics of the genetically engineered proteins to make them more effective – and with it rapid advances in the techniques of protein engineering itself. The third phase, which will probably come in the mid-1990s, seems likely to bring 'designer drugs' derived from the knowledge of protein structures and functions. Although rational drug design is a notion that has been around for a long time, it was limited in scope because the tools were not available. The tools are now available through protein engineering. (For a discussion of the potential of protein engineering, see Blundell et al., 1989.)

Ironically, these developments bring pharmacology full circle, for this new generation of 'designer drugs' will almost certainly be synthesised chemically, rather than cultured in bioreactors. But the route by which they are derived will be wholly different from the chemical drugs of the 1950s and 1960s, and will depend crucially on the understanding, stimulated by biotechnology, of human biochemistry and the functioning of the immune system. In this respect, what is happening to pharmaceuticals is part and parcel of the wholesale revolution taking place within the biomedical sector. The general trend will be towards disease diagnosis (immuno-tests, gene probes, biosensors) and prevention (vaccines) as much as cure.

One of the exciting new developments in pharmaceutical biotechnology stems from the convergence of the two main streams of development – recombinant DNA and monoclonal antibodies. *Antibody engineering* enables scientists to use the specificity of antibodies but to doctor them in such a way that they acquire also other valuable characteristics. Such chimeric antibodies have considerable potential in cancer therapy and other treatments where drug targeting is essential. (See G. P. Winter, 1989.) The abzyme (a cross between an antibody and enzyme) also has major potential industrial applications.

The new biotechnology has also brought a revolution in *vaccine development*, with a new and surer route to vaccines through recombinant DNA techniques. Diseases which had earlier eluded vaccines, such as hepatitis B (a disease endemic in Asian countries), now have vaccines under development. Unfortunately this is not yet true of AIDS (Acquired Immune Deficiency Syndrome), although the techniques developed through biotechnology have brought understanding of how the disease affects people. Being a retrovirus (a virus which attacks the gene and alters the genetic code) it is not easy to develop a vaccine, and the main effort at present is to develop drugs which inhibit the spread of the disease. (For a full discussion of applications to vaccines, see K. Murray et al., 1989.)

The new protein drugs also pose challenges in relation to drug delivery and targeting. These large-molecule drugs cannot be absorbed orally without being degraded, nor are they easily absorbed through artery walls, hence injection is no solution. Delivery problems for both drugs and vaccines have led to unexpected delays in getting them to market, and have aroused considerable interest in drug-delivery technologies.

Agriculture and Forestry

In the long run it seems likely that biotechnology will have its greatest impact on plant and animal agriculture and on food production. (For a full analysis of the impact of biotechnology on agriculture, see OTA, 1988, and Flavell, 1989.)

The application of rDNA methods to plants came more slowly than to bacteria, and it was not until the early 1980s that transgenic plants became a possibility. Even today, many applications are still in the development stage. But the range of potential application is enormous. Plant geneticists are now able to create plants which are resistant to diseases, insects, herbicides or harsh environmental condi- tions, thus raising the possibility, for example, of mass cultivation in desert areas or on salt marshes. One of the key developments has been the use of genetic engineering to engender resistance to a particular herbicide. This has led to many of the large chemical companies, traditionally manufacturers of herbicides, buying up seed companies in order to breed hybrid varieties of plant which are immune to their own brands of herbicide. To date, these techniques have been more successful with dicotyledons such as tomatoes or tobacco plants than with monocotyledons such as maize or rice, but it is unlikely to be long before they can be applied just as successfully to these staple crops. This will have major potential impact on yields and production techniques.

Microbial pesticides and herbicides have been developed both by using colonising (and engineered) bacteria to breed immunity in the plants themselves, and by engineering bacteria or fungal elements which inhibit or divert attack. For example, scientists with the British Forestry Commission have engineered a fungal spray which kills the larvae of the pine moth beetle and hence inhibits attack. In the long run such developments will have a major impact on the lucrative herbicide and pesticide markets of the big chemical firms. Indeed, anticipating these developments these firms are themselves in the forefront of research into the new techniques. (For more detailed discussion of biopesticides, see Lisansky, 1989.)

Plant tissue culture remains a vitally important area. A very large number of crops, from lettuce, through orchids, to Christmas trees, are now routinely raised through tissue culture rather than by seed. But the technique has been carried further. Methods for fusing plant cells enable cells from wild (and disease-resistant) varieties to be fused with cultivated varieties of a plant, and the hybrid cells can then be regenerated into fertile plants with the desired disease resistance. As with other types of plant genetic engineering, these techniques have been more successful with dicotyledons than with the monocot cereal crops.

The effect of biotechnology on *animal husbandry* has been influenced both by breeding methods and by developments in pharmaceuticals (the use of vaccines, growth hormones etc.). (See Cross, 1989, for a full discussion.) Public resistance may well limit the extent to which transgenic animals are bred, but in vitro fertilisation and implantation is now commonplace and enables far greater control over stock size and quality (e.g. the development of low-fat animals) than hitherto.

Another controversial area is the use of *growth and other hormones to enhance animal development*. The EC, for example, bans the use of growth hormones in cattle production, and it has also resisted the use of BST (bovine somatotropin), a naturally occurring protein in cows which enhances milk yield. The latter raises interesting issues, since, unlike growth hormones which can produce gross and ungainly animals, BST has no effect on the animal other than to increase milk yield so that, for farmers, it should mean keeping and feeding fewer cows (i.e., lower costs) for the same amount of milk. However, given the chronic problems of surplus in dairy products in the EC, a development which actually increases yields was hardly welcome! There is also concern about the long-term effects of continuing dosage of the hormone both on the animals and upon humans consuming the milk.

As with humans, the new biotechnology is bringing *new and improved vaccines* for such diseases as foot and mouth disease, scours, swine fever, etc. There is also study of proteins which are natural appetite modifiers such as the opiates and the endorphins, which should result in a better understanding, for both animals and humans, of the biological mechanisms that create appetite.

Finally *animals themselves are being used as 'bioreactors'* to produce rare proteins. For example, TPA (see table 11.3) has been engineered into the mammary glands of mice so that the protein is produced simply by milking the animal. (It does, of course, have to be isolated and purified.) Experiments along similar lines are being carried out using sheep and goats.

Fine Chemicals

Fine chemicals are used extensively in process industries such as *food processing and textile finishing*. (See Sharp, 1985, for a fuller discussion.) Some of the products used already derive from a biological route, via extraction from plant or animal material. Rennin, for example, the enzyme used for cheese making, is still largely derived from the stomachs of calves; spices such as chili are extracted from plants. Others are derived via fermentation. Citric acid, for example, is derived by fermentation from a molasses feedstock; antibiotics (which really are speciality chemicals) are produced by fermentation from a sugar or cereal-based feedstock. The amino acids, for example glutamic acid (monosodium glutamate – a food additive) or lysine (an animal feed addition) are produced in bulk by fermentation methods.

Enzymes are protein molecules which catalyse chemical reactions. There is a wide range of *industrial enzymes* already in use, such as subtilisin (the biological agent added to detergents) and glucose isomerase (used for turning starch-based materials into high fructose syrup). Potentially, enzymes could be used far more extensively than at present to manufacture speciality chemicals, but, as with the human genome, the problem is lack of knowledge about enzymes which exist and how they function. There are also problems in coping with unwanted by-products of the catalyst/fermentation route.

Potentially all such products, enzymes and the other fine chemicals, can benefit from developments in biotechnology:

  i as a result of improved methods of fermentation which increase yield and decrease wastage;
 ii by using rDNA techniques to improve strains and hence increase yields;
iii by the application of new methods of biotransformation which provide short cuts and new routes to existing products;
 iv by using the techniques of protein or molecular engineering to modify existing chemicals, or develop new ones with enhanced characteristics.

Until recently there was little interest in this sector, but new developments in biotransformations (the route by which organic substrates are modified by enzyme action) have reawakened interest. Japanese firms are leading the way, but others are now following their lead. Protein engineering and the development of abzymes have increased interest, as has the increasing concern for the environment and low-pollution alternatives to traditional chemical routes. Much fundamental work, however, remains to be done in this area.

Commodity Chemcals and Biomass Energy

The bulk production of *ethanol, acetone and butanol* from biomass feedstocks offers a potential route for biotechnology into the production of bulk chemicals. Although countries such as Brazil and the US have had considerable success in the development and use of *biomass alcohol* as a petroleum substitute, it is still not judged to be economically viable, and is justified largely on the basis of Brazil's shortage of foreign exchange and the need to find uses for US farm surpluses.

(The US have in fact scaled down drastically their gasohol project.) If the economics do not justify this route to a directly-used fuel, this is doubly true of the biomass route to chemical feedstocks. While protein engineering may increase yields and offer a technically feasible method of breaking down ligno-cellulose, the route is only likely to become viable when a severe shortage of fossil-fuel energy forces prices up very substantially. (The arguments set out in Spinks, 1982, hold as well today as they did on that date.)

Food and Drink Industries

The potential application of biotechnology to the food and drink industries is substantial; the impact as yet is relatively small although there is now (1990) increasing interest in this area. There are three main ways in which the industry will be affected.

i *New ways of producing old products* – for example, the extraction of essential oils from plant tissue cultured in the factory rather than from plants culled from the field. If put into practice extensively, this could hit exports of many Third World products, ranging from coffee and cocoa to the essential oils and spices.

ii *New products* – for example, mycoprotein, a single cell, fibrous protein product, developed originally by Rank Hovis and McDougall in the UK, but now manufactured and marketed by ICI as a high-quality, vegetarian protein.

iii *Improved techniques of production* – for example, quicker acting yeasts to cut down baking/brewing times; the use of biosensors to monitor and control process methods.

Waste Management and Pollution Control

The anaerobic digestion of waste materials is already widely applied to the treatment of sewage and animal slurries. The wider application of biological methods to the problems of *waste management*, and in particular *hazardous waste disposal* is at present limited, although the potential is great. In practice nearly every substance is biologically degradable. It is a matter of identifying the appropriate microorganism, and the metabolic pathway (i.e., route) by which it attacks the pollutant. Its characteristics can, if necessary, be altered by genetic engineering to make it more effective. Unfortunately, however, since it is usually cheaper to use landfill sites than to treat the pollutant, this rich application of biotechnology is as yet little used. The OTA (1988) concluded that only a tough regulatory programme would change the situation in the US. Some countries in Western Europe, for example Holland, are much further advanced. But it remains an area that requires public money and public action if the full potential is to be realised.

Mining and Mineral Ore Leaching

The microbial *de-sulphurisation of coal* and the *leaching of heavy metals* from solid wastes are both potential applications of biotechnology. In North America, something like 10 per cent of copper comes from leached ore. As yet the de-sulphurisation of coal is still very much in the R&D stage. The OECD concludes '[Both] will probably develop at a more sluggish pace. The reason for this . . . lies [in] the economics of raw materials – processing costs, capital costs and technological constraints which will work against biotechnology solutions until the cost/technology parameters shift' (OECD, 1989, p. 28).

NOTE

[1]   Monoclonal antibodies are cells derived from fusing antibody-producing cells from the spleen with myeloma (tumour) cells which multiply themselves fast, thus enabling the production of large quantities of pure antibody.

REFERENCES

Blundell, T. L. and 16 others (1989). Protein Engineering and Design, in W. G. Potter (ed.) (1989).
Cross, B. A. (1989). Animal Biotechnology, in W. G. Potter (ed.) (1989).
Flavell, R. B. (1989). Plant Biotechnology and its Application to Agriculture, in W. G. Potter (ed.) (1989).
Lisansky, S. (1989). Biopesticides: the next revolution. *Chemistry and Industry*, No. 15, 7 August 1989, pp. 478–482.
Murray, K., Stahl, S. and Aston-Rickardt, P. G. (1989). Genetic engineering applied to the development of vaccines, in W. G. Potter (ed.) (1989).
OECD (1989). *Biotechnology: Economic and Wider Impacts*. OECD, Paris.
OTA (1986). *Technology, Public Policy and the Changing Structure of American Agriculture*. US Congress, Office of Technology Asessment.
OTA (1988). *New Developments in Biotechnology, 4: US Investment in Biotechnology*. US Congress, Office of Technology Assessment.
Potter, W. G. (ed.) 1989. *Biotechnology: Spinks eight years on*. The Royal Society, London.
Sharp, M. (1985). The New Biotechnology: European Governments in Search of a Strategy. Sussex European Papers, No. 15, obtainable from SPRU, University of Sussex, BN1 9RF.
Spinks, A. (1982). Alternatives to Fossil Petrol. *Chemistry and Industry*, February.
Winter, G. P. (1989). Antibody Engineering, in W. G. Potter (ed.) (1989).

12

The Promise of Biotechnology for Vaccines

Phyllis Freeman and Anthony Robbins

The application of biotechnology to prevention of infectious diseases can become one of the most significant uses of the new science of molecular biology by reducing disease, disabilities and deaths around the world. This potential was recognized in the 1970s by biomedical scientists and modest steps have been taken toward realization of this potential. In order to appreciate the advantages biotechnology offers to public health worldwide one must understand the role vaccines have played in disease prevention to date and the role genetic technologies play in strategies for vaccine disease prevention in the future.

The Value of Immunization in Public Health, 1798–1980

The allure of vaccines and of their use in programmes of immunization is best illustrated by the global triumph over smallpox. For more than three thousand years smallpox panicked populations worldwide: the fatality rate was high and severe disfigurement and blindness were common. It was the leading cause of death in children under ten in 18th century England when Edward Jenner, an English country doctor, invented the first vaccine and put it to use in humans in 1798.[1]

Jenner's dream of global smallpox eradication took 175 years to fulfil. Political and economic impediments frustrated more rapid interruption of the spread of smallpox. Advances in epidemiology, virology, immunology and other fields enhanced the safety and efficacy of smallpox vaccines, their production and the delivery to humans during those years, but the scientific advances alone were not sufficient to ensure the triumph over even that one disease.

At the close of World War I smallpox was still endemic in many affluent countries as well as in the developing world. Unfortunately it was not until after World War II that immunization programmes gained sufficient public acceptance and governmental support to result in elimination of smallpox from most industrial nations. By 1967 extraordinary worldwide cooperation coalesced in the World Health Organization's Intensified Smallpox Eradication Programme which targeted 31 countries where endemic smallpox was still causing millions of deaths each year. After only one decade of global cooperation smallpox was eradicated from the world.[2]

In the 180 years between the invention of that first human vaccine and the recognition that biotechnology could enhance vaccines for disease prevention, approximately twenty additional vaccines were developed. Half of those remain in general use, but many are not optimal in terms of effectiveness and safety.[3]

Of the vaccines invented since Jenner's seminal discovery and until the first vaccine produced using recombinant DNA arrived on the market in 1986, it is those which prevent diseases in young children which have yielded the greatest gains in public health. Four out of every five children in the world are born in developing countries and many die or are permanently disabled by diseases potentially preventable with the development and use of new and improved vaccines. For example, acute respiratory infections and diarrhoeal diseases cause the greatest number of child deaths in the developing regions; and some tropical diseases caused by parasites, especially malaria, take an immense toll in young lives.[4] While a much lower percentage die from thesediseases in industrial countries where treatments are more readily available, the costs of the illnesses, disabilities and deaths are great and should become largely unnecessary as new and improved vaccines become available for use.

Vaccines: An Affordable Approach to Improve Health in the Third World

The international success in eradicating smallpox emboldened the World Health Organization to begin a global campaign against 6 childhood killer diseases for which vaccines were available. Those vaccines were prepared in industrial countries where the manufacturers had already recovered their research and development investments and were willing to sell the vaccines to the United Nations for use in developing countries at near the cost of production. These unprecedented low prices were offered as a gesture of good will or as a strategy for enhancing the sales of other products in the same markets. By distributing vaccines against tuberculosis, whooping cough, diphtheria, tetanus, polio and measles, some of which had already been in widespread use in industrial countries for many years, the World Health Organization and UNICEF have helped countries in Asia, Africa and Latin America to improve vastly their childhood immunization programmes.[5]

When this Expanded Programme on Immunization began, in 1974, only 5% of the world's youth had been fully immunized against these 6 diseases. By 1990 it is expected that over 70% will have received all of the recommended doses during the first year of life. UNICEF has annouced that by the start of 1989 1.5 million lives were being saved each year by these vaccines.[6]

The same delivery systems which are now reaching approximately 125 million children annually with the six older vaccines could be used to deliver new and improved ones to the Third World as soon as they become available – but only if the vaccine prices are affordable. These countries, or the United Nations programmes acting on their behalf, pay an average of $ .05 (US) per dose or $ .50 for all the recommended doses. That is only a fraction of the prices charged for the same vaccines in the North American and European markets.[7]

Although the health of adults can also be improved by the use of vaccines, no society has used adult preparations to the same extent as those for children. The new technology offers immense health advantages to adults, but even prosperous industrial nations pay enormous costs for treatment rather than focusing on prevention through lower-cost immunization programmes against pneumococcal pneumonia and influenza.[8]

The Application of Biotechnology to Vaccine Disease Prevention

Modern molecular biology can be expected to strengthen the public-health impact of vaccines and immunization programmes around the world. Just as the efforts of the United Nations have revolutionized the capacity to procure affordable vaccines and to deliver them to individuals in the most remote of communities, the new science has revolutionized the study of infection. If this new knowledge is con-scientiously applied to the communicable disease problems of the entire world, there can be vast improvements in health.

The new techniques of molecular biology usher in an era of syste-matic study of infectious disease and of achieving protective im-munity for humans; ushered out is the reliance on trial and error which characterized vaccine making from Jenner's day to the 1970s. Using the sophisticated techniques of biotechnology, such as mono-clonal antibodies, researchers can begin to satisfy centuries of curios-ity about debilitating diseases and their prevention: What is the genetic structure of the 'pathogen' or disease-causing organism? How does the pathogen enter the human 'host' and how does the disease occur when it does? What elements of the offending organism are 'antigens'; that is, which elements trigger immune response? And which of those responses protects against serious illness? Using the antigen, what are the other elements necessary to be included in designing a safe and effective vaccine? How can immunity to each disease be sustained over many years so that individuals do not need to be located for 'booster' doses of vaccine after the original innoculation? How can quality and efficiency be maximized in industrial-scale production of vaccines? And how can multiple antigens be combined in one vaccine to protect against a number of diseases, avoiding repeat visits for immunization?

The traditional methods for making vaccines required scientists to 'culture' or grow the pathogenic organism outside of humans in order to study it, and then to design and produce the product as best they could. Since there was success in culturing some bacterial and viral pathogens, but never with a parasitic organism, all vaccines to date combat bacterial or viral diseases.[9]

Where vaccines were developed, there were two approaches to design. One involved using an 'attenuated' or weakened strain of live pathogens which, when introduced into the human host, produced immunity instead of disease. These have been called 'live' vaccines; smallpox, oral polio and measles vaccines have been made in this way. The other approach relied on 'killed' or deactivated pathogens, or fragments, to produce immunity. The vaccines against whooping cough, inactivated polio, diphtheria and tetanus are examples of killed vaccines. The designs of the older vaccines are not ideal because there was no method to permit selection of only that material which would provide the most effective and safest vaccine without risk of side effects.[10]

Biotechnology can be used to identify and characterize the precise structure of any organism and to screen all elements to determine whic portion is the antigen which will stimulate a protective immune response. Once the antigen has been precisely identified there are several ways to design a vaccine:

1. Scientists can systematically attenuate or alter the pathogen at the molecular level so that it maintains its immunogenic qualities but is deprived of those which make it toxic.[11] This should reduce complications such as those associated with the existing oral polio vaccine. While it has saved countless lives, it has also caused a few cases of paralytic polio. This occurred when the polio organism,

attenuated with earlier and less precise methods, reverted to its virulent form. New techniques will permit genetic engineering of antigens designed for maximum effectiveness while reducing the risk of injury.

2. Scientists can insert the genetic information which 'codes for', or controls the production of, the desired antigen (i.e. against Hepatitis B) into a foreign cell system such as a yeast or bacteria which grows very efficiently. These new recombinant organisms are then grown in large batches to produce the needed amount of antigen. This is a modern adaptation of the design of 'killed' vaccines.[12] The new vaccines made in this way should be safer because they exclude altogether the unneeded parts of the pathogen. Moreover, industrial-scale production can be as simple as the fermentation of beer. The current process is more delicate and less predictable. The recombinant Hepatitis B vaccine licensed in 1986 has become the prototype for work of this type.

3. Scientists can use chemical synthesis to construct antigens containing only the desired elements as determined using genetic technology. These antigens are made up of chains of linked amino acids (called polypeptides) or of linked sugars (polysaccharides).[13] Although this approach offers precision, the vaccines may be weakly immunogenic. To strengthen their protective effects the vaccine designer selects an 'adjuvant' or a chemical to 'conjugate' or combine with the antigen to strengthen the immune stimulus. The stronger immune response may also last longer. Adjuvants also need to be studied carefully to ensure that those selected to enhance immunity do not increase the risk of side effects. This approach is a second way to build on the concepts used in making the older 'killed' vaccines.

Conjugated vaccines offer an advance in immunization of infants. For example, an *Hemophilus influenzae B* vaccine that is ineffective in children under two years of age can be conjugated to produce a vaccine effective in infants, the group most in danger of death or disability. Similarly, the number one killer of infants, pneumococcal pneumonia, may be prevented by a vaccine being designed in this way. Although a traditional 'killed' vaccine has existed for adults for some years, a vaccine effective in children may be one of the greatest breakthroughs in public health worldwide.

4. Scientists can make one vaccine to protect against multiple diseases by excising the genetic materials coding for each of several

different protective antigens and inserting them into a 'vector', which is an organism used as a carrier.[14]

This work is in an experimental stage and has not been approved for application to humans. The first test was completed using vaccinia, the virus of smallpox vaccinations, into which was inserted the genetic material for a protective rabies antigen. The result was a benign vaccination that protected dogs against a killer disease that also threatens humans. Because the vaccinia genome is large enough to accept genetic coding for additional antigens, scientists anticipate that infection with a single virus such as vaccinia can confer immunity to multiple pathogens. More recently researchers have been able to insert genetic material for several antigens into bacteria such as salmonella or E. coli to produce experimental live vaccines against diseases such as typhoid and cholera.[15]

For gastrointestinal diseases, the vector organism is chosen so that it will be one which reproduces in the gut of the human host, thereby maintaining immunity against all of the diseases whose genetic coding for antigens have been inserted.

5. Just as antigens can be encoded in a live vector vaccine to protect against a number of diseases simultaneously, immunogenic antigens can be combined into a single 'cocktail' for simultaneous administration. This has been done for years with DPT (diphtheria, pertussis, tetanus) vaccine. More recently, scientific work which supports family planning programmes has provided insight into new technologies for 'sustained release' of antigens.[16] Long exposure to the antigen achieves a booster effect and these vaccines may have longer lasting immunity than the traditional killed vaccines.

More efficient vaccine production is also the result of biotechnology. The new designs permit the desired antigen to be the major product of the production system so that little extraneous matter needs to be removed in expensive purification steps; and carrier organisms or cell systems are chosen so that they grow very rapidly. Taken together, all of these technical improvements can have a profound impact on how many diseases can be prevented using vaccines, on how many people can receive lifelong benefits of immunity, and on the reduction of the risk of vaccine injuries.

The challenge today will be to ensure that this new science is used to its greatest public-health advantage without being impeded by econmic and political barriers which are not in the broadest interests of the

human community. The commercial incentives for optimal investment in research, development, production, distribution and use of
vaccines around the world have proved inadequate.[17] The United
Nations has made immunizing all of the world's children with six
older and now inexpensive vaccines one of its highest priorities. A
similar commitment will be required to ensure that the new technology is harnessed to apply new and improved vaccines against many
more diseases as soon as possible. As the new technology will require
new investments in research and development, special efforts will be
necessary to see that adequate investment is targeted to the products
which can reach those areas of the world where disease incidence and
prevalence is highest and where the consequences of such illnesses
are so great, owing to the lack of medical care.

Approaches to Overcoming Barriers to Development of New Vaccines where Public Health rather than Profit is the Primary Gain

Several approaches to speed development of new vaccines are being
tested. The United States Congress passed a planning law in 1986, the
National Vaccine Program,[18] requiring the government to make a
national plan of top priority diseases which could be controlled by
immunization if new or improved vaccines were developed. The
government is to concentrate public resources on promoting development and testing of vaccines against those diseases where scientists
believe the probability of success is reasonably high and on encouraging industry to participate in development, production and pricing
to meet the national public-health goals. In part, this law was
modelled on experience in the Netherlands, where cooperation between public health institutes and biologics manufacturers has
allowed the public health authorities to identify which products
will not be developed by the commercial sector, and to target public efforts on needed vaccines which have been neglected by the
industry.

The Task Force for Child Survival, an organization sponsored by
four United Nations agencies and one foundation, initiated a programme in 1989 to encourage both public and private vaccine manufacturers to develop new vaccines which would be of great value in all
of the developing regions of the world. Their goal is to reinforce the
commitment of many developing nations to preventive health
through the Expanded Programme on Immunization by increasing
the number of diseases it is equipped to control. The method involves
paying public institutes or private firms to do the development work

with funds from international donor agencies so that the final product can be available for developing countries at or near the cost of production. The Dutch and Nordic public health institutes have formed a consortium to work on vaccines of high priority for developing countries which are not attracting investment from commercial firms seeking hard-currency markets and profits.[19]

The Pan American Health Organization is exploring the possibility of establishing a regional system of new vaccine development in Latin America. It would allow for participation of all countries in the region while being centred in two of those countries which have dedicated significant resources to immunizing their populations and to research on infectious diseases, development and production of vaccines. It would be the first programme to address diseases of regional significance, such as Chagas Disease and amebiasis, where the research is on complex, parasitic organisms which are not well understood, and where the market is far too small to interest multinational commercial firms.[20]

Previously the public-health programmes of the developing countries have had no option other than to wait for firms in the industrial world to produce vaccines for their own populations and then to wait the years until the price is reduced to an affordable level. No single developing country has sufficient resources to design and produce all of the new vaccines that would be of great benefit to the health of the population.

A regional approach, then, is meant to attract international investment to supplement what the Latin American region can dedicate collectively. The combined contributions would support world class science; and, in doing so, empower the regional scientific and public health communities to address their highest priority to infectious disease problems. The Dutch–Nordic consortium has discussed support for the regional initiative by transfer of technology in instances where the consortium may develop a vaccine of mutual interest. This willingness to share the new technology is based on a long-standing commitment of these countries to foreign assistance for Third World development.

Because some of the new techniques for research, development and production of vaccines are simultaneously more sophisticated and simpler than older ones, it is likely that state of the art vaccine work can advance in the Third World more easily in the coming decades than it has in previous ones. If this approach can succeed in Latin America, it could become a model for Asia and Africa as well.

NOTES

[1] F. Fenner, D. A. Henderson, I. Arita, Z. Jezek, and I. D. Ladnyi, *Smallpox and its Eradication* (World Health Organization, Geneva, 1988), p. 277.

[2] F. Fenner, *Smallpox and its Eradication*, p. 538.

[3] Kenneth S. Warren, New scientific opportunities and old obstacles in vaccine development, *National Academy of Sciences USA*, **83** (1986), pp. 9275 and 9276.

[4] Institute of Medicine, *New Vaccine Development: Establishing Priorities, Volume II* (National Academy Press, Washington DC, 1986), pp. 58–61.

[5] Anthony Robbins and Phyllis Freeman, Obstacles to Developing Vaccines for the Thirld World, *Scientific American*, vol. 256, no. 11 (1988), pp. 126–133.

[6] UNICEF, *The State of the World's Children* (Oxford University Press, Oxford, 1989), p. 4.

[7] World Health Organization, The Cold Chain Product Information Sheets, *EPI Technical Series, 1989/90* (World Health Organization, 1989), p. v.

[8] Institute of Medicine, *New Vaccine Development: Establishing Priorities, Volume I* (National Academy Press, Washington, DC, 1985), p. 56, and Jane Sisk Willems, Cost Effectiveness and Cost Benefit of Vaccination Against Pneumococcal Pneumonia, *Infection Control*, vol. 3, no. 4, 1982, p. 299.

[9] Ronald W. Ellis, New Technologies for Making Vaccines, in Stanley A. Plotkin and Edward A. Mortimer, Jr (eds), *Vaccines* (W. B. Saunders Philadelphia, 1988), pp. 568–575.

[10] J. L. Melnick, Virus Vaccines: Principles and Prospects, *Bulletin of the World Health Organization*, vol. 67, no. 2 (1989), p. 110, and Ronald W. Ellis, *Vaccines*, p. 568.

[11] J. L. Melnick, Virus Vaccines: Principles and Prospects, pp. 110 and 108.

[12] J. L. Melnick, Virus Vaccines: Principles and Prospects, pp. 110–111.

[13] World Health Organization, *Programme for Vaccine Development and Transdisease Vaccinology: Activities and Prospects 1989*, p. 74

[14] World Health Organization, *Programme for Vaccine Development and Transdisease Vaccinology*, pp. 6 and 74.

[15] World Health Organization, *Programme for Vaccine Development and Transdisease Vaccinology*, pp. 68–69.

[16] World Health Organization, *Programme for Vaccine Development and Transdisease Vaccinology*, p. 68.

[17] Anthony Robbins and Phyllis Freeman, The Future of the Vaccine Supply for the Developing World, *Protecting the World's Children: An Agenda for the 1990s* (The Task Force for Child Survival, Carter Presidential Center, Atlanta, Georgia; Conference Report, 10–12 March 1988, Talloires, France), pp. 161–163.

[18] 42 United States Code 300 aa et seq.

[19] Anthony Robbins, New Vaccine Products for the Third World, *The Lancet*, vol. 1, no. 8639, 1988, p. 679.

[20] Phyllis Freeman, Biotechnology Sparks Interest in a Regional System for Vaccines in Latin America: A Public Interest Initiative to Control Infectious Diseases and to Support Scientific and Technological Development, *Gene WATCH*, vol. 6, nos 2–3, 1989, pp. 6–7.

GUIDE TO FURTHER READING

Institute of Medicine. *New Vaccine Development: Establishing Priorities*, Volumes I and II (National Academy Press, Washington, DC, 1985).

Plotkin, Stanley A. and Mortimer, Edward A. Jr. *Vaccines* (W. B. Saunders, Philadelphia, 1988).

Robbins, Anthony and Freeman, Phyllis, Obstacles to Developing Vaccines for the Third World, *Scientific American*, vol. 256, no. 11 (1988).

Warren, Kenneth S. New scientific opportunities and old obstacles in vaccine development, *National Academy of Sciences USA*, **83** (1986).

13

The Genome Race is
On the Road

Clare Robinson

The race to map the human genome has begun. However, as a vast international research project, the Human Genome Project (HGP) is not so much a race between research groups (though fame and possibly fortune will reward the few), as a race against time. The motives of the researchers taking part vary – for some, the goal is a complete understanding of the role and origins of all genomic components whereas, for others, it is the prospect of being able to disentangle the complex causes of human genetic disease.

The Birth of HGP

In terms of scale and international co-ordination, it is certainly a unique project in biology. Although the HGP is now recognized as an international undertaking, the concept of the HGP originated, and its early development occurred, in the USA. (An excellent and highly readable account of the background, birth and maturation of the Project has been produced by Robert Shapiro in the recently published book *The Human Blueprint – The Race to Unlock the Secrets of Our Genetic Script*, St. Martin's Press, 1991.) As with many human dreams, the idea preceded the means of achieving it. Perhaps a catalyst for turning the dream into reality was publication in 1986 (*Science*, 231, 1055–1056) of an editorial by Renatto Dulbecco. He argued that to be able to understand and combat cancer, identification of all the genes involved was necessary, and although such information could be obtained piecemeal, knowledge of the entire genome sequence would be even more useful. Developing the idea further, he called for a national effort, similar to the effort toconquer space, pointing out that many areas of biology would benefit. Dulbecco's editorial furthered ideas which had arisen independently at a meeting in 1984 organized by the US Department of Energy (DOE), and another meeting called at the University of California at Santa Cruz, in 1985. Support for these ideas was forthcoming from the DOE, which already had experience in managing large projects, and also from NIH, which has long played a major role in supporting research into aspects of human health and biology. Endorsement of these initiatives by the US Congress resulted in start-up and organizational finance being provided to both NIH and DOE for 1988–1990.

Although without official status or power, the 'Genome I' meeting, convened in October 1989 in San Diego, resulted in 1 October 1990 being designated as the official start of the HGP. With US$3 billion to be appropriated by Congress over the next 15 years to support the HGP, and the new National Center for Human Genome Research, NIH being set up, the Project was on a firm footing in the USA.

Extension of the HGP to the international scene was promoted by the appearance of HUGO (Human Genome Organization). Although conceived in 1988 (named in advance by Sydney Brenner at a Cold Spring Harbor meeting in April, with the first meeting of the international committee in Switzerland later that year), it was not until 1990 that funding from the Howard Hughes Medical Institute (USA) and the Wellcome Trust (UK) permitted the setting up of offices in Bethesda and London. The role of HUGO is to co-ordinate the efforts of all countries who are, or become, involved with the HGP – likened by Robert Shapiro to a United Nations for the HGP. Collaboration and free exchange of data should avoid redundancy of effort.

Concerns that genome projects will have long-term implications, damage as well as benefits for biomedical and other research, have been voiced.[1] In addition, there are the social, legal and ethical issues of genome research. These include such questions as: will the misuse of data on genetic susceptibility to disease, which the project will eventually yield, impinge upon the freedom, opportunities and rights of the individual?; should such vast amounts of money be invested in a project whose potential benefits will not be seen for some time, when other areas of research whose potential health benefits are more immediate are neglected?; and will the HGP divert resources from all other areas of biological research, resulting in those areas being starved not only of finance, but also of trained researchers, well into the future? These, and other issues are on the agenda for both immediate and long-term study as part of the HGP,[1] as well as receiving discussion in the public domain. However, the project is already well under way, and with increasing international support and investment, has gained momentum of its own, such that it is no longer a question of 'if?', but 'how fast?'

The Participants

The HGP is an international project, involving co-ordination of many nationally and internationally administered official genome programmes in the USA, USSR, Japan, the European Community (EC) and the UK, France, and Italy. Major progress will depend on co-operation and the pooling of resources and the free exchange of data – a source of some contention in the past. As the project grows, the need for co-ordination to maximize efficiency increases, and co-ordinating roles are (or will be) played by CEPH (Centre d'Étude du Polymorphisme Humain), HHMI (Howard Hughes Medical Institute), UNESCO (United Nations Educational, Scientific and Cultural Organization) and HUGO.[2]

The HGP Milestones

The goals of the international HGP have evolved considerably since the concept was first proposed, as a result of technological advances and, perhaps even more significantly, extensive debate as to the priorities and organizational strategies that should be adopted. Its ultimate goal is the determination of the nucleotide sequence of the human genome and the genomes of model organisms. Achieving these goals necessitates highly organized progress on diverse research fronts, with targeted milestones scheduled over the next 14 years.[1] There is pressure to elucidate the sequence as rapidly and efficiently as possible while minimizing costs, and improving the technology available is a focal point.

The first phase of the project (1991–1995) concentrates on developing improved technologies for mapping and sequencing, and informatics systems for handling the vast amounts of data that will be generated over the course of the project. Low-resolution genetic (2–5 cM) [cM = centimorgan] and physical maps of the whole human genome, and completion of several pilot sequencing projects (both human and model genomes), should also be achieved by 1995. Genetic and physical maps of simple model genomes should be completed, including a high-resolution map (1–2 cM) of the mouse genome.

The second phase (1995–2000) should see further improvement in technology, which will enable production of refined physical and genetic (1–2 cM) maps of the human genome, and larger-scale human sequencing projects and completion of sequencing simple model genomes.

By 2005, the sequencing of the human genome (with the exception of tandem repeat regions) is scheduled to have been completed. During this final stage, a major increase in terms of understanding human disease should result from analysis of human polymorphisms, and knowledge of the complete sequence of medically important areas of the human genome and corresponding regions of model genomes. By 2005, the integration of databases to create a unified information sourc that can correlate sequence and map to known biological characteristics should have been achieved.

Whether these goals can be achieved on time will depend very much on the rate at which the required technology develops, as well as on co-operation between, and financial commitment from, all participants.

What does HGP Involve?

Mapping – finding the way

Maps are linear representations that describe the organization of a set of landmarks, using a defined system of measurement based on co-ordinates.

Physical maps can either be cytogenetic or molecular in basis. Cytogenetic maps order genetic loci based on their relative position and order along a chromosome, often in relation to the banding patterns obtained after differential staining. Fluorescence *in situ* hybridization (FISH) (see J. Korenberg, *TIBTECH*, vol. 10, Jan/Feb 1992, pp. 27–32) is a major source of input data for this type of map. Molecularly based physical maps use landmarks based on sequence features such as restriction-endonuclease cleavage sites, sequence-tagged sites (STSs)[3] or single-copy probes (SCPs).[4] STSs, short, single-copy DNA sequences that can be detected using the polymerase chain reaction (PCR), have been advocated as providing 'a common language between all types of mapping':[3] a region of DNA can be mapped by determining the order of a series of STSs and measuring the distances between them. Long-range physical maps are usually constructed by measuring the length of large DNA fragments which carry a DNA marker using pulsed-field gel electrophoresis (PFGE) separation. These fragments are generated using rare-cutter restriction endonucleases (i.e. enzymes that cleave DNA only infrequently). Cloned DNA in the form of yeast artificial chromosomes (YACs), cosmids or phage vectors carrying genomic inserts can then be characterized in terms of their content of physical markers such that their ordered relationship is established and groups of overlapping clones or 'contigs' are generated. 'STS-content mapping' of clones involves screening by PCR to detect which clones contain particular STSs, and then using the presence of other STSs to order and define the overlaps between clones.[5] The map must be in a form which provides clear and easy access for the entire mapping community, and it seems likely that the idea of using STSs,[3] which is compatible with strategies whereby probes and primer pairs are aligned along YAC contigs (see below), will be adopted.

Genetic linkage maps represent the genetic distance between markers: linkage maps base distance on centimorgan (cM) units, a measure of the frequency of recombination between the specified genes. Genetic distance does not have a constant relationship with physical distance, and, whereas physical distances are additive, genetic distances are not. Although the order of loci in physical and genetic maps will be the same, there is no easy way of converting from one to the other. A map unit of 1 cM indicates a 1% chance of recombination, but recombination frequencies vary with chromosomal region and sex, as well as with the genome size.

The initial phase of the HGP (1991–1995) has focused on establishing human maps – both physical and genetic – of landmarks such as polymorphisms, genes and specific DNA sequences. The process of establishing these maps in conjunction with sequencing efforts will yield the start of a composite DNA sequence.

Mapping and sequencing – converging routes to the same destination

Sequencing and mapping large genomes are inextricably linked, whether a 'top-down' or 'bottom-up' approach is considered. The top-down strategy starts at the level of intact genomes, and proceeds, for example, by the separation by PFGE of large DNA fragments, the physical and genetic linkage of DNA markers, and the construction of long-range maps. The construction of these maps leads to the generation of cloned regions of DNA, which can then be sequenced. The bottom-up strategy, on the other hand, starts at the level of nucleotide sequence in a large number of random cosmid or bacteriophage clones, and proceeds by assembling longer-range sequence by identifying overlapping sets of cloned sequences (contigs).

Quite a range of techniques have been developed for linking individual clones. Contigs can be assembled simply by identifying overlap between clones at the level of sequence. Alternatively, overlapping clones can be identified by 'finger-printing' methods. These can depend on identifying common restriction enzyme cleavage patterns in the overlaps, or identifying sequence overlap through the use of hybridization probes (e.g. repetitive sequences, STSs, synthetic oligonucleotides or whole clones). Repetitive screening of large libraries of clones with many different probes can be a tedious and time-consuming process and a range of methods have been proposed for arranging the clones at high density in two-dimensional arrays to form a matrix prior to screening. For example: the ordering of phage or cosmid clones by detecting identifying sequences with oligonucleotide probes;[6] or multiplexing[7] – the simultaneous analysis with many probes of a whole library of cosmid clones (the pairwise comparison of data generated by the mixed probes can be decoded using algorithms to predict the order and linkage of all clones in the collection into contigs); the sequential screening of complex libraries with 50–100 oligonucleotide probes, and computational analysis of the similarities of hybridization characteristics of each clone, to order the clones.[8] These techniques are amenable to automation, which should help to speed up considerably the process of ordering and mapping cones.[9]

In addition to the ordering of random libraries, mapping also involves the directed search from one clone in order to clone the sequence corresponding to immediately adjacent insert DNA, i.e. 'chromosome walking'. Two major approaches can be used for walking from one clone to the next: (1) screening clone collections using end-clones of vector-inserts as hybridization probes, or (2) by PCR techniques that remove the need for end-cloning (end-sequencing of inserts to design PCR primers, inverse PCR, Alu–Alu PCR, and Alu–vector PCR [see Glossary; and A. Rosenthal, *TIBTECH*, vol. 10, Jan/Feb 1992, pp. 44–48]).

A problem which persists is that whether a top-down or bottom-up approach provides the starting point for a project, there is a gap in terms of scale between the resolution that can be obtained by genetic-mapping techniques and the maximum length of contig sequence that can be assembled – as attributed to David Botstein,[5] the gap between 'the cosmid and the centimorgan'.

The standard recombinant-DNA cloning technology was limited until recently to small plasmid, cosmid and viral constructs – able to carry exogenous DNA inserts of not more than 50 kb. Although such systems are suitable for the manipulation and analysis of genes and small gene clusters where the information is tightly packed, most genes from higher organisms span great distances of DNA. Standard techniques may be used for the cloning of such genetic material in a large number of overlapping clones. The subsequent use of such clones, however, is unwieldy, error-prone and time-consuming. Cloning systems which enable greater lengths of sequence to be included in fewer clones improve efficiency and accuracy as well as the continuity of the final map.

The development of YAC vectors, which are capable of carrying much larger inserts and can be assembled into contigs spanning megabase lengths, but can also be manipulated by conventional molecular biology techniques to obtain the nucleotide sequence, seems to be closing the gap. The centimorgan represents the resolution obtainable by linkage-mapping studies, and on average corresponds to approximately 1 Mb of the human genome. Until the development of YACs, contigs derived from phage or cosmid clones usually only spanned 100–200 kb; individual YACs can carry several hundred kilobases of insert, such that YAC contigs approach the megabase range (see R. Anand, *TIBTECH*, vol. 10, Jan/Feb 1992, pp. 35–40). Are YACs the answer to all cloning needs then? Probably not, since for many manipulations, such as sequencing, there is a need to subclone into vectors such as phage or cosmids to obtain more manageable-sized fragments. In addition, YAC-based mapping is not without its problems: multiple fragments of chromosomal DNA may be co-cloned within the same YAC, and certain chromosomal regions appear to be inherently unclonable, yielding clones which are deleted for some segments of DNA.[10]

Genome-sequencing projects will probably proceed by integrating the capabilities of a range of vector systems (see P. Little, *TIBTECH*, vol. 10, Jan/Feb 1992, pp. 33–35). Mapping genomic spans in the megabase-plus range will continue to involve the use of PFGE and rare-cutter enzymes, though other new techniques which can link cytogenetics and molecular biology are being developed. An example is the use of laser 'microtechnology' for cleaving whole chromosomes visible under the microscope, and then physically manipulating the resulting fragments, enabling the cloning of megabase-sized fragments from defined chromosomal regions (see K. O. Greulich, *TIBTECH*, vol. 10, Jan/Feb 1992, pp. 48–51).

Sequencing technology

Sequencing technologies will need to improve dramatically in speed and capacity to handle the inevitable increase in DNA-sequencing activity associated with the world-wide genome initiative. Current methodologies are just not suited to large-scale sequencing projects. Until about five years ago, practical DNA-sequencing techniques were limited to the manual radioactive methods described by Sanger,[11] and by Maxam and Gilbert.[12] The introduction, in 1986, of a fluorescence-based modification of Sanger dideoxy sequencing enabled the DNA sequencing steps to be automated and the introduction of computerized analysis of the gel-based information: these were subsequently developed as in commercially available systems.

Motivation for developing novel sequencing approaches is strong – sequencing is a tedious and slow process. The two basic approaches to speeding up the progress are increased automation (robotics) and developing alternative technologies or new chemistries. New technologies, such as solid-phase sequencing systems (see M. Uhlén, *TIBTECH*, vol. 10, Jan/Feb 1992, pp. 52–55) offer the opportunity for semi-automation. Radically different new approaches, which are based on novel sequencing reactions as well as altered analytical methods, such as the flow-cytometry sequencing described by J. Harding (*TIBTECH*, vol. 10, Jan/Feb 1992, pp. 55–57) may not be implemented for some time. The introduction of capillary electrophoresis and other 'microsequencing' technology may permit automation to follow a different route and thereby facilitate increased throughput.[13] Although reduced sample volumes save on costs by reducing the amount of template and reagents required for sequencing, automation of such sequencing approaches may not prove to be straightforward. Many current 'automated' procedures do not extend beyond gel-reading equipment; attempts to increase automation and integrate different stages of the process are hampered by the sequencing chemistry, whichwas developed for small-scale projects. Thus, for advances in automation to succeed, they will probably need to be developed largely in conjunction with new sequencing technologies.

A range of techniques has been proposed for DNA sequencing using hybridization techniques (depending on the observation that mismatched and mismatch-free hybridization of oligonucleotides of 8–20 bases can be distinguished):[14] sequencing by hybridization (SBH),[15] fragmentation sequencing (FS)[16] and oligonucleotide hybridization sequencing (OHS).[17] All these techniques are based on the idea that the sequence of long fragments of DNA or even entire genomes could be built up from overlapping shorter sequences; all obviate the need for the enzymatic or chemical reactions and electrophoresis steps which are part of traditional sequencing methodologies.

An even more ambitious idea is the direct sequencing of DNA under the microscope. Recent advances in scanning tunnelling microscopy (STM) have enabled atomic resolution images of DNA to be obtained,[18] and, coupled with single-molecule spectroscopy, could possibly be used for DNA sequencing.[19]

So What Should be Sequenced?

With the HGP well underway, the focus is turning to improving efficiency and maximizing returns (in terms of biologically and clinically relevant data) on the investment. Careful budgeting and international organization (see C. Cantor, *TIBTECH*, vol. 10, Jan/Feb 1992, pp. 6–8) to avoid duplication of effort is one target area. The increased automation of most aspects of genome mapping and sequencing is essential to ensure a reasonable rate of progress, since the majority of procedures call for handling of vast numbers of component samples, whether at the level of initial sequence analysis, repetitive screening, or the management of ordered reference libraries of clones.[9] Use of the vast amounts of data being generated necessitates efficient means of storing, communicating, analysing and crossreferencing. Current technology cannot cope, and new systems are being devised (see C. Fields, pp. 58–61, and G. Cameron, pp. 61–66, in *TIBTECH*, vol. 10, Jan/Feb 1992).

Another consideration is that the technology being applied to the HGP five years from now will make today's methods look primitive. It therefore makes sense to target those regions which will yield information of greatest immediate value (such as coding and regulatory regions), and leave the remaining sequence to a later date when it can, hopefully, be obtained with less effort. Various strategies have been proposed for identifying and isolating coding sequences from complex mammalian genomic DNA (e.g. exon amplification, where by exons are isolated from cloned genomic DNA by selecting for functional 5′ and 3′ splice sites,[20] and cDNA cloning (see C. Venter et al., *TIBTECH*, vol. 10, Jan/Feb 1992, pp. 8–11). Although major advances in developing machine-learning and neural-network techniques for identifying functionally significant regions within 'naive' sequence are being made (see R. Mural et al., *TIBTECH*, vol. 10, Jan/Feb 1992, pp. 66–69), the techniques are not yet sufficiently advanced to replace a selective approach by 'blanket' sequencing.

Each milestone along the route of the HGP which is passed will increase our understanding of gene function and of the basis of genetic disease. Not only will the knowledge gained alter our perspective of the human evolutionary genetic inheritance, but the technological and biological advances made as part of the HGP initiative over the next 15 years will foster even closer links between the diverse research disciplines of biotechnology.

A Sequencing Glossary

Alu sequence Member of a family of interspersed repetitive dimeric DNA sequences (~ 300 bp long) in the human genome. Alu sequences form the major family of human short interspersed repeat sequences and are dispersed throughout the genome, with an average of 4 kb between copies.

Alu-PCR The use of PCR primers conforming to two Alu sequences in inverted orientation permits direct amplification of sequences between the Alu repeats, from complex backgrounds (such as human chromosomes in hybrid cells), as well as from YAC, cosmid or lambda clones – this technique is known as Alu–Alu PCR. A related technique, Alu–vector PCR, also uses PCR primers, one derived from an Alu repeat and one derived from vector sequence. (See PCR.)

cDNA Single-stranded complementary DNA. cDNA is synthesized from an mRNA template by reverse transcriptase.

cDNA cloning Molecular cloning of t coding sequnce of a gene from its transcript – does not contain any introns (non-coding sequences).

Centimorgan (cM) Measure of genetic distance (the distance that separates two genes between which there is a 1% chance of recombination) dependent on the size of the genome (e.g. *Arabidopsis*: 1 cM = 139 kb; Human: 1 cM = 1108 kb).

Centromere Site of chromosomal attachment to the spindle, required for accurate distribution of chromatids to daughter cells.

Chromosome banding Differential staining of metaphase chromosomes by a variety of techniques. Aids in chromosomal identification.

Chromosome flow sorting A method for sorting individual chromosomes based on quantification of laser-induced fluorescence of stained chromosomes.

Chromosome walking A procedure for cloning an ordered array of overlapping contiguous regions of chromosomes, based on the systematic isolation of successive clones by using th sequence of the end of one clone to probe by hybridization for the adjacent, overlapping clone.

Cloning vector Any DNA molecule capable of autonomous replication within a host cell into which exogenous DNA sequences can be inserted.

Contig A group of cloned DNA sequences which are contiguous.

Cosmid A hybrid bacteriophage cloning vector used for cloning long DNA fragments. Characterized by possessing the lambda *cos* site, which is required for packaging into the lambda capsid. After introduction into the host *E. coli*, the vector replicates as a plasmid. Cosmids are useful for constructing a gene library because only hybrid plasmids with large DNA inserts are packaged.

Cot value Parameter used in analysis of renaturation of DNA genomes (original DNA concentration × time). Highly repetitive sequences renature at low Cot values, unique sequences renature at high Cot values.

DNA library A store of cloned DNA in recombinant cloning vectors containing chromosome-specific fragments. May be a cDNA- or genome library.

Genetic mapping Production of a representation of the relative positioning and genetic distance separating genes, based on the frequency of recombination.

Genome mapping Production of an ordered set of overlapping clones that cover the entire genome.

Hybridization probe A small labelled nucleic acid molecule used to detect complementary sequences through base pairing (hybridization).

HTF islands '*Hpa* II-tiny-fragments' islands (also termed CpG islands). Single-copy, unmethylated loci in vertebrate genomes, containing a high density of *Hpa* II restriction enzyme sites, in which the dinucleotide CpG is abundant – associated with transcribed regions (i.e. genes).

Intron Any intervening sequence in eukaryotic genes that interrupts the coding sequence (exon) and which is transcribed but processed out of the precursor RNA to yield the mature RNA – does not therefore code for a protein product.

Inverse PCR The use of end-sequences in outward orientation as PCR primers, useful for amplifying unknown sequences adjacent to known sequences. (See Alu-PCR.)

Junk DNA In eukaryotes; any DNA sequences that have no apparent function.

LINE (Long Interspersed Nucleotide Element) – Found in the chromosomal DNA of eukaryotes (~6–7 kb in length, and ~10^4 copies in mammalian genomes).

Map Representation of the relative positions of genes or restriction sites and the distance between them. (See Genetic mapping.)

Map distance The distance, in terms of percentage recombination, between linked genes. Map distance is measured in centimorgans (see Centimorgan).

ORF (Open Reading Frame) – A stretch of nucleotide sequence with an initiation codon at one end, a series of triplet codons and a termination codon at the other end: potentially capable of coding for an as yet unidentified peptide or protein.

PCR   (Polymerase Chain Reaction) – A technique for the enzymatic *in vitro* amplification of specific nucleotide sequences between two convergent primers that hybridize to the opposite strands. The product of strand synthesis using one primer acts as template for synthesis using the other primer; repeated cycles of denaturation, primer annealing and strand synthesis result in an exponential increase in copies of the sequence bounded by the primers. Inverse PCR acts by the same mechanism to amplify DNA sequences flanking a known sequence by use of primers oriented in the reverse orientation. (See Alu-PCR.)

Primer   A short oligonucleotide which pairs with a complementary strand of DNA, providing a free 3′ OH terminus for extension of the other strand.

Restriction mapping   A method for rapid mapping of large segments of DNA by identifying the positions of restriction enzyme cleavage sites, and any insertions or deletions which alter the pattern of cleavage relative to a standard sequence.

RFLP   (Restriction Fragment Length Polymorphism) – Variation in the length of restriction fragments due to the insertion or deletion of restriction sites or intervening sequence, or rearrangements which affect the length of sequence between sites. Can be used to construct genetic linkage maps (RFLP maps) to follow inheritance of specific mutations and genetic diseases. Linkage of an RFLP with a specific gene permits subsequent identification and ordering of other RFLPs that straddle the gene by observing recombination characteristics.

RFLP marker   Any marker resulting in changes in the length of genomic DNA produced by digestion with specific restriction enzymes.

Shuttle vector   Any multifunctional vector which can replicate in two or more organisms and can be used to transfer genes between organisms.

Somatic cell hybrid clone panel   A panel of hybrid cell clones used for mapping human genes. Each clone contains a unique combination of the 24 human chromosomes. By correlating the presence/absence of a particular human gene to be mapped with the clones in the panel, the chromosome location of that gene may be assigned. Mapping to a specific region of the chromosome is possible when the panel of hybrid clones contains different segments of a particular chromosome.

STSs   (Sequence Tagged Sites) – Short, single-copy DNA sequences that characterize mapping landmarks on the genome and can be detected by PCR. Advocated as 'common language' of physical mapping projects [Refs: Green, E. D. and Olson, M. V. (1990) *Science*, **250**, 94–98; Olson, M., Hood, L., Cantor, C., and Botstein, D. (1989) *Science*, **245**, 1434–1435]. A region of a genome can be mapped by determining the order of a series of STSs.

Subcloning A procedure whereby smaller DNA fragments are cloned from a large DNA fragment insert which has already been cloned.

Targeted gene transfer The directed transfer of gene sequences to a specific site in the genome by homologous recombination between sequences in the vector and at the site of insertion in the genome.

Telomere The sequence/structure at the molecular ends of eukaryotic linear chromosomes that stabilizes the chromosome, prevents fusion with other sequences and permits chromosome replication without loss of chromosomal sequence.

REFERENCES

1 Yager, T. D., Nickerson, D. A., and Hood, L. E. (1991) *Trends Biochem, Sci.*, **16**, 454–461.

2 Watson, J. D. and Cook-Deegan, R. M. (1991) *FASEB J.* **5**, 8–11.

3 Olson, M., Hood, L., Cantor, C., and Botstein, D. (1989) *Science*, **245**, 1434–1435.

4 Barillot, E., Dausset, J., and Cohen, D. (1991) *Proc. Natl Acad. Sci. USA*, **88**, 3917–3921.

5 Green, E. D. and Olson, M. V. (1990) *Science*, **250**, 94–98.

6 Poutska, A., Pohl, T., Barlow, D. P., Zehetner, G., Craig, A., Michiels, F., Ehrlich, E., Frischauf, A. M., and Lehrach, H. (1986) *Cold Spring Harbor Symp. Quant. Biol*, **51**, 131–139.

7 Evans, G. A. and Lewis, K. A. (1989) *Proc. Natl Acad. Sci. USA*, **86**, 5030–5034.

8 Michiels, F., Craig, A. G., Zehetner, G., Smith, G. P., and Lehrach, H. (1990) *CABIOS*, **3**, 203–210.

9 Martin, W. J. and Walmsley, R. M. (1990) *Bio/Technology*, **8**, 1258–1262.

10 Schlessinger, D. (1990) *Trends Genet.*, **6**, 248–258.

11 Sanger, F., Nicklen, S., and Coulson, A. R. (1977) *Proc. Natl Acad. Sci. USA*, **74**, 5463–5467.

12 Maxam, A. and Gilbert, W. (1977) *Proc. Natl Acad. Sci. USA*, **74**, 560–564.

13 Swerdlow, H. and Gesteland, R. (1990) *Nucleic Acids Res.*, **18**, 1415–1419.

14 Wallace, R. B., Shaffer, J., Murphy, R. H., Hirose, T., and Itakura, K. (1979) *Nucleic Acids Res.*, **16**, 3543–3557.

15 Drmanac, R., Laqbat, I., Brukner, I., and Crkvenjakov, R. (1989) *Genomics*, **4**, 114–128.

16 Bains, W. and Smith, G. C. (1988) *J. Theor. Biol.*, **135**, 303–307.

17 Khrapko, K. R., Lysov, Y. P., Khorlyn, A. A., Shik, V. V., Florentiev, V. L., and Mirzabekov, A. D. (1989) *FEBS Lett.*, **256**, 118–122.

18　Beebe, T. P., Wilson, T. E., Ogletree, O. F., Katz, J. E., Balhorn, R., Salmeron, M. B., and Siekhaus, W. J. (1989) *Science*, **243**, 370.

19　Weaver, J. M. R., Walpita, L. M., and Wickramasinghe, K. (1989) *Nature*, **342**, 738.

20　Buckler, A. J., Chang, D. D., Graw, S. L., Brook, J. D., Haber, D. A., Sharp, P. A., and Housman, D. E. (1991) *Proc. Natl Acad. Sci. USA*, **88**, 4005–4009.

14

Reproductive Technologies: Whose Revolution?

Wendy Harcourt

The rapid developments in the field of reproductive and genetic engineering are generating widespread and diverse interest, opinion and controversy. This chapter examines the new innovations of biotechnology in reproductive technologies by exploring the tension between medical science offering women and society greater choice in reproductive practice and the ethical, political and economic aspects of research into human heredity. With these new ways of facilitating human reproduction, as Edward Yoxen points out,[1] our past set of prescriptions for describing and practising reproduction is now being challenged and needs to be replaced with another.

In the following discussion I look critically at the new innovations in genetic engineering in order to suggest questions which, in my view, should be accompanying the public debate on medical interventions in reproduction. My argument is that if this new medical knowledge will enable us to make decisions about our own environment we must open medical science to public discussion based on ethical and political as well as scientific and economic concerns.

The background to this more informed public discussion should be twofold. One is a greater public knowledge of the techniques themselves – the hazards as well as the potentialities – and the other is the need for medical decisions to take into account the public's response. Before moving on to some of the questions which could be raised in a more open discussion I will first briefly outline some of the reproductive techniques based on biotechnology and indicate something of the public's response.

New Technologies, New Possibilities

Since the 1970s biotechnological innovations in the medical field of reproduction have seen important breakthroughs particularly in the area of genetic manipulation. To quote Annemieke Roobeek, techniques such as DNA code restructuring have been heralded as a 'powerful tool in the struggle to control nature'.[2] These possibilities have set up a new industry of diagnostics. Diagnostic techniques based on monoclonal antibodies make it possible to carry out simple tests to detect disease at an early stage. More important to our discussion here, antenatal diagnosis has produced reliable techniques to monitor pregnancy for an increased risk of congenital abnormality; two of the best known being amniocentesis and chronionic villi sampling (CVS). Chromosomal metabolism and DNA examinations can be carried out within a few days, at an early stage of pregnancy, enabling abnormalities to be detected. Gene replacement therapy is also being developed to change a congenitally inherited defect.

These techniques are primarily focused on improving the health of the foetus, other reproductive techniques are changing definitions of biological parenthood. In vitro fertilisation (IVF) now allows women who would have previously been classified as infertile to have children. The creation of 'test tube babies', sperm banks and surrogate motherhood have changed the whole premise of parenthood as necessarily based on a biological relationship. The introduction of antenatal diagnosis mentioned above has also introduced another meaning to choice in parenthood and created a new set of concerns for future parents. Mainly used to detect Down's syndrome and neural tube defects, amniocentesis means that parents now face the question of what action to take on finding abnormalities. The techniques have created a set of moral and psychological dilemmas about whether to terminate the pregnancy, or whether one is always morally obliged to take an antenatal test. In making an antenatal test part of the routine procedure for a 'normal' pregnancy, pregnancy becomes further bound to medical procedure, which if unaccompanied by increasing women's knowledge of the tests will further reduce women's control over their pregnancy, further medicalising motherhood.

Sex predetermination as a result of antenatal diagnosis has also led to the parents' once undreamt of choice of the sex of their infant. A technique which has had somewhat distressing results in some countries. (See below.)

The responsibility of genetic choice is affecting even premarital arrangements. In seeking to prevent diseases such as thalassaemia, sickle cell anaemia and Huntington's Chorea, all caused by single cell defects, scientists have actively discouraged couples within known chromosome-deficient social groups from partnering in order to reduce the possibility of the disease.[3]

In looking at greater reproductive choice afforded by biotechnology we can also include contraceptive devices which have been widely available since the 1970s – the pill, injectable contraceptives such as Depo provera and Norplant. These choices have given a greater freedom for women to fulfil their potential outside of motherhood as well as inside the family, and allow for better spacing of children and improved maternal health.

All of these techniques have raised questions from varied perspectives. The major question from all sides has been how does one justify such techniques? From the economic perspective, the new possibilities of these techniques have not escaped industry. Medical and pharmaceutical applications of biotechnology controlled by the big pharmaceutical–medical complexes have come to assume a great importance in decisions about which techniques to develop – favouring the more expensive rather than the simpler preventative techniques.[4] The commercialisation of surrogacy, sperm banks and in vitro fertilisation by private sponsors, doctors and individuals also raises questions about how to ensure that what is socially rather than economically beneficial is being adopted.

On another level the decision to develop for example gene therapy, or to make amniocentesis compulsory for groups with a high risk of chromosome abnormality has also its economic considerations. If an embryo with a major defect is discovered in good time the cost to the state, family or insurance, is only the test and simple abortion. The expense to the state (or the public), however, of bearing the costs of a handicapped child, can be calculated at many times that amount.

This sort of calculation, however, immediately raises a set of questions which cannot be answered only on economic grounds: are we morally justified to interfere with some one's genetic heredity? Who sets the standards? Who will monitor experiments with human life? At what level is it a state or an individual choice?

As these questions indicate, reproductive technology is as much a social as a scientific phenomenon, which has moved many people to try and shape a response to what, after all, concerns one of the most fundamental of human processes.

The Public Response

We can roughly divide these responses into three camps. First is the group who welcome the innovations as heralding a technological era (revolution?) for future generations. Apart from offering infertile women and men the chance to have a child, techniques such as gene therapy also promise a healthier, more 'normalised' social gene pool; a space-age 'control over nature'.

The second group rejects the whole concept of genetic manipulation as an 'unnatural' choice. Very strong ethical and moral objections are raised from religious and conservative perspectives. (Who are we to play God and define what is normal and who can live?) Others object to the interference in 'natural' biological functions *per se*. (How do we presume to create life outside of nature and outside of the traditional family?) Groups such as elements of the Catholic Church and the 'right to life' campaign in the US emphatically oppose the notion of aborting foetuses or any form of embryo experimentation or gene manipulation. Surrogate mothering has also raised public outcry as well as legal questions.

A third group, who have been very active in questioning, scientifically as well as politically, genetic engineering and reproductive technology are feminist researchers. Working in Europe, the US and Australia, this group has set up its own research investigations, publishing a collection of their findings in books[5] and journals such as *Reproductive Toxicology* and *Reproductive and Genetic Engineering Journal of International Feminist Analysis* as well as creating networks such as Women's Global Network of Reproductive Rights and International Women's Health Coalition. Their work strongly questions the application of reproductive techniques in relation to the control it imposes rather than the choice it offers women. Their refutation of the success of the techniques (emphasising the human cost to the woman undergoing the tests) and stress on the use (abuse) of the female body underline the sense of powerlessness and invasion which many women feel. The basic feminist critique is that the techniques do little to empower women's control over their own lives and bodies. They employ phrases such as 'test tube woman' and 'walking wombs' to emphasise the objectification and invasion of women's bodies by medical science, which is depicted as a form of patriarchal oppression.

Individual Choice?

In the following discussion I question the use of reproductive technology from a slightly different angle than these sets of responses. My aim is not so much to judge the techniques themselves but rather to raise a series of questions about biomedical reproductive technology as it is practised today, in order to suggest a trajectory into the future. My question then is who decides what for whom and how? This I will look at from two perspectives – of the individual and of society. In both of these perspectives I take medicine as a social practice embedded in a complex ethical, political and social domain which cannot escape political and ethical questions.

Reproductive technologies have given individual women a potentially greater control over their own fertility. The 'pill' heralded the sexual revolution as well as an easy way to space families, and provoked challenges to many religious and cultural norms. The rider to this was the emphasis on responsibility for pregnancy remaining with the woman, who was now more 'available' (it would be interesting to investigate the reasons why similar research into men's reproductive pills has not enjoyed the same success), and the side effects – weight gain, greater risk of cancer, hormonal imbalances and, with long-term use, infertility – need to be taken into consideration when assessing the pill's success.[6] Another point which Yoxen raises in relation to this last concern, the great demand for in vitro fertilisation techniques to help infertility, has largely arisen from the generation who were first exposed to contraceptives and the new freedoms, suggesting either that this generation is more informed or that the incidence of infertility is higher. Greater freedom of sexual activity without pregnancy as a necessary corollary may not have given women greater control from a social perspective nor necessarily improved their long-term health.

Biomedical techniques of conception, on the other side, have allowed once infertile parents to chose children, as well as those not in heterosexual partnerships to have biological children. Here critics strongly question the amount of information available to the individuals undergoing treatment. For example, the number of procedures necessary – from screening to implantation and successive trials – to undergo IVF are not widely known (nor are the psychological pressure or physical discomfort). The poor success rate of around 13% could also be more widely taken into account.[7]

Antenatal diagnosis and genetic intervention now largely recognised as safe techniques do carry risks of which the women undergoing the tests are not always informed. Trials carried out in the US, the UK and Canada have judged that the foetal loss after amniocentesis is around 1%. This is slightly greater than the risk of having a Down's syndrome child until the woman reaches her forties.[8] Studies show a slight increase in the number of babies with respiratory troubles after amniocentesis.

The other major reason for antenatal tests is to detect spina bifida (when the column of the tissue around the spine fails to close) or anencephaly (where a portion of the brain is missing). Diagnostic tests are available which are done routinely in the UK to test high levels of protein in the foetal kidney prevalent in these conditions, whereas in the US there is more resistance to these tests because of the doubt about guarantee of medical expertise and counselling. This raises the serious concern about how to provide adequate counselling for women who have to face abortion after antenatal investigations. For those needing to undertake the tests, the waiting involved and the delay before women can act creates a period of uncertainty which in Barbara Katz Rothman's phrase becomes a 'tentative pregnancy'.[9] Many feel that this should lead to a questioning of the whole medical procedure.

Another concern is that, without adequate information on the purpose, nature and risks of amniocentesis, and the emotional complications, women are being ased to agree to termination before undergoing the analysis. This raises questions about the terms on which medical assistance is being offered. In seeking to reduce the number of handicapped children being born, stemming from genetic conditions and other causes, antenatal diagnosis has an important role to play. But so does better education, better nutrition, more appropriate antenatal care and higher standards in obstetrics. As well as offering the techniques, a priority should be to give people the time, the information, the assistance and the freedom to decide for themselves what use they want to make of the antenatal diagnosis. Giving people more information and counselling about the possible procedures allows them to choose for themselves what do to if a genetic defect does arise. It should not be assumed that people have to chose abortion or that, if they do so, it is not a serious life choice which needs careful attention from a psychological perspective.

Chronionic villus sampling, which permits investigation after 8 weeks, thus reducing the time of 'tentative pregnancy', raises a similar set of questions. First the test, which involves taking a sample of foetal blood tissues via the cervix, involves the risk of losing the pregnancy. Of over 3,000 cases, the pregnancy had to be terminated in 10% while 4.1% lost the foetus from miscarriage following the CVS.

The new tools for reading DNA have also opened up the possibility of experimentation with embroys using techniques ranging from transfer of embryos in in vitro fertilisation to gene screening to gene therapy. One technique, somatic cell modification, is the attempt to correct any malfunction by changing or supplementing the genes within them. The intention is to 'correct' an inherited trait. Manipulating cell genes is common in the life sciences in the 20th century but the question here is whether it is morally acceptable to move from animal and plant to human beings. A clear legal position or very clear directions to scientists after public debate needs to be put in place, rather than this being left to the dubious ethics of the scientifically curious. Clear directions would also guard against entrepreneurial interests that seek to capitalise on secret desires of potential parents – Huxley's

Brave New World where one pays for a genetically designed child. The point here is not that private enterprise is necessarily worse or better than public intervention but that open public dialogue takes place which allows standards to be set taking into account the ethical complexity of the matter with which we are dealing.

The commercialisation of parenthood, though not having yet reached the space-age quality of Huxley's fictional world, has found a niche in surrogacy. With all of the new technologies available surrogacy can today be completely removed from the simple (ancient) practice of sexual intercourse and transfer of baby. Yoxen discusses in detail how these increasing technological possibilities have also led to an increasing number of difficulties, with agencies to handle clients, and not enough legislation to cover the status of the contracts between the surrogate and the couple.

He wonders at the legal problems which could arise if an IVF clinic and surrogacy agency were to join forces – what would be the parental rights of the surrogate who carried an implanted foetus if she were to change her mind once the child had been carried to term? Is she just the vessel or do the sperm and ovum donors give up their rights once the embryo has been implanted? What are the ethics involved in contracting motherhood – is it to be seen as a form of prostitution? What about the active attempts by the agencies to convince the surrogates to 'give the child up' – are we talking about another form of organised use of women's bodies for others' gains? These are all questions which are yet to be taken on legally and openly.

Social Control?

In the following discussion I move from a focus on the individual, to take up the social implications of reproductive technology. This is a particularly important shift to the discussion of sex predetermination which antenatal tests can offer. Most cultures express preference for boys over girls. In the West reliable sex predetermination would not lead to an overwhelming male population, though given the popular preference for first-born sons (both women and men) it could lead to a considerable weakening of women's self-esteem and a changing social balance if society was made up almost exclusively of first-born sons and younger sisters. Tempting as it is to ponder on this, more insidious is the pattern which is beginning to emerge in China and India where preference for sons is very strong. Whereas traditionally the sister's health has been neglected in favour of the brother's, with higher incidence of early death, disease and malnutrition among girls, the pre-knowledge of the child's sex with antenatal tests has led to a sharp increase in the abortion of female foetuses. One study in India showed that where 700 women requested amniocentesis, 450 were informed that the foetuses where female. Of this 450, 430 were aborted. As Yoxen points out, these figures have a statistically improbable high female/male ratio. This suggests that of these tests 20% were incorrect due to faulty techniques which mistook maternal cells for foetal cells.[10] The adoption of amniocentesis to monitor femaleness could be seen as a misuse of medical technologies which is beginning to have disconcerting results on a wider social scale, as recent studies show.[11]

In raising the questions of adapting Western medical techniques to other cultures we are looking at reproductive technology as a set of practices which involve humn rights, population management, economic interests of multinational companies and state investment policies in public health and education.

Three general questions can be raised in relation to social responsibilities of biomedical interventions: how do different countries interpret and use technologies? Is the public able to make a choice about the use of technologies? Are short-term gains being made at the expense of longer-term goals?

In particular we can ask these questions in relation to the export of these technologies to the South. First, not all the latest innovations are available for the South. As Roobeek points out, much money is invested in expensive drugs which are only affordable by the very affluent in the North.[12] Often the choice of techniques exported to the South is made not so much because it is more suitable or adaptable; on the contrary, there are usually more questionable motives. Technology rejected by developed countries is dumped on the Third World markets (sometimes as aid).[13]

Further, it is not uncommon for devices unacceptable in the North to be tested on a public in the South. A public which has relatively less clout and awareness to enable it to refuse.[14] Or the ascribed needs of the South (for example, the need to drastically curb the birth rate) is perceived as requiring more stringent measures not acceptable in the North and often more dangerous, which the Third World public are not in a position to question.

There are examples of developing countries' governments adopting techniques which are applied as measures of social control as much as of the supposed population management (such as the enforced sterilisation in the 1970s in India). Contraceptive devices such as Norplant and other injectable contraceptives have been banned in the North but are widely used in the South with the encouragement of governments following the advice of some multilateral agencies.[15]

The coercion used and the way in which costly techniques are rec-ommended brings into question a number of problems. First, is the problem of local corruption and inadequately monitored techniques (unfortunately not an infrequent problem in the development indus-try). But perhaps more serious is the problem of reproductive techno-logies being heralded as part of the 'wonder' science of Western medicine. Over-zealous technological intervention and abuse of the public through continued use of banned drugs has been possible through the uncritical acceptance of medicine as part of 'develop-ment' and 'progress'. A well set up hospital for neonatal care can be enthusiastically funded by governments as a symbol of development and technological know-how, whereas, as recent studies show,[16] less money channelled into ensuring that girls have a better education would undoubtedly increase women's ability to monitor their fertility and family's well-being and more securely guarantee improved infant survival.

Another issue to examine when asking who decides what for whom is whether the Western model of science increases the knowledge and skills of the local users of the technology. The model of expert in-structing beneficiary (or doctor instructing patient) fits well with the prevailing practice of developing countries adopting developed coun-tries' models of industrialisation and scientific progress regardless of earlier customs.

In the same way that those interested in development are now beginning to see that sustainable development (development which guarantees the rights of future generations by a development growth which improves the quality of human life) needs to preserve the local environment and culture and allow for the existence of future gener-ations by relying on local expertise, knowledge and ability as well as external expert know-how, the medical model of Western-trained doctor applying techniques oblivious of local traditions needs to be challenged.

This challenge is not so much about the validity of the technologies as about their context. At both the individual and the social level we need to pose questions about the social location and meaning of reproduction when assessing the viability of introducing biomedical techniques.

Medical innovations should be introduced sensitively with the ob-jective of giving women access to knowledge of their own bodies, control of their fertility and safer birthing practices. The technological methods of control over reproduction and population increase clearly offer an important and necessary development for both the individual women who wish to control their fertility, and society, when we look at ways to achieve economic and environmental sustainability on a global level. The change needed is to establish this new knowledge of women's bodies, fertility and childbirth within parameters which those most closely affected can actively shape and carry out.

In balanced development of medical research, funds used to control fertility, to develop contraceptive techniques, foetal transfer and artificial wombs, need to be assessed in relation to funding available for adequate testing outside laboratories, basic knowledge of available family-planning techniques and wider adequate nutrition and health. All are areas which would enable means of improved reproductive technology that would be just as effective and, in the long run, less costly in economic and human terms.

Which Revolution?

The development of reproductive technologies should not be seen so much as a revolution, but as a combination of the prevailing 20th century scientific medical approach with increasingly complex technologies. From this perspective the revolution may not necessarily benefit those whom it proposes to benefit. A revolution, then, would nt be a super-development of reproductive technologies as 'man' triumphs over nature; rather it would be a humbler and more wide-sweeping revolution of changing the process of medical research and practice in order to allow the subjects and consumers of these technologies to determine their use. Money devoted to the improvement of medical technologies should be channelled into giving women access to better education, better public health and awareness of their own bodies' fertility and functioning.

To conclude slightly tongue in cheek, one further, even more revolutionary, step would be to turn the focus away from female reproduction to techniques for greater choice and control of *men's* reproduction.

NOTES

[1] Edward Yoxen, *Unnatural Selection? coming to terms with the new genetics* (London: Heinemann, 1986), p. 6. In my opinion Yoxen has written probably the most complete and certainly one of the more balanced summaries of biotechnological innovations to date. I have therefore referred liberally to Yoxen's work throughout the chapter.

[2] Annemieke Roobeek, *Biotechnology: a challenge full of promise and pitfalls* (Socialist Group European Parliament, Belgium, December 1989), p. 11.

[3] Yoxen, *Unnatural Selection*, p. 135.

[4] See Annemieke Roobeek, *Biotechnology*, p. 19.

[5] For example, Michelle Stanworth, in M. Stanworth (ed.), *Reproductive Technologies, Gender, Motherhood and Medicine* (London: Polity Press, 1987); Brighton Women and Science Group (eds), *Alice Through the Microscope* (London: Virago, 1980); H. Roberts (ed.), *Women Health and Reproduction* (London: Routledge and Kegan Paul, 1981); G. Corea, R. Duelli, R. Klein (et al.), *Man-Made Women: How New Reproductive Technologies Affect Women* (London: Hutchinson, 1985); R. Arditti, R. Duelli, R. Klein and S. Minden (eds), *Test-Tube Women: What Future for Motherhood?* (London: Pandora Press, 1984); R. and R. Rowland, Hormonal cocktails: women as test-sites for fertility drugs, *Women's Studies International Forum*, vol. 12, no. 3 (1989), pp. 333–348.

[6] For examples of the side effects the pill and other drugs have on women patients, see G. Corea et al., *Man-Made Women: How New Reproductive Technologies Affect Women.*

[7] Edward Yoxen reports that from 58 IVF teams, which had together actively intervened in 9,641 ovulatory cycles, 1,209 viable pregnancies were achieved. Yoxen, *Unnatural Selection*, pp. 69–70.

[8] Yoxen, *Unnatural Selection*, p. 125.

[9] Quoted in Yoxen, *Unnatural Selection*, p. 127.

[10] Yoxen, *Unnatural Selection*, p. 111; see also, William Lavely, China's One Child Policy, in *Gender and Society*, vol. 2, no. 2, June 1988, pp. 241–242.

[11] See Gopalakrishna Kumar, Gender, differential mortality and development: the experience of Kerala, in *Cambridge Journal of Economics*, vol. 13, no. 4 (December 1989), pp. 517–539.

[12] Roobeek, *Biotechnology*, p. 17.

[13] See Our Health, *Echo*, Issue 9–10; pp. 10–17.

[14] This is difficult to document but from talking to members of family-planning clinics in Jakarta, Suva and Manilla, and in reports of networks such as the Women's Reproductive Rights Network in Gabriela, Philippines, as well as health workers in Northern agencies such as International Women's Health Coalition, it appears that this is accepted practice.

[15] Examples, again from my own research, are of the Fijian, Philippines and Indonesian governments. See issues of *Populi* Journal of the United Nations Population Fund, for clear statements of encouragement. UNFPA and Finnida (Norplant is produced in Finland) are strong supporters of Norplant which is banned in almost all countries in the developed world with the exception of Finland).

[16] Nafis Sadik, *State of the World Population Report*, 1989, UNFPA.

15

The Biotechniques in Agricultural Research

Office of Technology Assessment

New biotechnologies have the potential to modify plants so that they can resist insects and disease, grow in harsh environments, provide their own nitrogen fertilizer, or be more nutritious. Technical barriers, however, still exist. In particular, widespread success in applications for multigenic traits (such as salt tolerance or stress resistance) will for the present remain elusive, perhaps decades away. Nevertheless, the newer technologies can potentially lower costs and accelerate the rate, precision, reliability, and scope of improvements beyond that possible by traditional plant breeding.

Two broad classes of biotechniques – cell culture and recombinant DNA – are likely to have an impact on the production of new plant varieties. Plant tissue and cell culture date from the turn of the century, but were only minimally exploited until the late 1950s. Successful in vitro cultivation of plant cells and related culturing techniques underlie today's gene transfer techniques and subsequent regeneration of altered, whole plants. Plant tissue and cell culture are also critical tools fo increasing fundamental knowledge through basic research. The history of genetic engineering and a detailed description of the principles of recombinant DNA technology are discussed in an earlier OTA report. In general, the fundamentals of genetic engineering are similar for microbial, animal, and plant applications, but developing some new approaches for plant systems has been necessary.

The endpoints of crop improvement using biotechnology are those of traditional breeding: increased yield, improved qualitative traits, and reduced labour and production costs. New products not previously associated with classical methods also appear possible. Box 15.1 briefly describes some of the new biotechniques exploited to achieve these aims. Comprehensive descriptions of strategies de-signed to transfer foreign genes to plants and plant cells have been published elsewhere.

Box 15.1 Techniques used in plant biotechnology

Plant Tissue and Cell Culture. Plant cultures can be started from single cells, or pieces of plant tissue. Cultures are grown on solid or in liquid media. Several species of plants, including alfalfa, blueberry, carrot, corn, rice, soybean, sunflower, tobacco, tomato, and wheat, can be cultured in vitro.

Plant Regeneration. Regenerating intact, viable organisms from single cells, protoplasts, or tissue is unique to plants and pivotal to successful genetic engineering of crop species. (To produce a protoplast, scientists use enzymes to digest away the plant cell wall.) Although genes can be transferred and examined in laboratory cultures, ultimate success is achieved only if the culture can be regenerated and the characteristic expressed in the whole plant. Figure 15.1 illustrates steps involved in regenerating plants in vitro.

Protoplast Fusion. Protoplasts from different parent cells are artificially fused to form a single hybrid cell with the genetic material from each parent. Protoplast fusions are useful for transferring multigenic traits or for fusing cells from plants that cannot be crossed sexually, thus permitting the exchange of genetic information beyond natural breeding barriers. Successful gene transfer via protoplast fusion depends on the ability to regenerate a mature plant from the fusion product.

***Agrobacterium tumefaciens* plasmid.** One of the most widely used and probably the best characterized system for transferring foreign genes into plant cells is *Ti* plasmid-mediated transfer. The technique involves a plasmid vector (*Ti* plasmid) isolated from *Agrobacterium tumefaciens*, a naturally occuring soil-borne bacteria that can introduce genetic information stably into certain plant cells in nature. Using recombinant DNA technology, the plasmid has been modified to increase its efficacy in the laboratory.

Transformation (Direct DNA Uptake). Certain chemical or electrical treatments allow direct uptake and incorporation of foreign DNA into plant protoplasts – a process called transformation. Since hundreds of thousands of cells can be simultaneously treated, transformation is a relatively easy technique. Cells expressing the desired trait can be regenerated and tested further.

Microinjection. Using a special apparatus, fine glass micropipettes, and a microscope, DNA is directly introduced into individual cells or cell nuclei (in plants, protoplasts are usually used). The process is more labour-intensive than transformation, requiring a trained worker. Although fewer cells can be injected with DNA than in mass transformation, a higher frequency of successful uptake and incorporation of the foreign genetic material can be achieved – up to 14 per cent of injected cells.

Virus-Mediated Transfer. Virus-mediated transfer of DNA has played a critical role in nonplant applications of biotechnology. But, in large part due to an underdeveloped knowledge base, viral vectors for plant systems generally have not been exploited. Cauliflower mosaic virus has been used with some success in turnips, and Brome mosaic virus in barley. Developing generic virus-mediated transfer systems could accelerate progress in plant biotechnology.

DNA Shotgun. One novel approach uses gunpowder to deliver DNA into plant cells. The DNA to be transferred is put onto the surface of four-micrometer tungsten particles and propelled into a plant cell by a specially designed gun. While an innovative approach, it is unclear whether it will prove to be a routine method for gene transfer in monocots.

Source: Office of Technology Assessment, 1988.

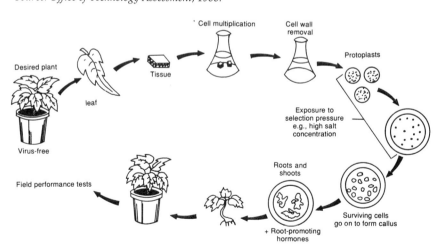

Figure 15.1 Plant propagation: from single cells to whole plants: the process of plant regeneration from single cells in culture
Source: Office of Technology Assessment, 1988.

Applications of the Techniques

The new biotechniques are useful for investigating diverse problems and plant types. For example, plant tissue and cell culture is an important technique for breeders. It can be used for screening, at the cellular level, potentially useful traits. As many as ten million cell aggregates can be cultured in a single 250 ml flask (less than 1 cup). This can be compared with a space requirement of 10 to 100 acres if individual test plants were put into the field.

Several species of plants can be clonally regenerated to produce genetically identical copies. The process is widely used for a range of commercial applications, including forestry and horticulture (e.g., producing strawberry, apple, plum, and peach plants). Several crop species, such as asparagus, cabbage, citrus, sunflower, carrot, alfalfa, tomatoes and tobacco are also routinely regeerated. Although monocotyledonous plants, such as the cereals, have been more difficult to regenerate, rapid progress is being made with these as well.

Plant regeneration is a powerful tool not only for increasing the numbers of propagated materials, but also for reducing the time required to select for genetically interesting traits. Furthermore, under certain conditions, genetic variants arise during the culturing process (somoclonal variation). Somoclonal variation can uncover new, useful variants and again reduce the time spent selecting genetically interesting traits.

Many important agricultural applications of biotechnology depend on regenerating whole plants from protoplasts. Protoplast fusion has been applied successfully in several plants, including the potato. In this instance, cells from wild and cultivated potato plants were fused to transfer the viral resistance of the wild species. The hybrid cells were regenerated into fertile plants that expressed the desired virus-resistant characteristic. Virus-free potato cells can now be cultured in vitro, and virus-free plants regenerated; the yield of these plants has increased substantially. Culturing virus-free plant cells is particularly important in certain horticulturally important species, including ornamentals and certain vegetable crops. As is the case with single cell or tissue regeneration, protoplasts of the monocotyledonous subclass of plants, such as cereals, have been much more difficult to regenerate than protoplasts of the other major plant subclass, dicotyledonous plants, such as tobacco and tomato.

Ti vectors are especially useful for genetically engineering dicotyledonous plants, such as tobacco, tomato, potato, and sunflower. For example, *Ti*-mediated transfer has been used to engineer virus-resistant tobacco plants and insect-tolerant tomato plants. The technique is less useful for gene transfer in monocots (which include important cereal crops). Increasingly, however, technical hurdles identified as barriers only a few years ago are being cleared. Recent success using the *Ti* vector for corn (a monocot) has been reported, with continued progress for monocots anticipated. Furthermore, direct DNA transformation apparently allows gene transfer in several cereals (monocots), including rice, wheat, and maize, with an efficiency approaching comparability to the frequency of *Ti*-mediated gene transfer in dicots.

New applications and new techniques, such as the 'DNA plant shotgun', are continuously arising. Table 15.1 describes a few recent applications of biotechnology to plant agriculture.

Impact of Biotechniques on Agricultural Research Investment

In part, the advent of genetic engineering and related biotechniques has, itself, altered the shape and scope of U.S. agricultural research investment decisions. In particular, the emerging technologies present fundamental challenges and opportunities for the public component of U.S. agricultural research. Basic science advocates charged that the USDA-led system had not been on the cutting edge of science nor had it been paying enough attention to basic research, stimulating an evaluation of the system that continues today.

Table 15.1 Some recent applications of biotechnology to plant agriculture

---

Rice: Whole rice plants can be regenerated from single-cell protoplasts; recent advances that improve the efficiency of the process are important to progress in applying genetic engineering to cereals in general.

Maize: The *Ti* vector was recently used to transfer the maize streak virus into corn plants, a monocotyledonous member of the grass family. The study is a landmark because the *Ti* plasmid is probably the best characterized plant vector and an efficient gene transfer mechanism, but monocots had been refractory to its use. Successful plant regeneration of maize protoplasts also was reported recently.

Rye: Using a syringe, DNA was injected into rye floral tillers. The new genetic material was introduced into the germ cells of this monocot, and some recovered seeds grew into normal plants that expressed the foreign gene. This simple strategy, which does not require plant regeneration from protoplasts, could be useful in other cereals.

Orange: Orange juice-sac cells have been removed from mature fruit and maintained in tissue culture. The cells produce juice chemically similar to that squeezed from tree-grown fruit. Such laboratory cultivation could advance trait selection and speed up varietal development, although laboratory-produced juice is not on the immediate horizon.

Tomato: A gene that confers a type of insect tolerance was recently transferred via the *Ti* system to tomato plants. The tolerance is also expressed in progeny plants. Since over $400 million per year is spent to control this type of pest, constructing insect tolerant transgenic plants of this sort is of great interest to the agricultural community. See also fig. 15.1.

Source: Office of Technology Assessment, 1988.

Some have argued that the biotechnologies have led to private sector, proprietary-dominated research efforts. Others, however, point out that increased private sector research investment resulting from the biotechnology boom has uniquely contributed to the fundamental knowledge base and resulted in a positive economic impact. And, through increasing alliances between companies and universities, industy involvement has also resulted in resources for new ideas, with potential to further enhance economic return through accelerated technology transfer.

Biotechnology has also stimulated greater interest in agricultural research by the nontraditional agricultural research community. Today, agricultural applications command greater interest within the general research hierarchy. While some believe this shift is valuable, others fear that research directed to address regional and local problems could suffer and that 'have' and 'have not' institutions will result.

In addition to the effect of biotechnology on research investment decisions, concern has been raised about biotechnology's influence on investment in human capital: namely, a decline in the number of full-time equivalents (FTEs) in traditional plant breeding at the expense of increasing numbers of FTEs in molecular biology. Improvements in varieties with the new biotechniques will be hollow achievements if there is a shortage of traditional plant breeders who conduct the complementary field research that is essential to develop varieties for use by farmers. Some reports indicate a 15 to 30 per cent decrease in university-based plant breeders and an increase of about one-third in molecular biologists between 1982 and 1985. This trend might, in part, reflect the glamour image of plant molecular biology coupled with industrial demands for plant breeders.

A continuing industry demand for trained plant breeders might be an attractive argument for those making career decisions and ensure an adequate supply of plant breeders. However, a large majority of graduate students in the plant sciences still want to work in molecular biology, and siphoning university plant breeders to industry could leave a teaching void for those who want to learn conventional breeding. At present, some argue that a balance in supply seems to have been (or is being) struck. Others within industry and academia assert a lack of plant breeders exists. Regardless, evidence for both sides is largely anecdotal, and accurate accounting would be useful for forecasting and planning the direction of plant agricultural research.

The impact of the biotechnologies on the direction of agricultural research has not, however, occurred in a vacuum. Intellectual property issues and who funds projects also are important factors. For example, the concern about the exchange of plant-breeding materials just mentioned has been generated both by the research thrust using the biotechnologies and by interpretation of patent law. The biotechniques have also contributed to an evolution in the investment emphasis (i.e., the types of projects funded) of private and public sources.

16

The Realities and Challenges of Plant Biotechnology

Indra K. Vasil

Plant biotechnology has often been touted overzealously and prematurely as a cure-all for many agricultural problems, and a panacea for the genetic manipulation and improvement of plants.[2] There has been a great deal of hype and rhetoric, but little substance. This general lack of objectivity, combined with excessive speculation, has led to heightened expectations. Worldwide, there are about 500 companies and 150 research organizations that have at least some interest or activity in agricultural biotechnology, and substantial amounts of money have been invested in research. Despite all this the sobering fact remains that thus far not a single major product of plant biotechnology – either as improved plant or plant product – is produced commercially any where in the world.

The lack of real and clear benefits, as well as the difficulties and delays in implementing biotechnological objectives, has forced a reassessment of the potential and the problems of plant biotechnology, leading to what should be more objective and realistic goals and timetables.

Plant biotechnology, like its biomedical counterpart, has two important but critical interacting technical components – cell culture and molecular biology. Interest and investment in plant molecular biology is rather recent but is already paying rich dividends in the understanding of the molecular basis of plant growth and development.[3,4] Cell culture techniques, on the contrary, have been available and continuously improved since the early 1930s.[5,6] Many of these procedures are simple and can be practised without substantial investment in facilities and infrastructure for the production and multiplication of improved plants. These constitute the short-term or immediate applications (five years) of this technology. Technologies that are based on integrating cell culture and molecular genetics should bear fruit in the intermediate term (5–10 years). Plant biotech's major impact on agricultural production, however, must await significant scientific advances in the understanding of plant growth and development – which are not expected until beyond the year 2000.

A Look at the Short Term

Clonal propagation

Plants can be regenerated *in vitro* from cultured meristems (micropropagation), or by the formation of somatic embryos or adventitious buds.[7] The most widely used and commercially successful application of plant biotechnology is the rapid and large-scale clonal multiplication of plants by meristem culture.[6,8] Scores of companies use this process to mass-produce many ornamental plants, and increasingly for fruit, vegetable, medicinal, and tree species. Current methods of micropropagation are time-consuming and rely heavily on manual labour. The resulting high costs and low profitability have restricted the market to high-value species only. Therefore, automated methods of micropropagation canreduce production costs substantially and help extend the technology to a wider variety of crop plants and expand the market.[9]

Perhaps the most useful application of this technology can be for *reforestation*. It has been difficult to regenerate both hardwood and softwood species in culture, but there have been significant advances during the past decade in methods for propagating tree species *in vitro*. Reforestation on a worldwide scale will require billions of plantlets. Integrating tissue culture methods with automation can provide a reliable and efficient way to produce an unlimited supply of high-quality planting material.

Virus-free plants

Meristem cultures contain either no viruses or a very low titer. Excision and culture of the shoot meristem with only one or two leaf primordia, therefore, often physically eliminates any virus.[10] Such virus-free meristems can be rapidly multiplied in culture by micropropagation. This simple procedure is being widely used in many countries to obtain virus-free potato, sugarcane, cassava, garlic, banana, strawberry, and many species of ornamental plants and vegetables.[11] The practical benefits of this technology are substantial: virus infections can result in crop losses of more than 50 per cent.

Embryo rescue

Agronomically useful hybrids – even between closely related species – are often unattainable because of sexual incompatibility. In many instances fertilization takes place, but the developing embryos abort at various stages of development. The technique of embryo rescue, which involves the excision and culture of the young hybrid embryos, has proved to be very useful in growing such embryos to maturity and in obtaining hybrid plants.[12,13] Furthermore, genetic and field evaluation of the hybrids can be accelerated by generating multiple clones of the embryos in culture.

Haploids

Scientists have recovered haploid plants from cultured anthers, microspores, and ovaries of more than 100 species. Haploid gametophytic cells can be stimulated to form either embryos directly or callus tissues that can be induced to differentiate shoot buds.[14,15] The doubling of chromosome numbers in such haploid plants readily yields homozygous diploid and fertile plants which are useful in breeding programmes. This method has produced a few new cultivars of rice, wheat, maize, tobacco, oil seed rape, and some other species, but their commercial success has been limited – partly because these efforts have not been coordinated with and integrated into established breeding programmes.

Mutant selection

It is well recognized that mutations occur during cell division and differentiation *in vivo*.[16,17] Meristem cells, which serve as a 'germ line', are generally immune to such genetic changes. In the normal life cycle of a plant the mutant somatic cells are eliminated during sexual reproduction and are not passed on to the progeny. Such mutant cells, however, have an excellent opportunity to divide and multiply (as do non-mutant cells) when plant tissues are placed in culture. Imposing selective pressures on cultured cells can result in the preferential growth of mutant cells, establishing mutant cell lines from which whole plants are recoverable. The procedure is useful only for single-gene traits, such as resistance/tolerance to certain pathogenic toxins,[18] herbicides,[19] anti-metabolities,[20] heavy metals,[21] and for creating mutants that overproduce useful amino acids.[22]

There have been many attempts to use similar experimental strategies to select mutant cell lines and plants that are resistant to a variety of stresses (such as salt, aluminium, drought, and frost) or mutants that affect agronomically important traits such as yield and growth. None of these attempts has been successful, largely because each of these traits is complex and multigenic in nature. There are no known methods that allow the simultaneous selection of desirable mutations in a number of genes. Furthermore, it is likely that the genes involved are functional only in specific tissues or organs during specific phases of plant development; it may be impossible to select for these mutations in cultured cells.

There has also been considerable speculation about the potential uses of somaclonal variation in plant breeding and improvement.[23,24] The variation found in cell cultures results either from perpetuating preexisting variations in the plant's differentiated somatic cells,[16,17,25] or from new mutations generated during growth in culture.[26] Very little of the variation found in cell cultures is actually recovered in regenerated plants, and even then almost all of it is either useless or similar to the variation recovered after sexual crossing or mutagenesis. Despite the many potential uses claimed for somaclonal variation, and substantial efforts by scores of individuals, the fact remains that thus far there is not a single example of any significantly important new variety of any major crop species developed as a result of somaclonal variation.

Secondary metabolites

Growing plant cells in large bioreactors has long been considered an ideal way to produce pharmaceutically interesting secondary metabolites. Indeed, there have been significant technical advances over the past 10 years in the synthesis of secondary metabolites by dispersed plant-cell cultures grown in large-volume automated fermentors.[27-29] Nevertheless, the high cost of production, coupled with the limited market size of many of these products, have severely restricted commercial applications. Shikonin is thus far the only product manufactured in this manner. Recent advances in bioreactor design and technology, however, coupled with experiments on elicitors and cellular immobilization, may result in the production of additional useful products in the foreseeable future.

Somatic hybrids and cybrids

Protoplast fusion has long been proposed as a novel and important method for producing hybrid plants that cannot be obtained by sexual means. Although the literature describes many somatic hybrids,[30] few of these are actually useful. Of those, for example, *Brassica naponigra*, produced by the fusion of protoplasts of *B. napus* plus *B. nigra*, is resistant to *Phoma lingam*,[31] and the somatic hybrids of *Solanum tuberosum* plus *S. brevidens* are resistant to certain virus diseases.[32] In general, it is unlikely that useful and fertile somatic hybrid plants can be obtained by a simple additive combination of the complete genomes of two related or unrelated parents. The recent work on the production of asymmetric hybrids, therefore, is more promising.[33] Perhaps the best use of protoplast fusion will be in the production of cybrids which contain the nuclear and cytoplasmic genome of one parent and only the cytoplasmic genome of the second parent. These can be of particular advantage in the transfer of cytoplasmic male sterility, an important trait in plant breeding.[34, 35]

Intermediate-Term Applications

Applications of plant biotechnology that are being actively developed today and will be commercially available within the next 5–10 years are based on the transfer, stable integration, and expression of useful genes in crop species. Several methods are now available to transform cultured tissues, cells, and protoplasts: via the *Ti* plasmid of *Agrobacterium tumefaciens* (as well as the *Ri* plasmids of *A. rhizogenes*); the direct delivery of DNA into protoplasts by polyethylene glycol or electroporation; and introducing DNA into intact cells by microinjection or accelerated particles coated with DNA.

At the present time only three genes of agricultural importance are available and have been successfully transferred to and integrated in a few crop species. These genes confer resistance to several modern and ecologically safe herbicides,[36] a number of destructive insects,[37,38] and viruses.[39,40] Transgenic plants containing such genes have been produced in soybean, potato, tomato, tobacco, and oil seed rape. Considering the fact that enormous plant losses, prior to as well as postharvest, are caused by weeds, insects, and viruses, the savings effected by such transgenic plants will be substantial. Information about the yield and quality of the transgenic plants and their products is not generally available, but these factors will undoubtedly play a significant role in determining their practical use. Most of the transformed plants listed above are now undergoing extensive laboratory and field trials for regulatory purposes, and are expected to enter the agricultural marketplace within the next 5–10 years. Nevertheless, there are issues that raise serious concerns – about excessive use of herbicides with the availability of herbicide-resistant plants, and the evolution of new and more virulent viral strains and insect pests by mutations.

The production of transgenic plants listed above was rather rapid and simple considering the fact that the first transgenic plants were reported only in 1983. This was because protocols for transforming and regenerating these particular plants were already available. Nevertheless the production of transgenic plants of important crop species, such as the cereals and grain legumes, is still rather difficult because the regeneration and transformation methods are far from routine. Transgenic maize, rice, and soybean plants containing kanamycin-resistance genes have been obtained, and it is expected that in these species plants with agronomically useful genes can be produced in the near future. These results now need to be extended to other important crops such as wheat, sugarcane, sugarbeet and cassava. Continued efforts are needed to improve plant regeneration from cultured cells and the efficiency of genetic transformation in important crop species.

The Future

The benefits of plant biotech's many powerful and novel procedures will be realized only after we achieve a more complete understanding of plant growth and development, and of the structure, function, and expression of agronomically important genes. Several of the inherent problems are clear. First, most of the agronomically important genes have neither been identified at the molecular level nor cloned. Second, most of the important agronomic characters are under multigenic control. Third, current transformation methodologies allow the integration of only a few foreign genes, generally no more than one or two. Fourth, the introduced genes are integrated in the host genome – often in multiple copies – at random and multiple sites. While this has not proved to be a problem with characters that are controlled by single genes, such as herbicide tolerance, insect resistance, and virus cross-protection, integration of more than one copy of several genes at multiple sites will undoubtedly disrupt and adversely affect the activity of other vital genes. Furthermore, position effects and other factors will require that the genes for multigenic traits be integrated at specific loci. (Work on site-specific integration of introduced genes – gene targeting – is just beginning.[41])

The recent interest and advances in mapping the genomes of important crop species through rstriction fragment length polymorphism (RFLP) analysis, and the use of polymerase chain reaction (PCR) technology for amplifying small DNA fragments, are welcome steps in the right direction. Information obtained from such studies will not only be useful in conventional breeding programmes, but will also be crucial in the identification and cloning of important genes and in the understanding of their interrelationships at the molecular level.

The study of developmentally regulated genes in plants is rather recent. Information on genes involved in reproductive processes, including those regulating the synthesis and storage of proteins, starch and fats in seeds, and fruit development and maturation, would be crucial in controlling and manipulating flowering and fruit and seed production. Understanding the temporal and spatial control of the activity of such genes will be required to manipulate the quality (composition) and quantity of seed storage products, including starch, proteins, and fats and oils. Anti-sense RNA technology, which allows one to deliberately delay or decrease gene expression,[42] may be used to regulate other genes in a temporal manner. And ribozyme ('gene shears') technology also holds great promise for the control of gene expression in transgenic plants.

Many cultivated as well as wild plant species show natural resistance to insects, fungi, bacteria, viruses, and environmental stresses. Concerted efforts are needed to identify the genes conferring such resistances – to develop effective biocontrol methods[43] and to exploit the natural genetic variability of plants. And understanding the molecular basis of pathogenesis may also allow the 'disarming' of plant pathogens to make them less damaging to crops.[44] The production of ice-minus bacteria, which prevent and minimize freezing damage,[45] by minimal molecular manipulation, could very well be used as a model for future work. In the long term, such engineered organisms may prove to be not only most useful, but also socially and environmentally acceptable.

Transgenic plants could also be used as 'natural bioreactors' for producing large amounts of biologically active peptides in the form of chimeric seed proteins – including neuropeptides, blood factors, and growth hormones.[46–48]

Plant biotechnology in the future will certainly supplement and complement conventional breeding in plant improvement. A close interaction and collaboration between biotechnologists and plant breeders – who unfortunately are wrongly viewed by some as antagonists and competitors – will be essential in transferring many of the successes in the laboratories to the field. While it is already clear that some important benefits of plant biotechnology will be available in the next decade, an objective assessment of long-term prospects suggests that 'its contributions will be evolutionary, not revolutionary, in nature'.[49]

REFERENCES

Note: In the citations listed below, *CCSCGP* stands for *Cell Culture and Somatic Cell Genetics of Plants* (Academic Press, New York). *Volume 1: Laboratory Procedures and Their Applications*, I. K. Vasil (ed.), 1984. *Volume 2: Cell Growth, Nutrition, Cytodifferentiation, and Cryopreservation*, I. K. Vasil (ed.), 1985. *Volume 3: Plant Regeneration and Genetic Variability*, I. K. Vasil (ed.), 1986. *Volume 4: Cell Culture in Phytochemistry*, F. Constabel and I. K. Vasil (eds), 1987. *Volume 5: Phytochemicals in Plant Cell Cultures*, F. Constabel and I. K. Vasil (eds), 1988. *Volume 6: Molecular Biology of Plant Nuclear Genes*, J. Schell and I. K. Vasil (eds), 1989.

1   Borlaug, N. E. and Dowswell, C. R. 1988. World revolution in agriculture, pp. 5–14, in *1988 Book of the Year* (Encyclopedia Britannica, Chicago).
2   Vasil, I. K. 1988. The contributions and prospects of plant biotechnology – an assessment, pp. 15–19, in *Plant Cell Biotechnology*. M. S. S. Pais, F. Mavituna, and J. M. Novais (eds) (Springer-Verlag, Heidelberg).
3   *The Molecular Basis of Plant Development*. 1989. R. Goldberg (ed.) (Alan R. Liss, New York).
4   *CCSCGP, V. 6.*
5   Gautheret, R. J. 1985. History of plant tissue and cell culture: a personal account, pp. 1–59 in *CCSCGP, V. 2.*
6   *CCSCGP, V. 3.*
7   Vasil, I. K. and Vasil, V. 1980. Clonal propagation. *Int. Rev. Cytol. Supp.*, **11A**: 145–173.
8   *Tissue Culture as a Plant Production System for Horticultural Crops*. 1986. R. H. Zimmerman, R. J. Griesbach, F. A. Hammerschlag, and R. H. Lawson (eds) (Martinus Nijhoff Publishers, Dordrecht).
9   Levin, R., Gaba, V., Tal, B., Hirsch, S., DeNola, D., and Vasil, I. K. 1988. Automated plant tissue culture for mass propagation. *Bio/Technology*, **6**: 1035–1040.
10   Klartha, K. K. 1984. Elimination of viruses, pp. 577–585 in *CCSCGP, V. 1.*
11   Wang, P. J. and Hu, C. Y. 1980. Regeneration of virus-free plants through *in vitro* culture, pp. 61–99, in *Advances in Biochemical Engineering, Volume 18. Plant Cell Cultures II*. A. Fiechter (ed.) (Springer-Verlag, Heidelberg).
12   Collins, G. B. and Grosser, J. W. 1984. Culture of embryos, pp. 241–257 in *CCSCGP, V. 1.*
13   Raghaven, V. 1986. Variability through wide crosses and embryo rescue, pp. 613–633 in *CCSCGP, V. 3.*
14   Nitzsche, W. and Wenzel, G. 1977. Haploids in plant breeding. *Z. Pflanzcht. Supp.* **8**: 1–101.
15   *Haploids of Higher Plants* in vitro. 1986. H. Hu and H. Yang (eds) (Springer-Verlag, Heidelberg).
16   D'Amato, F. 1985. Cytogenetics of plant cell and tissue cultures and their regenerants. *CRC Crit. Rev. Pl. Sci.* **3**: 73–112.

17   Vasil, I. K. 1988. Progress in the regeneration and genetic manipulation of cereal crops. *Bio/Technology*, **6**: 397–402.

18   Sacristan, M. D. 1986. Isolation and characterization of mutant cell lines and plants: disease resistance, pp. 513–525 in *CCSCGP, V. 3*.

19   Chaleff, R. S. 1986. Isolation and characterization of mutant cell lines and plants: herbicide-resistant mutants, pp. 499–512 in *CCSCGP, V. 3*.

20   Bourgin, J. P. 1986. Isolation and characterization of mutant cell lines and plants: auxotrophs and other conditional lethal mutants, pp. 475–498 in *CCSCGP, V. 3*.

21   Misra, S. and Gedamu, L. 1989. Heavy metal tolerant transgenic *Brassica napus L.* and *Nicotiana tabacum L.* plants. *Theoret. App. Genet.*, **78**: 161–168.

22   Hibberd, K. A., Anderson, P. A., and Barker, M. 1986. Tryptophan over-producer mutants of cereal crops. United States Patent No. 4,581,847.

23   Larkin, P. J. and Scowcroft, W. R. 1981. Somaclonal variation – a novel source of variability from cell cultures for plant improvement. *Theoret. Appl. Genet.*, **60**: 197–214.

24   Evans, D. A., Sharp, W. R., and Medina-Filho, H. P. 1984. Somaclonal and gametoclonal variation. *Amer. J. Bot.*, **71**: 759–774.

25   Swedlund, B. and Vasil, I. K. 1985. Cytogenetic characterization of embryogenic callus and regenerated plants of *Pennisetum americanum L.* *Theoret. Appl. Genet.*, **69**: 575–581.

26   Karp, A. and Bright, S. W. J. 1985. On the causes and origins of somaclonal variation. *Oxford Surv. Pl. Molec. Cell Biol.*, **2**: 199–234.

27   CCSCGP, V. 4.

28   CCSCGP, V. 5.

29   *Plant and Animal Cells: Process Possibilities*. 1987. C. Webb and F. Mavituna (eds) (Ellis Horwood, Chichester).

30   Gleba, Y. Y. and Sytnik, K. M. 1984. *Protoplast Fusion: Genetic Engineering in Higher Plants* (Springer-Verlag, Heidelberg).

31   Sjodin, C. and Glimelius, K. 1989. *Brassica naponigra*, a somatic hybrid resistant to *Phoma lingam*. *Theoret. Appl. Genet.* **77**: 651–656.

32   Gibson, R. W., Jones, M. G. K., and Fish, N. 1988. Resistance to potato leaf roll virus and potato virus Y in somatic hybrids between dihaploid *Solanum tuberosum* and *S. brevidens*. *Theoret. Appl. Genet.*, **76**: 113–117.

33   Sacristan. M. D., Gerdemann-Knorck, M., and Schieder, O. 1989. Incorporation of hygromycin resistance in *Brassica nigra* and its transfer to *B. napus* through asymmetric protoplast fusion. *Theoret. Appl. Genet.*, **78**: 194–200.

34   Kyozuka, J., Kaneda, T., and Shimamoto, K. 1989. Production of cytoplasmic male sterile rice (*Oryza sativa L.*) by cell fusion. *Bio/Technology*, **7**: 1171–1174.

35   Izhar, S. and Zelcher, A. 1986. Protoplast fusion and generation of cybrids for transfer of cytoplasmic male sterility, pp. 589–599 in *CCSCGP, V. 3*.

36   Padgette, S. R., della-Cioppa, G., Shah, D. M., Fraley, R. T., and Kishore, G. M. 1989. Selective herbicide tolerance through protein engineering, pp. 441–476 in *CCSCGP, V. 6*.

37 Vaeck, M., Reynaerts, A., and Hofte, H. 1989. Protein engineering in plants: expression of *Bacillus thuringiensis* insecticidal protein genes, pp. 425–439 in *CCSCGP, V. 6.*

38 Delannay, X., La Vallee, B. J., Proksch, R. K., Fuchs, R. L., Sims, S. R., Greenplate, J. T., Marrone, P. G., Dodson, R. B., Augustine, J. J., Layton, J. G., and Fischoff, D. A. 1989. Field performance of transgenic tomato plants expressing the *Bacillus thuringiensis* var. *kurstaki* insect control protein. *Bio/Technology,* **7:** 1265–1269.

39 Hemenway, C., Tumer, N. E., Powell, P. A., and Beachy, R. N. 1989. Genetic engineering of plants for viral disease resistance, pp. 405–423 in *CCSCGP, V. 6.*

40 Stark, D. M. and Beachy, R. N. 1989. Protection against potyvirus infection in transgenic plants: evidence for broad spectrum resistance. *Bio/Technology,* **7:** 1257–1262.

41 Paszkowski, J., Baur, M., Bogucki, A., and Potrykus, I. 1988. Gene targeting in plants. *EMBO J.,* **7:** 4021–4026.

42 Kramer, M., Sheehy, R. E., and Hiatt, W. R. 1989. Progress towards the genetic engineering of tomato fruit softening. *TIB-TECH,* **7:** 191–194.

43 *Innovative Approaches to Plant Disease Control.* 1987. I. Chet (ed.) (John Wiley, New York).

44 Timberlake, W. E. and Marshall, M. A. 1989. Genetic engineering of filamentous fungi. *Science,* **244:** 1313–1317.

45 Lindow, S. E., Panopoulos, N. J., and McFarland, B. L. 1989. Genetic engineering of bacteria from managed and natural habitats. *Science,* **244:** 1300–1306.

46 Gasser, C. S. and Fraley, R. T. 1989. Genetically engineered plants for crop improvement. *Science,* **244:** 1293–1299.

47 Vandekerckhove, J., Van Damme, J., van Lijsebettens, M., Botterman, J., de Block, M., Vandewiele, M., de Clercq, A., Leemans, J., van Montagu, M., and Krebbers, E. 1989. Enkephalins produced in transgenic plants using modified 2S seed storage proteins. *Bio/Technology,* **7:** 929–932.

48 Hiatt, A., Cafferkey, R., and Bowdish, K. 1989. Production of antibodies in transgenic plants. *Nature,* **342:** 76–78.

49 Duvick, D. N. 1989. Research collaboration and technology transfer: the public and private sectors in developing countries and the international seed companies, pp. 21–32 in *Strengthening Collaboration in Biotechnology: International Agricultural Research and the Private Sector.* J. I. Cohen (ed.) (Agency for International Development, Washington, DC).

17

Herbicides and Biotechnology: a Threat to Sustainable Agriculture?

Jack Doyle

Introduction

In recent years, biotechnology has become the pivotal technology in agricultural universities and related institutions. Many biotechnology centres have been set up to develop new techniques that will boost agriculture. Yet, biotechnology may not produce all the benefits these biotechnology centres are hoping for.

The problem with the new biotechnology centres is that they are targeting 'more of the same' for America's farms and rural environment. That is, they are pushing biotechnology to bring more high-tech products to agriculture; products that will extend the high-yield, high-capital and high-energy system of production. As has been well documented in recent years, this system has had undesirable side effects, many resulting from fertilizer and pesticide pollution, the poisoning of farmworkers and farm families, and groundwater contamination.

Sustainable agriculture offers another kind of agricultural production that promises to produce adequate agricultural returns without locking farmers into an expensive and environmentally damaging array of new farm inputs. This system relies chiefly on working with nature, rather than trying to conquer it.

Sustainable agriculture, however, is in danger of being sucked into the 'black hole' of biotechnology. Some scientists and representatives of industry are now asserting that biotechnology *is* sustainable agriculture. But making that claim could be stretching the truth and ruining a promising alternative in the process.

Biotechnology, as presently structured, is *not* sustainable agriculture because it is not 'systems biology'. As will be illustrated in this chapter on herbicides, genetic engineering offers agriculture no more than a fix-it approach that does not yield a truly integrated understanding of the biological systems that make crop and animal agriculture possible in the first place.

It is in this light that one should see the concerns with biotechnology among many groups in society. These concerns have much to do with how this new knowledge will be used, who will have access to the technology, and what kinds of research will be undertaken. This is why the current emphasis on using genetic engineering to make crops herbicide resistant is a bad idea, and exactly the wrong way to use this new technology.

Herbicides: a rejuvenation by biotechnology?

Herbicides are increasingly suspect in the environment: many are being found in groundwater globally, some are known carcinogens, and most have not been tested adequately for their long-term health effects on wildlife, farmers and farmworkers, or consumers. In addition, an increasing number of weed species are now genetically resistant to herbicides, increasing the cost of using them, and questioning their effectiveness when compared with other methods.

But despite the new concerns over herbicide safety and efficacy, chemical and pharmaceutical manufacturers and new biotechnology companies are enlisting the powers of bioengineering to enhance and broaden the use of herbicides in agriculture: that is, they are planning to give crops that may be damaged or killed by herbicides the genes that will make them resistant to the chemical's killing action or its presence in the soil.

Industry officials and scientists say the making of herbicide-resistant crops will only help herbicides do their job more efficiently, and in a manner that will be better targeted and safer for the environment, wildlife, and public health. Others disagree, seeing a darker side to this marriage of genetics and chemistry. The critics charge that herbicide-resistant crops – and the economic investment behind them – will only increase and prolong the use of herbicides rather than reduce or eliminate it.

But how did we get to this point? How did we develop such a dependence on herbicides in agriculture?

Herbicide History

During the last 100 years or more, farmers, agricultural scientists, and the agrichemical industry have waged war against weeds with all manner of substances and strategies – from fire to cultivators. Chemical weed killers made their debut rather crudely in the 1850s; sodium chloride, sulfuric acid and some petroleum products were initially used.

The modern herbicide era actually began in 1941 in the botany laboratories of the University of Chicago. Scientists at the time were conducting research on plant hormones and synthetically produced chemicals that regulated various kinds of beneficial plant growth. These researchers were often frustrated in their work by the killing action of the growth regulators on the plants. But one scientist envisioned that these chemicals might work to kill weeds if purposely applied in toxic doses. And sure enough, the researchers soon discovered that certain of these compounds would indeed kill weeds.

In Washington, this work soon came to the attention of the National Academy of Sciences' war research committee and the U.S. Army. The Army was then becoming interested in biological warfare research, and shortly recruited the leader of the University of Chicago botany laboratory to begin secret military research on herbicides at Fort Detrick in Maryland. Between 1944 and 1945, the Army tested the effects of more than 1,000 different chemical compounds on living plants at Fort Detrick. The use of herbicides in warfare was seriously contemplated at that time, but never unleashed.[1]

With the war's end, herbicides soon made their way to agricultural experiment stations, private companies, and farmers. In 1945, the American Chemical Paint Company (later named AmChem, which subsequently became part of Union Carbide) began selling the first systemic herbicide – 2,4-D, under the brandname Weedone.[2]

Two years later, 30 different preparations of herbicides containing 2,4-D were being sold in the U.S. By 1949, about 23 million acres of agricultural land were treated with herbicides of all kinds. Ten years later, that figure rose to 59 million acres. Still, by 1962, only 14 per cent of the $2 billion that farmers spent to control weeds was being spent to buy chemical herbicides. For some corporations, herbicides have become their single most lucrative product. In recent years, Monsanto has sold over $1 billion worth of herbicides worldwide in one year, nearly half of which derived from one popular herbicide named Roundup. When Eli Lilly's Treflan was riding high in the U.S. and other markets during the late 1970s, it alone comprised as much as 12 per cent of the company's total earnings.[3] Today, herbicide use in the U.S. and many other countries is exploding. In 1981, U.S. farmers applied 625 million pounds of herbicides to their fields, which constitutes a 175 per cent increase since 1958. Herbicides now comprise more than 60 per cent of all pesticides in U.S. agriculture. Sales in the

U.S. alone now come to more than $2 billion annually; worldwide they are $4.5 billion.[4]

No-Till Boosts Herbicide Sales

A significant increase in herbicide sales began in the mid-1970s with the increasing popularity of minimum and no-till agriculture, a system of crop farming in which chemicals are substituted for cultivation as the primary means of weed control. For example, some Chevron literature of a decade ago explains: 'When you switch to no-tillage farming, you switch to an often total reliance on chemicals for control of unwanted vegetation.'[5] With low-tillage methods there is reduced soil erosion, lower energy consumption, less runoff of fertilizer and pesticides, and increased moisture retention in the soil. Critics argue, however, that minimum-till crop residues increase pest and disease problems, resulting in greater pesticide use, increased costs for farmers, and additional groundwater pollution from pesticides.

Nevertheless, some predictions indicate that as much as 85% of all cultivated cropland will be in some form of conservation tillage by the year 2000, and that means more herbicides.[6] Not surprisingly, some chemical and pharmaceutical companies are gearing up their sales force and changing their sales pitch, casting themselves as conservationists.

Biotechnology's New Harvest of Biotechnology

In recent years, the seed and biotechnology industries have begun to research ways to protect seed and crops from the harsher side of herbicides: they have begun to adapt crops and seed to the likelihood of herbicide-treated soils.

One of the first products marketed to help crops tolerate high herbicide environments was 'herbicide safened' seed. Ciba–Geigy, the Swiss pharmaceutical company, introduced one such product in 1979. Ciba–Geigy and its subsidiary, Funk Seeds International, called their product a herbicide antidote and named it Concep. Concep is a chemical coating applied to sorghum seeds to protect them from any damaging effects of herbicides such as Dual, Milocep and Bicep, all of which Ciba–Geigy sells.

Another problem with herbicides is that some of these chemicals kill crops, or linger too long in the soil, damaging crops that may follow in rotation. This phenomenon is known to farmers and herbicide makers as 'residue carryover'.

In Illinois, for example, about 10 million acres of corn are treated each year with the herbicide atrazine. Corn can tolerate atrazine because it contains certain enzymes which detoxify the chemical inside the plant. Thus, corn is naturally resistant to atrazine's lethal effects. But this is not the case with soybeans, a crop often used in rotation with corn. And that can be a problem when atrazine is still present in the soil. A farmer using atrazine on corn, and then following that corn crop with a soybean crop, is likely to have atrazine-damaged soybeans and reduced yield because of atrazine soil residues. Other crops, such as alfalfa and small grains, are also sensitive to atrazine residues.

The same kind of problem exists with even the newer generation of low-dose herbicides, such as Du Pont's wheat herbicide, Glean. Only wheat, it seems, 'knows' how to detoxify Glean, or metabolize it into harmless (as far as we know) by-products. Crops such as soybeans, sugarbeets, sunflowers, and corn that might follow wheat in rotation will be damaged by residues of Glean remaining in the soil.

One solution to this problem has become clear with gene-splicing research: find the specific genes that enable a given crop to resist the ill effects of certain herbicides and move those genes into the sensitive crops.

First Research Reports and Early Corporate Programmes

During the late 1970s, some genetic engineers began to investigate the genetic and biochemical mechanisms in weeds that made them resistant to certain herbicides. By 1982, the first scientific papers were presented which suggested that such resistance could be genetically engineered into important crop plants, such as the soybean. By this time, chemical and pharmaceutical companies had also begun their own programmes of herbicide-resistance research.

In July 1982, *Chemical Week* wrote of a 'slow but steady push', among herbicide makers toward the 'genetic manipulation of corn, soybeans, and other crops to make them more resistant to herbicides'. 'The theory is', explained the magazine, 'that farmers would then be willing to use even more of the weed killers, safe in the knowledge that their crop won't be damaged.'[7]

In November 1982, Calgene, a California-based biotechnology company, announced that it had cloned a gene for resistance to the world's most widely used herbicide, Monsanto's Roundup, chemically named glyphosate. Two years later, Calgene had contracts to develop herbicide-resistant crops with at least three major corporations: Kemira Oy, Finland's largest chemical company; Nestlé, the world's largest food company; and Rhone–Poulenc, a major French agrichemical company.[8]

By this time, Du Pont had developed tobacco plants in the laboratory with resistance to its herbicide Glean – plants that were 100 times more resistant to Glean than were normal plants. The company also cloned the gene for Glean resistance. Ciba–Geigy, meanwhile, was working with atrazine. Ciba–Geigy was hoping to develop soybean varieties resistant to its old herbicide atrazine, and began work with a model system in tobacco, which it field tested in North Carolina in 1986.

According to George Kidd, an analyst with the L. William Teweles Company, a seed and biotechnology consulting firm, the development of atrazine-resistant soybeans would mean, for example, that atrazine use by farmers each year would double or triple over what it was in the mid-1980s, increasing sales by about $120 million annually. The same consulting firm predicts that the use of genetic engineering to develop herbicide-resistant crop plants will bring about a complete restructuring of the $2.4 billion-per-year U.S. herbicide market. This firm reports that some chemical companies are developing crops resistant to their herbicides 'in the hope of selling the seed and chemical as a pair'. Other companies are seeking herbicide resistance, according to Teweles, 'as a way of gaining market share lost after a well-known herbicide has declined in price and popularity'. In other words, the old herbicide will be sold in combination with a new seed resistant to it. The Teweles' consultants also see herbicide-resistant crops as creating 'a complementary demand for both chemical and seed'.[9]

Since the early 1980s, a whole host of chemical, energy, pharmaceutical and biotechnology companies have begun research on herbicide resistance. In 1988 there were at least 33 companies involved.

In addition to this commercial activity, public funds from the U.S. Department of Energy, the National Science Foundation and the U.S. Department of Agriculture have been used to support herbicide-resistance research at a number of universities and agricultural experiment stations.

Herbicide Toxicity and Safety

While most herbicides are generally not regarded to be as toxic as the chlorinated hydrocarbon insecticides used widely in the 1960s, they do have side effects – side effects we are only now beginning to understand.

And while it may be true that some of the new low-dose herbicides may indeed be safer than those of the past, this does not mean they are toxic-free or incapable of being misused. American Cyanamid's new family of herbicides, the imidazolinones, for example, have low mammalian toxicity; they are not toxic-free. Other herbicides such as DuPont's Glean and Monsanto's Roundup are claimed to be among the safest chemical compounds developed by industry so far, but not much is known about the long-term effects of herbicides in the environment or how they break down. Moreover, recent discoveries of herbicides such as alachlor, atrazine, and paraquat showing up in groundwater in the U.S. and other counties, have raised new questions about herbicide safety.

In addition, many existing herbicides, though less toxic than their predecessors, have not been adequately tested for their possible mutagenic, carcinogenic and teratogenic effects. According to some scientists, such information exists for only a handful of the 150 herbicide compounds presently in use.[10]

The Herbicide Safety Record

We know that herbicides have caused some horrendous problems in the past. In 1980, Canada suspended the use of nitrofen, produced by Rohm & Haas under the tradename TOK, after laboratory tests showed cancers and birth malformations in animals. In 1982, for example, *Science* magazine published an article on the teratogenic effects of nitrofen – mice born without heads.[11] Shortly thereafter, Rohm & Haas pulled its TOK herbicide off the market because it said it didn't want to spend the money to do the required toxicity tests.

Between 1982 and 1985, EPA listed several popular herbicides for special review, indicating that testing on laboratory animals had shown these chemicals to be potential carcinogens.

In June 1984, citing high dietary and applicator health risks, EPA placed the DuPont soybean herbicide linuron in a 'restricted use' classification. DuPont data from 1980 showed the potential of linuron to cause benign, dose-related testicular tumours in rats and mice. After DuPont had satisfied some EPA concerns for applicator safety, the agency lifted the 'restricted use' classification on linuron in September 1984, but kept the herbicide in special review because some tests showed dietary cancer risks.[12]

In late November 1984, EPA announced a special review of Monsanto's herbicide alachlor, explaining that recent studies showed the chemical to cause at least four types of tumours when fed to laboratory animals. EPA banned the use of the chemical on potatoes as well as its application by aerial spraying. EPA also tightened the herbicide'slabelling requirements, attempting to protect the 600,000 farmers and field workers who face a cancer risk from alachlor of one in 10,000, according to the agency.[13]

Used widely on grains intended for human and livestock consumption, alachlor was, according to one EPA official at that time, likely to be present at some level in meat, milk, and poultry. The herbicide has also shown up in surface or groundwater in 10 states, mostly in the East and Midwest – in some places as high as 267 parts per billion – and human exposure through drinking water is likely, according to EPA.[14]

In April 1985, EPA moved on a third herbicide, cyanazine, produced by the Shell Chemical Company. Concerned that cyanazine might be contaminating groundwater and presenting a health risk to farmworkers, EPA announced a special review of the herbicide's safety. EPA noted that cyanazine was found to cause birth defects in laboratory animals.[15]

In its 1987 study, *Regulating Pesticides in Food*, the U.S. National Academy of Sciences identified several herbicides now targeted by biotechnologists – including glyphosate – as potential carcinogens.

Beyond their potential impact for human health, some herbicides also produce physiological changes in crop plants – such as reduced wax formation on leaves, and disruption of plant growth – that make the crops more susceptible to disease pathogens. The herbicide picloram increases the sugar output in the roots of wheat and corn, encouraging sugar-loving fungal pathogens, while 2,4-D reduces the sugar content of leaves, making them susceptible to other pathogens.[16]

Herbicide-Resistant Crops: The Side Effects

The first genetically-engineered herbicide-resistant crop strains reached the market in 1990. A number of herbicide-resistant or herbicide-tolerant crop strains have already been field tested. Thirteen of the 21 field tests for genetically engineered crops approved by the U.S. Department of Agriculture since late 1987 have been conducted by three chemical and pharmaceutical companies – Monsanto, Sandoz and Du Pont – and more than half of these were for herbicide-tolerant plants.

Not everybody, however, is thrilled about the prospect of herbicide-resistant crops. 'If we create food crops with herbicide resistance,' asks Sheldon Krimsky, a social scientist at Tufts University and former member of the NIH committee on recombinant DNA, 'are we not going to reinforce the use of herbicides? Are we not going to reinforce greater chemical use in food production at a time when people are increasingly questioning the agricultural use of chemicals?'

By spreading more herbicide around in the environment, it is possible that more weeds will become resistant to herbicides because of the additional selection pressure on those weeds.

Another problem is the possibility that weed species sexually compatible with crop species will cross with the engineered crops, potentially resulting in the transfer of a herbicide-resistant trait to one or more weed species.

Only The Tip of the Iceberg?

Some scientists say that their excursion into herbicide resistance will be short-lived, and that the important thing is what is learned in the process about the chloroplast genome or the basic molecular biology of the plant. Yet capital is being invested, new herbicide factories will surely be built, and high-tech agriculture will not turn around overnight. The direction of this research is unmistakable.

In the seed and plant genetics industry, research priorities are already being reshuffled to accommodate herbicide-related traits in ways that may divert scientific attention away from more socially important kinds of research. There is some concern that plant scientists may be spending too much of their time trying to come up with genetically engineered varieties that will accommodate the use of chemicals, rather than developing disease or insect resistance in those crops. Such priorities may well detract from, if not foreclose, other research that could lead us away from continued dependence on chemical pesticides.

But herbicide resistance in crops appears to be only the tip of the iceberg. Consider, for example, the following research:

- U.S. Forest Service researchers in Rhinelander, Wisconsin, are exploring ways to make forest trees like poplar and jack pine resistant to herbicides such as glyphosate and hexazinone.

- Others are working on chemical plant-growth regulators that will signal or prompt certain genes in crop plants to 'turn on' or 'turn off' at the right time – genes that control certain metabolic or growth traits which put the plant into dormancy, increase its flowering or pod set, increase stalk rigidity, or cause it to defoliate uniformly at a certain time.

- Similar kinds of research are being conducted in animal science, where a certain kind of chemically-laced feed will 'turn on' or activate the right promoter gene at the right time to begin a particular growth function, or to bring the animal into fertility, or to facilitate some other function.

- And still others are considering ways to give honeybees the genes to make them resistant to the side effects of certain insecticides. Such applications of biotechnology are troubling because they suggest a continuation of chemical toxity in agriculture, and the extension of the pesticide era. With more pesticide in the environment, public resources like groundwater or wildlife refuges will be threatened, not to mention farmers', and farmworkers' health, or that of consumers exposed to pesticides in tainted fruits and vegetables and other food products.

Such applications of biotechnology are also troubling because of the kind of 'product synergy' they pose for agriculture, extending the web of integrated products that may increase agriculture's production and capital costs, as well as its dependence and vulnerability.

What we should be doing with our new biological research talents is figuring out how to *eliminate* pesticide use rather than conjuring up new and exotic ways to extend and expand pesticide use through genetic engineering.

Pesticides and the Changing Political and Economic Environment

It is clear that in the last 6 years or so there has been a growing public concern over environmental degradation, food-safety issues, and public health generally. This concern is manifest in many ways; through public opinion, consumer reaction, concern in the investment community, changed business practices, and public policy. But in the matter of pesticides in the environment and food supply, there is increasing evidence of a political groundswell from many quarters which indicates a fundamental change with regard to society's willingness to accept the continued use of pesticides, particularly in the food system.

Then there is the matter of cost, and the changing economics of pesticide use. Once upon a time, pesticides were a cheap and effective way to kill pests. But that was before the cost of energy rose, and the emergence of pest resistance was widely understood.

According to the chemical companies it costs between $20 million and $50 million to develop a new pesticide. By comparison, the cost of developing a new crop variety that is resistant to disease or insect pathogens is considerably lower than $20 million, and the cost/benefit ratio of such crops has been found to be extremely high, something like 300 to 1. Industry officials say they have to screen more compounds today to find acceptable chemicals. Secondly, there is more testing required of pesticides; new compounds have to run a longer gauntlet of toxicity tests to screen for potential cancer and birth defects.

And then there is the matter of effectiveness in the field. As of 1984, for example, genetic resistance to one or more pesticides has developed in: 447 species of insects and mites; 100 species of plant pathogens; 55 species of weeds; 2 species of nematodes; and 5 species of rodents. More recent data on weeds, for example, suggest that there are now 100 species of weeds resistant to one or more herbicides.

So, what do we have after all of the chemical warfare on agricultural pests? Well, the amount of crop loss to disease and insects is still very high, despite the continued and increasing use of pesticides.

'Common Sense Biotechnology' and 'Biotech-knowledge'

What is needed today is a return to what we already know; to what I call 'common sense biology' and 'common sense agriculture'. This is based upon a thoughtful 'biotech-knowledge', which can make an important contribution to sustainable agriculture.

There is a lot of good biology already 'on the shelf', so to speak. And there is a lot of good, innovative work going on all over the world by researchers at universities and agricultural experiment stations. We have extensive information about soil tilth, plant breeding, crop rotations, intercropping, insect adaptability, agricultural diversification, and other fields – all of which has application today. This huge practical knowledge base is now ripe for rediscovery and new application, and it doesn't need genetic engineering to be made useful. Secondly, it ought to be possible to integrate the old biology with the insights of the new biotechnology. New, molecular-level insights into crop/pest interactons, or pest life cycles, should help to enhance the breeder's ability to 'build in' more durable, multigenic forms of disease and insect resistance.

For example, consider some common sense observations and practices that are available right now:

Improved Fertility Through Classical Genetics

Donald Barnes, a plant breeder at USDA's research centre at the University of Minnesota, has developed a new variety of alfalfa called Nitro that produces good livestock fodder and puts high amounts of nitrogen back into the soil. When ploughed back into the soil, Nitro puts 94 lbs of nitrogen into the ground compared with 59 lbs for the top standard variety. Although Barnes began working on this variety in the late 1970s 'to make American agriculture less susceptible to rising energy costs', Nitro represents the kind of genetic improvement in agriculture that can save growers money on the input side of their operation, and thus increase their profitability.

Thicker-Husked Corn and Insects

In some areas of the U.S. where sweet corn is grown, the corn varieties in use are so prone to insect damage that farmers have to spray them with pesticides 20 times in a growing season. Yet, there are alternatives to this kind of pest management. A variety of sweet corn that would offer a thicker-husked ear of corn would prevent worms from burrowing into the ear. Today, some such varieties are available on a limited basis and only in certain regions of the country. More of this kind of common sense genetics in varietal design could help dramatically reduce if not eliminate pesticides.

Manure Applications and Insects

According to research at the University of Minnesota, manure applications may reduce rootworm problems by increasing the population of mites that feed on the rootworm eggs. And some Minnesota farmers have observed a possible inhibition of alfalfa weevils by manure applications. At the University of Washington, researchers have found that corn borer infestations are much less on organic farms, which generally use green or animal manures, than they are on conventional farms.

Crop Rotations Help Control Insects

Adding legumes to rotations increases predators that prey on insects. In Wisconsin it has been discovered that mixing grasses with alfalfa aids in the control of insect pests of alfalfa.

Classically-Bred Strawberries

One example of common sense genetics in the public sector that has resulted in protability for the private sector is the work of strawberry breeders at USDA's Beltsville Agriculture Research Center near Washington. There, the work of people like Donald H. Scott and geneticist Gene Galletta has resulted in a long line of strawberry varieties including Earliglow (1975), Allstar, Tribute and Tristar (all in 1981), and Earliglow (1986). Tribute and Tristar, for example, have emerged as important eastern varieties after the Beltsville team successfully incorporated cold-tolerant traits from Rocky Mountain varieties. And they were also equipped with better disease and pest-resistant genes. This classical genetics work enabled one New York grower to gross about $50,000 an acre with Tristar in 1987 because he was able to ship his berries up until frost to Manhattan markets. This is the result of 20 years of patient plant breeding and testing.

Some Examples of 'Common Sense' Biology From the Private Sector

Ingene Biotechnology, Inc.

The Ingene Biotechnology company of Columbia, MD, has discovered a safe and effective way to kill nematodes, which now cause $3 billion in crop damage annually. Chitin, a protein formed in crab and shrimp shells, nourishes the growth of soil microbes such as fungi. Fungi, in turn, can convert chitin into an enzyme that destroys young nematodes and nematode eggs.

Safer, Inc.

Safer, Inc. of Wellesley, MA, has discovered that fatty acids extracted from animal tissues kill insects by degrading the outer membranes of their cells. These fatty acids aso appear to be useful as herbicides, and are harmless to mammals, birds, and fish.

Evans BioControl

Evans BioControl of Broomfield, Colorado, is using Nosema locustae, a natural spore that kills grasshoppers by destroying their fatty tissue. The spore is applied to tiny wheat bran specks which can then be sprayed on fields by conventional crop dusters. This biological agent, which was discovered in 1953, should be welcomed by western farmers and ranchers who now confront grasshoppers that have genetic resistance to many insecticides.

Herbicides, Weed control and Biotech-knowledge: An Outlook

In fact, there is a whole new programme of agricultural and biological research possible – one that incorporates disease and insect-resistance work; biological weed-management research; new fertility strategies; new research in apomictic hybrids and open-pollinated lines; a new crop and livestock diversification programme; agro- and microbial ecology work; and finally, a complete system-wide inventory and cataloguing of what there already is 'on the shelf' in our universities and agricultural experiment stations (e.g., discovered disease or insect resistances that have never made their way into commercial use), as well as abandoned old research that might now have new life, given our advances in molecular and cellular biology.

The better understanding of basic plant science that is possible with biotechnology could help in designing non-herbicide weed-control strategies. It could facilitate better crop rotation systems, aid in the domestication of new crops, or enhance the use of alleleopathic genes in crops. Used in these ways, biotechnology would also allow such crops to inhibit the growth of competing weeds through natural defence mechanisms.

Yet, as chemical herbicide development continues – now intensified by the use of biotechnology – non-chemical alternatives might be completely overshadowed unless a strong push is made to bring these alternatives to the fore. Such strategies – and even a few viable products – are available today, and other research shows promise. Let's give some examples:

In 1981, Abbott Laboratories of Chicago registered a fungus with EPA named Phytophthora palmivora. This particular fungus is a naturally-occurring organism, discovered in Florida soil, that induces a disease in the citrus strangler vine, or milkweed vine, that plagues citrus groves in Florida. Abbott sells this 'myco-herbicide', as it is called, under the tradename Devine. It is a highly effective, highly specific herbicide (attacking *only* milkweed vine), and does not harm citrus trees or fruit. In fact, this fungal herbicide is so effective that retreatment in some Florida groves may not be necessary for several years.

The Upjohn Company also produces a fungal herbicide that attacks the weed Northern Joint vetch, which is a problem in rice and soybean fields. Upjohn's product, trade name Collego, was registered with EPA in 1982.

Alan R. Putnam at the Pesticide Research Center at Michigan State, and William B. Duke of Cornell's Agronomy Department, noted the alleleopathic properties of cucumber in suppressing weed growth as early as 1974, and suggested the possibility of moving such traits into commercial crop varieties through conventional plant breeding. Today, gene splicing and biotechnology could, no doubt, enhance the range of possibilities for incorporating such alleleopathic traits into commercial crop varieties.

Other approaches to nonchemical weed control could proceed without the help of biotechnology. The successful application of biocontrol using plant-eating insects has been repeatedly demonstrated (see 'Biological Control of Weeds in the United States', *Journal of Pesticide Reform*, 5(3): 18–20). Cochineal insects imported from India in 1856 ate prickly pear cactus in Ceylon. In the 1950s and 1960s, a runaway ornamental lantana was partially controlled in Hawaii with insects from South and Central America.

Use of plant-eating insects to control weeds is currently being largely ignored in the U.S., however, in the rush to employ biotechnology and herbicides. Nevertheless, says the U.S. Department of Agriculture's Lloyd Andres, 'All weeds have natural enemies somewhere in the world.'

Unless all herbicide and pesticide alternatives are pursued with equal vigour – by either the commercial and/or public sector, using conventional or biotechnological techniques – the primary conclusion that begs to be drawn is that this technology will be used primarily to extend the pesticide era rather than end it. If that turns out to be true, it will indeed be unfortunate – for farmers, society and the business community – for there are genuine opportunities in applying this new genetic knowledge to free agriculture from the too-long dependence on toxic and environmentally-damaging pesticides.

NOTES

[1] Gale E. Peterson, The Discovery and Development of 2,4-D, in: *Agricultural History*, Vol. 41, July 1967, pp. 243–253.

[2] D. L. Klingman and G. C. Klingman, Focus on Herbicides, in: *Farm Chemicals*, December 1984, pp. 36–37.

[3] Maureen K. Hinkle, Problem with Conservation Tillage, in: *Journal of Water and Soil Conservation*, May–June 1983, pp. 201–206; Herbicides Follow the Current Trend to Low-till Farming, in: Chemical Week, 9 May 1984, pp. 12–13.

[4] The Hot Market in Herbicides, in: *Chemical Week*, 7 July 1982, pp. 36–40; see also note 3.

[5] Elyse Axell, The Toll of No-Till, in: *Soft Energy Notes*, Dec.–Jan. 1981, pp. 14–16.

[6] See note 3.

[7] The Hot Market in Herbicides, in *Chemical Week*, 7 July 1982, pp. 36–40.

[8] Calgene Inc. Announces First Genetically Engineered Gene for Herbicide Resistance, in: *Seed World*, November 1982; Vicki P. Glazer, Researchers Tackle Herbicide Resistance, in: *Biotechnology*, December 1983; David Valiulis, Plant Biotechnology at Calgene: After the Hype, There's Hope, in: *California Farmer*, 15 June 1984, pp. 6–7 and pp. 36–37.

[9] Herbicide Markets for Resistant Crop Plants, in: *Genetic Technology News*, April 1984, pp. 6–7; see also note 8.

[10] D. D. Kaufman and P. C. Kearney, Microbial Transformations in the Soil, in: L. J. Andus (ed.), *Herbicides*, New York: Academic Press, 1976, pp. 29–64.

[11] L. E. Gray et al., Prenatal Exposure to the Herbicide 2-4-dichlorophenal-p-nitrophenyl ether Destroys the Rodent Harderian Gland, in: *Science*, Vol. 215, 1982, pp. 293–294.

[12] Harry Meier, Quest for Pest-Resistant Crops Is Raising Ecological Concerns, in: *Wall Street Journal*, 30 August 1985, p. II-1.

[13] See for example, Andy Pasztor, EPA Plans to Impose Severe Restrictions on Use of Monsanto's Lasso Weed Killer, in: *Wall Street Journal*, 19 November 1983; Cass Peterson, EPA Tightens Restrictions on Widely Used Herbicide, in: *Washington Post*, 21 November 1984; Another Corn Herbicide Under Scrutiny, in: *Washington Post*, 4 April 1985, p. A-15.

[14] See note 13.

[15] See note 13.

[16] John P. Giere, K. M. Johnson and J. H. Perkins, A Closer Look at No-Till Farming, in: *Environment*, Vol. 22, No. 6, July–August 1980, pp. 15–41.

18

Applications of Biotechnology to Livestock Production

W. Jos Bijman

1. Introduction

Biotechnology can and will be applied to livestock production in many different ways. There is a broad range of research and development activities from the relatively simple technique of embryo transfer, which is already in use on a commercial scale, to the very advanced technique of genetic engineering, for which there is still a long way to go. Livestock agriculture is becoming more and more a high-tech business, where animals are raised with sophisticated compound feed, are continuously diagnosed for even the slightest sign of disease or stress, and are very precisely selected for their production task. The next step is that animals will be engineered to fit the production or environmental requirements.

Here we will discuss some of the main applications of biotechnology to animal production. Four groups of applications can be distinguished (see figure 18.1):

- reproduction and breeding;
  animal health;
- animal nutrition; and
- physiology of growth and lactation.

2. Reproduction and breeding

In order to increase the economic merit of farm livestock their reproduction is organised along the lines of breeding programmes. This process of genetic improvement of commercially important traits like growth, production, and disease resistance is based on quantitative

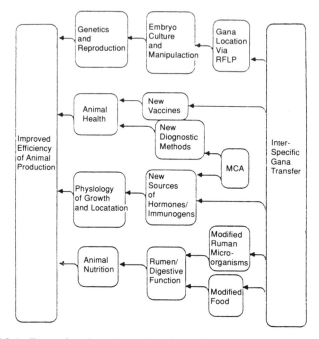

Figure 18.1 Biotechnology in animal production

Taken from: E. P. Cunningham, 'Animal production', in: Gabrielle J. Persley, *Agricultural Biotechnology. Opportunities for International Development* (Wallingford: CAB International, 1990), pp. 169–175.

genetic theories. By selecting the best animals as parents for the next generation the available genetic variance is used to enhance the average performance of the whole population. The annual improvement for economically important traits is 1 to 2%. These classical breeding programmes, however, have certain disadvantages. It is a slow process owing to the long generation intervals (1 to 2.5 years for the different farm species). Since most animals combine both good and bad traits, it takes many years of selection to obtain the right combination of desired traits.

With the application of biotechnology some of the restraints of classical animal breeding can be removed. New techniques working at cellular or molecular level greatly increase the possibility to manipulate farm livestock to improve their traits.

Biotechnology research for reproduction and breeding is carried out for several farm animals, like sheep, pigs, chicken and cattle. Since beef and dairy production together form the largest sector within the animal-production industry, we will here focus on application of biotechnology to cattle. Where appropriate we will also make comments on applications to other animals.

Embryo technology

The latest advances in embryo technology continue along the lines set by the introduction of embryo transfer in the 1970s. Embryo transfer basically consists of the recovery, storage and implantation of embryos. Cows normally only release one egg at a time to be fertilised. In order to obtain more than one fertilised egg, superovulation is induced with a hormone injection. In this way up to 30 fertilised embryos can be recovered. The average yield is 5 to 6 embryos, of which 75% are good for implantation into receiver cows. For embryo transfer over distance and time embryos can be frozen into liquid nitrogen, with relatively small effects on their viability.

The principal benefit of embryo transfer is the ability to produce more caves from a female animal than would be possible with normal reproduction, just as artificial insemination produces more offspring from a male animal. While normally a cow will give birth to four calves in an average lifetime, this can be readily increased to 25 calves. Increasing the reproductive rate of selected cows has important benefits for breeding programmes:

1. Genetically outstanding cows can contribute more to the breeding programme, particularly if their sons are being selected for use in artificial insemination.

2. The rate of genetic change can be enhanced if specially designed breeding schemes are set up which take advantage of the increased intensity of female selection combined with increased generation turnover.

Other (potential) advantages of embryo transfer technology are:

• International transport of breeding stock: transport of embryos is much cheaper than live animals; moreover it avoids the risk of importing endemic diseases.

• The transfer of embryos of specialised breeds (e.g. pure beef breed) of higher value into cows of lower potential (e.g. dairy cows not used for reproduction).

• The rapid expansion of rare genetic stocks, for example of a new breed.

- Avoiding the environmental shock to susceptible imported genotypes by having them born to dams of local breeds, rather than
importing them as live animals. This is particularly important when
farmers in tropical areas want to import breeding stock from temperate climate zones.

In Vitro Fertilisation (IVF)

A more advanced step in embryo technology is In Vitro Fertilisation
(IF). For IVF, eggs are recovered from the ovaries of a cow, are then
matured and fertilised outside the body, and further developed until
they are ready for implantation into receiver cows (after about a
week). Eggs can be obtained by surgical and non-surgical methods
from living cows, or from the ovaries of just slaughtered cows. If the
non-surgical method is used, it can be repeated many times. About 10
to 20 eggs at a time can be recovered, up to 250 a year.

IVF has several advantages above normal embryo transfer. First, a
large number of embryos can be recovered from one single female
animal. This will reduce the cost of embryos, and therefore make
embryo transfer techniques economically feasible on a larger scale. A
larger number of embryos from cows with important production
traits can increase the efficacy of breeding programmes. Secondly, the
availability of a large number of relatively cheap embryos will facilitate research in genetic engineering. Thirdly, IVF makes available
embryos in one or two cell stadium. Only this phase of development
is suitable for embryo manipulations like cloning.

Embryo cloning

A next step in embryo technology is the asexual reproduction through
embryo cloning and embryo splitting. With the method of embryo
splitting, which is cutting the embryo in two or more pieces, a maximum of four identical animals can be produced. Embryo cloning can,
theoretically, produce an unlimited number of identical individuals.

Embryo cloning is based on the technique of nuclear transplanta-
tion. An embryo in the blastocyste phase (64 cells, no differentiation
yet) is divided into individual cells. Each of these cells is fused with a
mature egg from which the nucleus is removed. After fusion with
these nucleus-free cells, each of the 64 cells of the original embryo
produces an identical copy (clone) of that embryo. In principle, out of
each clone 64 new clones can be obtained. However, the research on
embryo cloning so far has only been successful with the first generation of clones.

The purpose of cloning is to obtain a large number of identical embryos to become identical production animals. Also, these clones can be used for further research, for instance, on the sex and quality of embryos, and on genetic engineering.

It is not expected that clones will be used in cattle-breeding programmes. As clones are animals with identical genotype their use within breeding programmes would lead to a loss of genetic variance. Clones will mostly be used as production animals, for either dairy or beef production. The main emphasis in most (applied) research projects is on producing clones for more uniform beef production.

Sexing of embryos and of semen

Two methods are being developed for sex selection of embryos before implantation. One is using monoclonal antibodies, which bind to the H–Y antigen of male cells. This binding is made visible by a fluorescent flag tied to the antibody. The other method makes use of DNA-probes. Both methods have been successfully applied in experiments. Commercial applications of embryo sexing have not been reported yet, but are expected in a few years. Sexing of semen has not been accomplished, but research is going on.

Sexing of embryos (and eventually of semen) is important for the efficient application of embryo transfer and IVF techniques. As more embryos are implanted (either as clone or not) it becomes increasingly important to know in advance what sex they are, since there is only a limited number of receiver animals. On a dairy farm only a limited number of cows are needed to reproduce the herd. These cows will receive a female embryo. The other cows will receive a male embryo, since the calves born out of these embryos will only be used for beef production.

Genetic markers

So far we have discussed techniques applied in research at the cellular level. The application of these techniques can improve and speed up the efficacy of breeding programmes, and thereby indirectly improve the genetic traits of farm aimals. The most revolutionary biotechniques, however, work at the molecular level, by manipulation of the DNA-sequence. Before genetic engineering of an animal can actually take place one has to know its genetic code, or genome. The most important technique for gene mapping is the RFLP (Restriction Fragment Length Polymorphism) technique: DNA-probes (cloned DNA-fragments) are used as a genetic marker to identify chromosome segments with the same DNA-sequence. Gene mapping using the RFLP technique has several applications:

1. characterising commercially important genes or gene complexes;
2. establishing positive parent–offspring relationship (genetic fingerprint);
3. identifying heterozygous carriers of genetic disorders;
4. identifying genetic markers for performance traits;
5. identifying genes for genetic engineering.

Within breeding programmes genetic markers (application 4) will be a very useful tool, because the relationship between genetic variance at the molecular level and variance in production traits can be analysed.

Using RFLPs in the selection process is called 'Marker Assisted Selection' (MAS).

Research on genetic markers is carried out to study, among other things, genetic polymorphism of milk protein variants in dairy populations. Different kinds of caseins (milk proteins) show different characteristics in milk processing, for instance in cheese making. Once the genetic variants of the desired kinds of casein are known, cows can be selected on their genetic disposition to these variants.

Transgenic animals

Research on transgenic farm animals is carried out for several commercially important traits, like milk production and disease resistance of cows, and growth characteristics of pigs. The use of transgenic animals in breeding programmes will require a long process of development and testing. Transgenes must show at least 10% genetic improvement for common traits to be of practical interest.

Apart from common traits under selection, transgenes can be of interest for traits which presently are non-existent or which presently are difficult to select for. These new traits can be the production of new pharmaceuticals, the improvement of the quality of existing animal products, and improved disease resistance (see below for discussion of these new traits).

3. Animal Health

Monoclonal antibodies

The development of monoclonal antibodies (MABs) is probably the most interesting biotechnology invention after genetic engineering.

MABs are highly specific, they can be produced in large quantities, and they can be used in simple, rapid diagnostic tests. Beside in diagnostics, MABs can be used for prevention and treatment of viral and bacterial diseases, for elimination of pharmaceuticals and toxins, and for manipulation of physiological processes, like regulation of growth and reproduction.

MABs are currently used in diagnostic tests for pregnancy of cattle, for sex selection of cattle embryos, and for quality control of meat products. An example of the use of MABs for disease prevention is Genecol TM99, administered to newborn calves to prevent diarrhoea caused by infection with E. Coli bacteria. An example of MABs used for regulating physiological processes is Neutra-PMSG. This product, working against PMSG (Pregnant Mare Serum Gonadotrophin), is used to neutralise the residue of exogenous PSMG during a superovulation treatment. MABs can also be used in basic research on genetic disposition, for instance to disease resistance.

New vaccines

Infectious diseases form a major threat to livestock farming, in both developed and developing countries. Veterinary vaccines are an important tool in keeping the animals healthy. Biotechnology provides new techniques to develop and administer (new) vaccines.

One new technique is to express the genetic sequence coding for a specific antigen in a microorganism or mammalian cell. Large quantities of the antigen can then be harvested and used for vaccine production. This method of producing specific antigens for use in vaccines makes the difficult task of cultivation of large numbers of complex infectious agents obsolete. This process also provides purified materials which do not contain other toxic products or infectious agents. In practical tests, however, these vaccines do not always show the expected efficacy.

Another promising method is live vaccines, produced with recombinant-DNA technology. Genes coding for a certain antigen are inserted into a virus or bacterium, which will then act as a vector vaccine. Currently, the vaccinia (cowpox) virus is the most studied vector, but other viruses and bacteria can be used, too. The large size of the vaccinia virus also means that it can accommodate more than one foreign antigen gene without losing its ability to infect an animal, so giving the option of immunisation against several different antigen genes from the same or different infectious agent(s) in a single dose.

A recombinant DNA vaccine against diarrhoea of pigs is already available. Research is going on for, among others, foot-and-mouth disease and theileriosis (cattle diseases), African swine fever and Aujeszkys disease (swine diseases), Newcastle disease and coccidiosis (poultry diseases).

Transgenic animals

Transgenic farm animals with an improved disease resistance have not been reported, and it will take some time before they will be. However, research towards that goal is being carried out for several farm animals: cattle, sheep, pigs and chickens. The first step is to study the genes or gene complexes which are responsible for the imune system. The primary task of the immune system is the induction and regulation of protective immune responses against disease-causing pathogens such as bacteria, viruses and other microorganisms. It has been found that a gene complex called the Major Histocompatibility Complex (MHC) plays a central role in the immune system. Research on the MHC is carried out for cattle, pigs and chickens. Studies indicate the presence of associations with the BoLA, as the MHC of cattle is called, for several diseases: mastitis, bovine leucosis, parasite infections (worms, ticks), theileria parva infections and trypanosomiasis. Experiments on improving mastitis (udder infection) resistance of cows have recently started in the Netherlands. A gene coding for lactoferrine, a protein with antibacterial qualities, has been inserted into cattle embryos. The goal is a prolonged lactoferrine secretion in the mammary gland, and thereby a better mastitis resistance.

4. Animal Nutrition

Biotechnological research on animal nutrition can be categorised under three headings: improvement of feed quality, improvement of the digestibility of feed, and a decrease in the excretion of environmentally polluting matter. Most of the research is carried out for pigs and poultry as their nutrition requirements are largely met by (industrially produced) compound feed.

The quality of feed can be improved by enzymatic treatment, by adding inoculants to ensilaged fodder, and by decreasing the antinutritional factors in certain plants used as feed, like legumes. Enzymatic pre-treatment of feed components can improve the digestion of cellulose-rich feed by pigs and poultry. These single-stomached animals do not have the enzyme system like ruminants have to digest cellulose. With enzymatic pre-treatment a larger variety of raw materials, for instance byproducts from the food industry, can be used for compound feed. By adding inoculants (bacteria) and/or enzymes to ensilaged fodder the nutritional quality of the feed can be improved. Elimination of anti-nutritional factors, like anti-trypsin, lectin, and tannin in legumes, is direced at the improvement of protein absorption of pigs and poultry.

Amino acids, such as lysin, can be produced with biotechnological methods. By adding amino acids, which are essential for growth but can be lacking in raw materials, the compound feed can be better adjusted to the need of the animal. The required production level can thus be reached using feed with a lower protein content. The resulting improved nitrogen utilisation has important advantages from an environmental point of view. Environmental pollution also can be prevented by adding the biotechnologically produced enzyme phytase to compound feed. With phytase the available phosphate (a growth enhancer) will be better utilised by the animal, leading to a lower phosphate excretion.

5. Physiology of Growth and Lactation

Somatotropin

Most of the biotechnology research on growth and lactation of farm livestock is targeted at the growth hormone somatotropin. This protein occupies a central role in the regulation of growth, body composition and milk production in mammals. It exhibits regulatory effects on metabolism and determines how absorbed nutrients are partitioned to different parts of the body.

By isolating the gene responsible for somatotropin production and expressing it in bacteria it is now possible to produce large amounts of this protein through recombinant-DNA technology. Both recombinant bovine somatotropin (rBST) and recombinant porcine somatotropin (rPST) are now available for research purposes.

Exogenous administration of rPST has shown clear effects on production performance of fattening pigs: an increase in growth, a decrease in feed intake and therefore an improvement of feed conversion, an increase in carcass meat content and a decrease in carcass fat content. The milk production of lactating sows and the growth of nursing piglets can be significantly improved by daily treatment of lactating sows with rPST. However, administration to sows has a negative impact on fertility and health. A major constraint for use of rPST (besides official approval) is that no practical method for administration is available. The method used in research projects, i.e. daily intra-muscular injections, is not considered a practical method for intensive pig-production systems.

The administration of rBST to dairy cows during lactation has shown increases in milk output by 15–30%, and also increases in efficiency of milk production (the conversion of feed into milk). Here, too, the main technical problem to be resolved is that of an efficient delivery system, since current use requires regular injections. The commercial use of rBST has been approved in just a few countries, owing to a considerable discussion over the desirability of a milk-production increasing device in times of oversupply in Western livestock agriculture, and also over the safety of rBST milk for consumers.

Transgenic animals

The physiology of growth and lactation is also a major target of research on transgenic farm livestock. Most of this research is dealing with genes responsible for growth-hormone production. Results of experiments with transgenic pigs, probably the most advanced experiments as far as growth hormone is concerned, have been rather ambiguous, with improvements in growth and in feed conversion into protein but detrimental effects on the health of the pigs.

Another goal of current research on transgenic farm animals is to let them synthesise proteins of medical value. This would be particularly valuable for the production of proteins that cannot be synthesised in their active form by microorgansms, such as certain bloodclotting enzymes. These proteins might be produced either in blood or in milk under the control of regulatory elements from genes that are expressed in liver or mammary gland, respectively. The most appropriate production site seems the mammary gland, which allows excretion via the milk. In Edinburgh, transgenic sheep are reported to have secreted coagulation factor IX and alpha-1-antitrypsin in their milk.

The quantity and quality of existing animal products are also subject to transgene research. Improvement of wool production through genetic engineering is being studied in Australia. Research on (changing) the composition of milk is carried out in France, the USA, Germany and the Netherlands. Important (potential) applications of genetic engineering to changing the nutritional and technological qualities of milk are the following: (1) Reduction of lactose content of milk to promote its consumption by people intolerant of lactose and to facilitate the manufacture of lactose-free products. (2) Suppression of beta-lactoglobulin in bovine milk used for preparing humanised infant formulae. (3) Production of milk with high casein and organic phosphate contents, in order to improve cheese yield.

REFERENCES

W. Jos Bijman, *Biotechnologie in de zuivelproduktieketen: Implicaties voor de melk-veehouderij* (Biotechnology in the Dairy Chain of Production: Implications for Dairy Farming) (The Hague: NOTA, 1990, NOTA Werkdocument W19).

Nationale Raad voor Landbouwkundig Onderzoek (NRLO), *Studiedagen genetische manipulatie bij landbouwhuisdieren, rundvee, schapen, varkens en pluimvee* (Symposium on Genetic Manipulation in Farm Animals, Cattle, Sheep, Pigs and Poultry) (The Hague: NRLO, 1989, NRLO-rapport no. 89/15).

C. Smith, J. W. B. King and J. C. Mckay, *Exploiting New Technologies in Animal Breeding. Genetic Developments* (London: Oxford University Press, 1986).

K. Sejrsen, M. Vestergard, A. Neimann-Sorensen (eds), *Use of Somatotropin in Livestock Production* (London/New York: Elsevier Applied Science, 1989).

P. van der Wal, G. J. Nieuwhof and R. D. Politiek (eds), *Biotechnology for Control of Growth and Product Quality in Swine. Implications and Acceptability* (Wageningen: Pudoc, 1989).

19

Enzymes and the Food and Beverages Industry[1]

Z. Towalski and H. Rothman

Many sections of the food and beverage industry use enzymes in their production, the chapter on enzyme technology has already indicated how wide-ranging is this utilisation. This is in many cases the result of long-standing craft tradition which predates the scientific understanding of enzymes, as, for example, cheese making, beverage manufacture via alcoholic fermentation, baking and so forth.

We have summarised the sectoral usage of enzymes by the food and drink industry in table 19.1. Since it is not possible here to treat such diverse uses in detail, we will, therefore, concentrate our attention on describing the main enzymes used, and their role in a small selection of industries.

Table 19.1 Summary of enzyme usage by food industry (2)

Industry	Application	Enzyme
Baking and Milling	Reduction of dough viscosity, acceleration of fermentation process, increase in loaf volume, improvement of crumb score and softness, maintenance of freshness and softness	Amylase
	Improvement of dough texture, reduction of mixing time, increase in loaf volume	Protease
Beer	Mashing	Amylase
	Chill-proofing	Protease
	Improvement of fine filtration	Beta-Glucanase

Industry	Application	Enzyme
Cereals	Precooked baby foods, breakfast foods	Amylase
	Condiments	Protease
Chocolate, cocoa	Manufacture of syrups	Amylase
Coffee	Coffee bean fermentation	Pectinase
	Preparation of coffee concentrates	Pectinase Hemicellulase
Confection-ery, candy	Manufacture of soft-centred candies, and fondants	Invertase, pectinase
	Sugar recovery from scrap candy	Amylase
Corn syrup	Manufacture of high maltose syrups	Amylase
	Production of low D.E. syrups	Amylo-glucosidase
	Production of glucose from corn syrup	Glucose isomerase
	Converting corn syrup to fructose-containing product	
Dairy	Residual H_2O_2 removal from milk, after sterilisation by H_2O_2	Catalase
	Manufacture of protein hydrolysates	Protease
	Stabilisation of evaporated milk	Protease
	Production of whole milk concentrates, whey concentrates, ice cream and frozen desserts	Lactase
	Curdling milk	
Distilled beverages	Mashing	Amylase
Eggs, dried	Glucose removal	Glucose oxidase
Feeds, animal	Pig starter rations	Amylase, protease

Table 19.1 Continued.

Industry	Application	Enzyme
Flavours	Clarification (starch removal)	Amylase
	Oxygen removal	Glucose oxidase
Fruit juices	Clarification, preventing gelling of concentrates, improvement of juice extraction yield	Pectinases
	Oxygen removal	Glucose oxidase
Meat	Tenderisation	Protease
	Preparation of fish protein concentrates	Protease
Protein hydrolysates	Preparation of protein hydrolysates	Protease
Soft drinks	Stabilisation of citrus terpenes from light-catalysed oxidation	Glucose oxidase and catalase
Vegetables	Preparation of hydrolysates	Pectinase cellulase
	Liquifying purees and soups	Amylase
Wine	Clarification of must	Pectinase

Modified from: J. T. P. Böing, Enzyme Production, in G. Reed (ed.), *Prescott & Dunn's Industrial Microbiology* (London: Macmillan, 1982).

Important Food Enzymes

Proteases hydrolyse peptide bonds in proteins to yield shorter peptides and amino acids; they can be obtained from microbial, plant and animal sources. Animal sources are used for proteases such as: trypsin, chymotrypsin, rennet and pepsin.

Pepsin is obtained from cattle and pig stomachs. Rennet, the milk-coagulating enzyme produced by the fourth stomach of calves, can be extracted from minced stomach or from dried, ground stomach. Trysin and chymotrypsin are obtained from the pancreas glands of cattle or pigs.

Papain is the most extensively used plant protease, being the dried powdered latex of the fruit of the papaya tree, *Carica papaya.*

Glucose oxidase oxidises glucose to gluomates and H_2O_2, and is obtained from the microorganism *Aspergillus niger*.

Lipase splits glycerides and other esters of higher acids. The major commercial sources are *Rhizopus arrhizus* and *A. niger*.

α-Amylase splits starches into short oligonucleotides, and is commercially produced from microbial sources such as *Bacillus subtilis, B. licheni* and *A. niger*.

β-Amylase splits amylose to maltose, and is found in higher plants, especially grains; barley is the major commercial source.

Invertase digests sucrose into glucose and fructose, Saccharomyces sp. being the main source.

Glucose isomerase catalyses an equilibrium between glucose and fructose, and is obtained commercially from various microorganisms. Later chapters discuss both its role in sweetener production and its commercial diffusion.

Examples of Enzyme Innovations and Use

We shall briefly discuss four food and beverage sectors: brewing beer, flour modification, fruit juices, and meat tenderisation.

Brewing Beer

It is a historical fact that the brewing industry has been a major influence on enzyme science, as well as the site of much enzyme technology innovation.

Starch is the principal raw material for beer production, and it must first be converted into sugars since yeasts are unable to utilise starch directly. They are, of course, able to ferment the sugars into alcohol and carbon dioxide. During brewing starch is converted to sugars by enzymic process, which only became well understood scientifically in the second half of the twentieth century. When grain seeds germinate, large quantities of amylases develop; these are able to digest starch and convert much of it to maltose. Malt is produced by using drying to stop the germination process in its initial stages. Barley malt is the principal material for beer brewing, and it supplies the bulk of the enzymes and starches needed. The practice has evolved over centuries, and since the late 19th century many scientifically based innovations have been made.

Amongst those involving enzymes we will mention two. The first is the Amylo Process, which used maize starch saccharified into glucose by the action of an enzyme released by a fungus, *Alomyces rouxii*. The convertion of glucose to alcohol was achieved by a strain of *Rhizopus delemar*. It was a more efficient process than the more traditional yeast fermentation, and became popular, during the inter-war period, in the USA and France.

Our second example concerns the role of enzymes in the chill-proofing of beer. When beer is chilled it becomes turbid because of proteinaceous matter. Leo Wallerstein, in 1911, suggested the use of proteases to digest this protein. Modern chill-proofing of beer developed from Wallerstein's idea. The agents used in the process are mixtures of several enzymes, primarily pepsin and papain, with an inert filler such as sugar.

Flour Modification

Like the brewing industry, the starch industry has also been a major locus for enzyme innovation; here we will discuss one important area, flour modification.

Wheat flour contains α- and β-amylases and a protease, the amount of each depends on grain variety, and growing and harvesting conditions. The α-amylase content of flour increases in wet climates, because it is the enzyme which mobilises the starch reserves during germination. The proteases are particularly evident in newly ground flour, producing a sticky dough which is difficult to work with, and which produces a doughy loaf on baking. Aging the flour inactivates the proteases and makes the flour stronger (this effect can also be achieved artificially using chemical agents).

The β-amylase content of flour is also important since its action releases the fermentable sugar maltose and controls the amount of gassing in the bread, and therefore its rising. As we have indicated, flours rich in α-amylase are very sticky, this interferes with modern bakery equipment. This can be controlled in various ways. One is by mixing flours with high α-amylase content with flour with low α-amylase content to produce a balanced flour. Alternatively, low-level α-amylase flours have been preferred, to which is added malt flour, which contains α-amylase. However, this requires careful monitoring to ensure preferred levels. More recently, therefore, millers have resorted to using a fungal α-amylase. These have proved more suitable for baking than bacterial varieties since they are more easily inactivated by baking and prevent further enzyme action once the bread is worked.

Enzyme use in fruit-juice production

By the 1930s enzyme producing companies were diversifying their product ranges into each other's products. Many had small research and development departments which were looking for new enzyme applications.

In 1926 the fruit-juice industry was fairly static, but innovations into better processing and improved packaging for distribution, by bottling or canning the juice, made fruit juices more accessible to the public; then the consumption of fruit juice in the United States began to grow rapidly. Rohm and Haas identified a major market need that could be met by an enzyme. Willaman and Kertesz obtained a patent for the production of pectinases, enzymes that digest the micilaginous pectins found in fruit and vegetables. Rohm and Haas began the production of pectinases shortly after, and marketed an enzyme product, Pectinol. Similar products followed from other companies.

Enzymes for the meat-tenderisation industry

The tenderisation of meat with the aid of papaya juice was reported by early travellers to the tropical regions of Central and Southern America. Papain is the powdered latex of *Carica papaya*. Latex is obtained by making 3–4 scratches in the green fruit when it is still hanging on the tree. The liquid coagulates on the surface and is collected and dried. About 50% of the total mass is the protease. This product is stabilised by heating with hydrogen sulphides, sodium or potassium sulphites. The application of the papain produced is mainly for tenderisation of meat. Commercial products are marketed usually with a bacteriostat, for example, dilute alcohol. These preparations are applied to the meat prior to its cooking. Some slight but definite digestion occurs of the connective tissue and muscle fibres, which softens the meat. Recently the use of papain has been extended to injection of animals prior to slaughter. The major distributors of papain are its importers and as a result they only trade in this enzyme and constitute a market for it.

REFERENCES

N. Blakebrough and E. J. Parker (eds), *Enzymes and Food Processing* (London: Applied Science Publishers, 1981).
J. T. P. Böing, Enzyme Production, in G. Reed (ed.) *Prescott & Dunn's Industrial Microbiology* (London: Macmillan, 1982).

20

Metals Biotechnology: Trends and Implications

Alyson Warhurst

Introduction

With the depletion of easily accessible high-grade mineral concentrations many of the new ore deposits planned for development are low-grade, complex, multi-mineral bodies, such as those that typify the Andean mountain chain. These deposits were previously mined for individual metals – like copper or tin – but now it is becoming increasingly necessary to recover associated by-products, such as gold, silver, nickel and zinc, and to extract pollutant elements like sulphur, arsenic and bismuth. There are also pressures to stretch further existing mine capacity through recovering values previously not exploited, in ancient waste dumps and in the marginal ore that has to be removed as overburden in mine development. Furthermore, instability in base-metal markets is pushing many mining companies to search for gold. However, a large proportion of new gold reserves are found in inaccessible hard-rock deposits associated with iron and arsenic. The exploitation of all these complex and refractory (difficult to recover) deposits in the present unstable metal-market conditions poses a unique set of metallurgical problems which biotechnology is poised to resolve.

Biotechnology is not an invention of the 1970s, when the magc of manipulating recombinant DNA was unleashed. Rather, it incorporates hard-won advances in the long built-up science of molecular biology as well as a wealth of multi-disciplinary expertise from engineering, environmental sciences and chemistry – much of which is not at all new. In spite of promises of heralding a revolution in solving the world's health and resource problems, there is still a dearth of evidence about commercial applications. The diverse fields of biotechnology are evolving at different paces, and at different levels of sophistication in terms of the technologies they incorporate. Nonetheless, national economies do not need to wait for the bio-revolution to hit them unprepared. There are opportunities now to be seized to direct biotechnology advances towards their own development goals. Metals biotechnology is interesting in this respect. It is a technology which is already commercially applicable and viable. It involves the optimisation of naturally occurring biological processes, using a combination of inputs which are currently accessible and do not necessarily include genetic engineering techniques. It is also a technology which is highly relevant to developing countries, owing to the nature of their geological resources and mining histories.

Metals biotechnology employs micro-organisms in two ways: first, in leaching processes for extracting metals from low-grade ore and concentrates; and, second, for recovering accumulated dissolved metals through bio-absorption processes.

Bacterial leaching is a natural process which occurs at most mine sites and mineral waste dumps. It is, in effect, an *in-situ* solution mining process. Bacteria (of the Thiobacillus group) which thrive in acid mine water help to dissolve sulphide minerals, copper, zinc, nickel, gold metals, thus freeing the associated metals into an effluent from which they can be recovered. The bacteria act as oxidising agents and generate powerful leaching solutions by their growth reactions. Since the bacteria are living they require special conditions for optimal growth which include oxygen, an acid pH, and specific nutrients and elements. These requirements are site-specific and reflect the fact that each bacteria strain will have adapted through time to the vagaries of its mine habitat – the local ecosystem, the pH, acid dilution during the rainy season, etc. In addition, the characteristics of the leach system itself (i.e. of the mine, heap or tank) – its mineralogy, particle size, porosity and temperature profile – will also determine the efficiency of the biological reactions and thus the amount of and rate at which contained metals can be liberated.

It is this range of requirements which provides the scope for designing parameters to optimise the leach system for economic gain.[1] Optimisation measures include: improved system design through, for example, engineering optimal dimensions for the leach heap, mine or tank and its 'watering system'; second, improving solution management through the determination and implementation of process optima such as lixiviant application rate and mode, pH, and rest periods; and, third, improving bacteria activity through strain selection, employing mixed cultures, changing environmental parameters and, of course, genetic manipulation. The extent to which these measures are undertaken will vary depending on mine geology, the potential return to investment for the individual mine and the availability of investment resources, skills and experience, especially in the field of bacteria improvement.

Industrialisation has meant that metals are amongst the most commonly used raw materials. Mining, metal refining, the use of metals in manufacturing and the final disposition of manufactured products all result in metal losses. Metal wastes in turn not only pose environmental problems but also a critical loss of non-renewable resources, particularly in the case of gold and silver. The use of biotechnology to attract and accumulate such metal values represents a significant commercial opportunity. In North America several companies are commercialising 'bio-absorption' technologies using, for example, fungi, yeasts, algae, as well as combined microbial biomass in pellets (the AMT-Bioclaim process)[2]. Since this technology is in the experimental stage our chapter will concentrate on bacterial leaching.

Emerging Applications of Metals Biotechnology

The exciting feature of bacterial leaching is the range of metals and problems to which it can be applied, all of which are in different stages of development. These include:

Dump leaching

Metal is leached out of the old mining dumps by spraying on acid mine water containing bacteria and letting it percolate through the material, collecting copper on its way. The metal values are recovered from the effluent by precipitation on scrap-iron or more efficient techniques of solvent-extraction and electro-winning. Commercial applications, mainly for copper extraction, are found throughout the United States, Canada, Chile, Peru, Papua New Guinea and the Eastern bloc.

Investment and operating costs for dump leaching are generally low since the possibilities for optimisation are limited. Extraction, transportation and dumping costs would be foregone. Although the process is very flexible, given that ore grades are low (generally less than 0.5% copper) and recovery rates are unpredictable and sub-optimal (less than 40% of the contained metal), economies of scale can be important. But, generally, bacterial leaching can be used to extract metals from small dumps of less than 25 metric tonnes to huge dumps of thousands of millions of tonnes of material producing up to 23 metric tonnes of copper per day. If the process works efficiently, acid is self-generated, no purchased energy is required and the costs of solvent-extraction/electro-winning compare favourably with those for smelting and refining.

As yet there exists no alternative to bacterial leaching for the processing of low-grade material and since the technique is applied to previously unexploited resources the employment effects are necessarily positive. Environmental impacts are also positive. In fact, the metal ions and sulphuric acid produced by natural bacterial-leaching processes, through the action of percolating ground waters or rain, can generate pollutant effluents if allowed to enter water supplies. When this natural process is harnessed for economic gain and metals are recovered, chemical changes render the solutions non-pollutant. Also, the barren solution, once the metals have been recovered, can be recycled over the dump to begin the leaching process anew.

In-Situ leaching

Bacterial leaching can also be applied to leach out metal values from underground mineral concentrations. The solutions are piped or sprayed into pre-shattered rock and allowed to percolate through, collecting metals as they leach through the mineralised rock. Then the solutions are pumped up from their natural collecting point to be processed. There are a number of advantages involved in this method. First, it enables metal recovery from parts of mines which are too dangerous to be mined by people; second, it avoids the environmental damage of having to dump huge amounts of overburden and transport large volumes of ore for processing, which would be associated with open-pit mine development. The costs of *in-situ* mine development would be substantially reduced but metal recovery is more likely to be uneven, relative to conventional processes. There are two commercially successful examples of this technology in the United States, but, with new advances in metals biotechnology, *in-situ* leaching is likely to become a viable alternative to traditional open-pit development.

Heap leaching – copper and gold

Perhaps the most commercially promising application of biotechnology to metal extraction is the leaching of marginal copper ore during ongoing operations and of overburden from newly developed mines, in heaps designed, constructed and operated according to parameters for optimal bacterial activity at that mine. Metal recovery rates for heaps are more predictable and may range from 40% to 80%

depending on geology, environmental context and the degree of optimisation sought. For example, optimal heap dimensions could be determined to fulfil aeration and temperature requirements of specific species of bacteria; strain selection, nutrient addition and ecosystem design and control may be undertaken to stimulate bacterial activity. The chemical composition of the mine-water solutions may be changed to accelerate oxidation and inhibit insoluble precipitates forming within the heap. The investment and operating costs of an optimised heap-leaching system are more variable than those for a dump system, since they will depend on the extent to which R&D is undertaken. Other variables also need to be considered: the rock type and its natural particle size after blasting will determine the necessity for expensive preliminary crushing, the addition of extra acid or the design of long, narrow 'finger-heap' systems; and the hydrogeology and soil types will determine the need for costly ground-base preparation; the characteristics of local acidified mine water and a vibrant bacterial population will ultimately determine the efficiency of bacterial leaching for a given site. Investment costs for heap operations could therefore range betwen $5 million and $50 million. (Current mine investment costs are over $1000 million.) Employment effects are positive since bacterial leaching is a unique process for treating previously unviable ore. However, these effects may be masked by situations where a mine might be shedding its conventional mining labour force.

The potential of heap leaching also offers particular benefit for applications to refractory gold ores and tailings (processed waste). This has been the focus of innovative activity by Giant Bay Biotech Inc. of Canada, who have patented a process; and commercial applications to refractory gold are already under way in Brazil, South Africa and North America.

Concentrate leaching – Copper, zinc and gold

Energy-intensive and pollutant smelting operations are currently facing a growing range of regulations concerning the nature of their environmental emissions. For this reason it is likely that confined bacterial leaching systems may become an attractive alternative to smelting for the treatment of copper, gold and zinc concentrates in the future. Developing countries frequently export their minerals in concentrate form to European, Japanese and North American consumers. Through bacterial leaching they could obtain more value-added for their minerals, since both leaching and final metal recovery takes place at the mine site.

Since more controls can be imposed on confined systems, genetic engineering of the leaching bacteria may provide a breakthrough. Related research has focused on selective mineral leaching, speeding up oxidation functions, reducing toxicity effects, bacteria adaptability to saline waters and the self-generation of nitrogenous nutrients.[3] Significant advances have already been made in South Africa, Chile and the United States. Concentrate leaching has been the focus of innovative activity by Giant Bay Biotech Inc. and Coastech Research Inc. of Canada. The former company has developed the proprietary BIOTANKLEACH process as an alternative to the smelting and roasting of gol concentrates. It demonstrates very clear advantages over conventional processes.

Uranium leaching

Uranium has been recovered by bacterial leaching using mainly *in-situ* methods in underground mines (e.g. Elliot Lake, Ontario, Canada). The advantages of uranium leaching are: no rock is brought to the surface, eliminating the risks to miners' lives and saving on labour costs; solid waste pollution is prevented; and, liquid wastes can be minimised by recycling the leaching solutions. The process is ideal for low-grade and finely disseminated ores which cannot be treated economically by conventional methods. Disadvantages include: rates of extraction and metal recovery are difficult to predict, and pollution could occur through the seepage of leach solutions rich in uranium and associated ions.

Mixed sulphide leaching

Bacterial leaching techniques are also being developed for several types of problematic minerals including: mixed zinc and lead in disseminated form; nickel and cobalt sulphides; and arsenopyrite, thus resulting in a reduction of pollutant emissions which are normally associated with the processing of ores with arsenopyrite by-products.

Bacterial leaching of high-sulphur coal and oil shale

Current research suggests the possibility of microbial removal of the organic sulphur content of coal and oil shale. Since a high sulphur content is a significant constraint in coal development, if this process could be developed on a commercial scale the implications for coal-producing countries like India and Colombia are enormous.

Socio-economic Aspects of Metals Biotechnology

The opportunties biotechnology for mining offers developing countries are due to their unique mineral reserves and their long mining histories. The resource base of the developed countries influences the priorities of firms and institutions which then push biotechnology advances in the direction of solving problems in their own economic interest. Just as we see in agricultural biotechnology the targeting of frost resistance rather than drought resistance (developing-country priorities), in metals biotechnology, firms in North America are looking at recovery of gold and silver from inland waterways rather than metal recovery from old dumps (which in developing countries are rich in formerly marginal ore, which now commands a high commercial value). Since biotechnology advances in mining build on a wealth of site-specific disembodied technology and expertise, and since developed-country interests are not yet targeting Third World markets in this area, this opens a window of opportunity for mineral-dependent developing countries. Since the 1980s, efforts by Chilean and Peruvian mining companies to introduce biotechnology provide evidence of the success that seizing such opportunities early can bring.

An interesting exercise is to compare the North and South American cases. At present, over 15% of total United States copper production comes from bacterial dump leaching. That figure is increasing as many firms shut down high-cost mines and rely on bacterially leaching waste dumps for 100% of their coppe production at less than 50 cents per pound (compared with 80 cents a pound for conventionally mined copper) when prices are around 70 cents per pound. The significance of this is that these dumps, although they are quite profitable, have surprisingly low recovery rates varying between 5% and 40% of the metals contained over periods as long as 5–20 years. This is because the microbiological part of mineral leaching was not realised at the time of dump construction and efforts were not made subsequently to control and optimise the process. Since waste is usually placed in large mounds in valleys for dumping convenience, dump interiors may be starved of air, which the bacteria crucially require. Also, as a result of insufficient cooling, the temperature in the dump may rise making the environment inhospitable for bacterial activity. Furthermore, weathering and the weight pressures will facilitate 'fines' formation which prohibits solution percolation, thus preventing 'bacteria–substrate' interaction. Even now it is difficult to persuade conservative mine management that metal leaching in their dumps is due to biological as well as chemical reactions and that therein lies the scope for improving efficiency.

Indeed, by designing and operating optimal leach systems and improving bacterial activity and solution management, it is possible to obtain metal recovery rates of over 80% in periods of less than three years. This was demonstrated by the case of Chile, where a commercial operation involving a combined chemical/biological leaching plant is generating copper at rates of 90% recovery, in less than 6 months at costs under 40 cents per pound. In fact, the Chilean State mining company – CODELCO – is introducing bacterial leaching at most of its mines and is investing in optimising the technology.

The main implication of all this is that there exists little precedent in developed countries for optimising bacterial leaching operations. This helps to explain why bacterial leaching has not been introduced on a larger scale (and practically why it has been difficult to persuade managers of the technology's potential). High transport and dumping costs, environmental restrictions and a lack of new mine sites there reduce the possibilities of changing this situation in the future. There are therefore no general models upon which developing countries can base their leaching projects. And if their economic feasibility was in any way based upon the experience of developed countries in bacterial leaching, the potential for optimising the process would probably not be realised.

The potential fr optimised bacterial leaching operations is clearly much higher in developing countries, therefore the onus is on them to take the initiative. Dumps can be re-designed, leach heaps built at new mine sites and future projects may benefit from metal-concentrate leaching. This of course poses a huge challenge. Mining companies are rigidly structured and bacterial-leaching technology development requires a multi-disciplinary approach and constant and close cooperation between research and production from the outset. Furthermore, the developed countries are clearly a limited (though nonetheless important) source of this technology. And, the development of this technology involves the search for and assimilation of disembodied technology from diverse sources – industry, research institutes and universities – and its on-site integration with locally-derived technological inputs.

Biotechnology is often employed as an all-embracing term, and often as shorthand for genetic engineering. Applications in mining, however, are likely to encompass quite different principles from those in pharmaceuticals. This leads us to raise questions about the different policy implications for developing countries of 'open process' biotechnology applications (i.e. those for natural resource 'processing' in the open environment) as opposed to 'closed process' biotechnology applications (i.e. those for manufactured production undertaken in confined systems under precisely controlled conditions). The interesting issue here is that 'closed process' biotechnology applications are more likely to involve sophisticated genetic engineering techniques to produce miracle drugs and new products. But 'open process' applications involve a broader range of knowledge, and often optimisation of already occurring natural processes. Site-specific conditions, ecological contamination and the effects of the diversity of genetic resources are the changing variables in 'open process' biotechnology development.

Under these circumstances there are clear advantages to a planned approach to investment in the technological capabilities needed for developing projects in metals biotechnology. Given the complex and multi-disciplinary requirements involved, the investment in such knowledge-intensive technology might best be spread across firms and institutes, or even across countries. Indeed, this is precisely what Canada and Chile are doing at a national level and the Andean Pact countries are doing at a regional level.

The market for metals-biotechnology applications largely comprises traditional and existing mining companies – not new manufacturers. One of the main barriers to entry for new biotechnology companies has been the lack of ownership of a mine and the lack of a strategic alliance with a mining company. The small and medium-sized mining companies seem reticent in paying for perceived risky and untried technology and unwilling to establish relationships involving royalties on the metals extracted by biotechnology. The large mining companies seem to be undertaking highly secretive R&D programmes in-house, sub-contracting in from different sources only those elements missing from their own rapidly developing set of capabilities. They tend to contract consultants on an individual basis, rather than contract biotechnology firms on a turnkey basis.

Four approaches to the metal-biotechnology market are currently adopted:[4]

(a) R&D consultancy services and process-route design and testing (e.g. Coastech Research Inc., Warren Spring Laboratory (UK), CANMET (Canada), INTEC (Chile) and various universities).
(b) Provision of specialised equipment and engineering services (e.g. Davy McKee).
(c) Turnkey services, 'toll' processing and selling radical new proprietary technologies (e.g. Giant Bay Biotech).
(d) Bioprocessing of metals in owned mines, generating revenues through the sale of gold or copper (e.g. Aurotech, CODELCO).

In addition there are a number of mining companies worldwide that are applying bacterial-leaching technologies, e.g. Denison Mines in Canada; Kennecott, Duval and Inspiration in the United States; and mining companies in South Africa, Mexico, Brazil, Chile, India, Italy and Peru. Those involved in genetic engineering include: CODELCO, Rio-Tinto Zinc, Freeport Minerals Co., St Joe Minerals, Gencor, Minorco, General Electric, Inco and Idaho National Engineering Laboratory. In the area of metal removal/waste-water treatment, start-up companies are establishing their market position through patenting products (e.g. De Voe–Holbein and Bio-recovery Systems). Some companies have developed strong in-house research programmes (e.g. Coastech Research Ltd). Others have acquired their technology from university research or government-sponsored work (e.g. Bio-recovery Systems, De Voe–Holbein and Giant Bay). However, as a consequence of the capital burden involved, 1988 saw the demise of one metal-biotech company – the five-year-old Advanced Mineral Technology Incorporated – and 1989 threatens to herald the demise of another, Giant Bay Inc. But that has not meant the demise of the technology. Several other biotech firms and mining companies are moving in fast to buy up their patents, technology and expertise.

Policy and capability development in Canada and Chile

Canada's national policy on metals biotechnology arose out of a combination of two new areas of interest. First, within the Ministry of Energy, Mines and Resources, the Canadian Centre for Mineral and Energy Technology (CANMET) launched a programme in 1982 'to develop hydrometallurgical processes to extract and reover residual metal values from sulphide ores, specifically those portions of ore deposits that are usually left underground'. Secondly, within the national biotechnology networks (others being for health and agriculture) the Mineral Leaching and Mineral Recovery Network (BIOMINET) was set up in 1983 within CANMET with the following defined scope: 'Biominet is a network of research organisations, companies and associated agencies interested in applications of biological systems in mineral processing for the purpose of recovery of values and environmental control and protection. The client community is viewed as the mining industry including metallic, non-metallic and energy minerals.' (BIOMINET Newsletter, no. 13, 1988). These two institutional concerns have led to the prioritisation of metals biotechnology R&D in Canada, and, within that concern, to the prioritisation by joint industry–researcher committees of selected technologies: underground and *in-situ* leaching, gold leaching, uranium leaching, complex copper and zinc sulphide leaching and environmental mine-waste treatment.

CANMET's bio-hydrometallurgy programme involves a budget of C\$7 million from the Ministry and C\$4 million from the Industrial Research Assistance Programme in Biotechnology. It is used to support company R&D, university and in-house research. The programme has changed from starting off as 'all technology and government push' to its current phase of now being largely 'industry pull', with companies matching public funding commitments. This has largely been the result of a dynamic joint industry research advisory committee, its specialist working-group struture and the success of the annual BIOMINET conferences and networking activities.[5] The example and experience of Canadian government policy in metals biotechnology is clearly an interesting example for developing-country mineral producers, although a recent paper comments that long-term 100% government support is required to enable small and medium-sized firms to develop, consolidate and commercialise metals biotechnology.[6]

An illuminating human-resource development project in metals bio-technology was undertaken by Chilean institutions and enterprises during the 1980s, with support from both the Chilean government and UNDP.[7] The goal was to develop proprietary innovations both in genetic engineering of the leaching bacteria as well as in the industrial application of optimised leaching technology to dumps, heaps, concentrates and *in-situ* systems. Building on past diffuse capabilities in bacterial leaching, the project successfully brought together two universities, two research institutes – one state owned and one private – and the state mining company CODELCO, as well as a number of private companies which later developed independent links with the research institutions working in the project.

The organisation of this research and training programme was interesting in several important respects. *First*, the overall level of starting skills and knowledge was high. Two of the university researchers had doctorate degrees in biotechnology from United States universities. Micro-biological laboratory facilities were also fairly sophisticated. *Second*, the research institutions and universities involved had already developed close links with specific mining companies, so the prospects for future collaboration in training and industrial application were high. *Third*, company management was, from the outset, convinced of the commercial returns likely from developing metals bio-technology and was therefore willing to make the necessary commitments of personnel, time and resources. This was a key feature since training and the research programmes involved working as a multi-disciplinary team at the mine site. *Fourth*, the programme was organised as three mutually supportive research and development sub-projects. Training took place within them along the lines of programmed supervision, and was complemented by consultancies and seminars by international experts and the hosting and attending of international metals-biotechnology conferences. As a result of this nationwide effort at capability development in Chile several apparently profitable industrial projects are underway in both CODELCO and privately owned mines.

Conclusions and Policy Implications

Metals biotechnology, althoug not always recognised by name as such, is already used in mining to stretch existing capacity through the leaching of metals from old dumps and mines, and to recover gold from refractory ore. What is being argued here is that, as the potential of biotechnology unfolds within the changing context of the world mining industry, a new range of commercial applications is emerging. Some techniques are relevant for the immediate future, others for the longer term. But all of them have implications for the future organisation, efficiency and environmental impact of national mining activities.

The current period of unstable metal prices and depressed metal markets has tended to exert a marked downward pressure on production costs. These 'pull' factors have led to the closure of high-cost mines and have encouraged economies in the remainder. Harnessing the naturally occurring bacterial leaching process to extract copper from marginal ore has evidently been a profitable response in the North American and Chilean contexts. This is clearly reflected in their current operating costs, new investment plans and the substitution for conventional mining and smelting capacity of low-cost bacterially leached copper production. This trend is also reflected in emerging industrial interest in metals biotechnology. Several factors have emerged from our discussion to suggest that bacterial leaching offers particular advantages to developing-country producers. Their mineral deposits are higher grade, they have long mining histories, therefore their dumps might be expected to be rich in residual metal values. They also have fewer environmental restrictions on building new optimised heaps over larger surface areas and they own relatively less installed capacity in pollutant smelting and refining facilities.

At the same time the biotechnology revolution is providing the scientific impetus or 'push' to develop further the potential of the technology, through opening up new possibilities for manipulating and optimising bacterial activity. The potential of metals biotechnology will not be developed in a vacuum. It will be spurred on by advances made in other areas, particularly in 'open process' applications of biotechnology. There is currently a wealth of knowledge being generated in agricultural biotechnology about the effects of ecosystem contamination, bacterial resistance to chemical elements in soil and rocks, climatic influences, nitrogen fixation, pH and nutrient balancing, which are relevant to metals biotechnology. In addition, the new microelectronics-led waves of technical change, which are permeating many industries, will also promote biotechnology applications in mining through facilitating the control and monitoring of the leaching process, more efficient genetic experimentation and computerised leach-system design.

Our chapter suggests that four main characteristics of metals biotechnology warrant serious consideration regarding national policies or corporate strategies. These are:

First, there exists a considerable R&D effort in metals biotechnology throughout the world, which includes a growing participation by a number of developing-country centres of excellence. Similarly, there exists a growing pool of experience of the problems and potential of commercial applications in copper and gold. But for various reasons this technology is not being widely diffused. These reasons include: the historical tradition in developing-country mining companies of relying on technology inputs and expertise from the industrialised countries, who have different priorities regarding which technologies to develop; the conservatism within mining companies, which makes it difficult to incorporate microbiology within traditional mining, metallurgy and engineering activities; the fact that the potential benefits of commercial application are hidden through insufficient investment in measures to optimise the bacterial-leaching process. This situation is changing slowly as Chile and other countries, including Canada and the Soviet Union, makeiportant advances in this field. However, policy research is required in order to investigate what are these obstacles to diffusion and what would be the implications for the international mining industries and developing-country mineral producers of the widespread adoption of bacterial leaching.

Second, metals biotechnology, particularly where applied to extract residual metal values in existing mines, may provide developing countries with new ways to expand their mineral exports when prices are low or to react to unstable market conditions, thus safe-guarding mining employment. Another important positive productivity and employment effect associated with the introduction of metals biotechnology is that it also involves the adoption of new downstream technologies for metal processing at the mine site. Many developing countries sell their semi-processed minerals in concentrate form to foreign smelters either abroad or in their own countries, thereby incurring penalties for impurities and losing by-product values. Bacterially leached metals do not need smelting. Instead, marketable high-quality copper cathodes can be produced by electroplating copper cathodes directly from enriched leach solutions. This process of solvent-extraction/electro-winning must be undertaken at the mine site and by definition would also increase employment requirements. The productivity effects of metals biotechnology are therefore both labour and capital saving in nature.

Third, metals biotechnology is new and requires a close linkage of scientific R&D with industrial production and a fresh approach to training, on account of its knowledge-intensive characteristics. To ensure optimal mineral leaching, a collectively trained multi-disciplinary research team needs to be working with production personnel from the outset. Initial fieldwork inputs are required from geologists, hydro-geologists, geographers and micro-biologists. These results must then be interpreted during the 'process route' design stage by chemical and metallurgical engineers, who also need to work alongside miners and mining engineers. The latter will ultimately be responsible for the extraction and dumping of mined material in accordance with the parameters developed by chemists, metallurgists and micro-biologists during research. So far there has not been an effective 'welding' of the micro-biological and engineering bodies of knowledge and experience to optimise the potential of emerging applications. Multi-disciplinary training by an appropriate mix of experts, at the mine site as well as in the laboratory, is therefore necessary. Efforts by Chile and Canada in this direction provide useful lessons about how such training might best be organised to accelerate skill acquisition.

Fourth, metals biotechnology offers environmentally sound alternatives to conventional mining activities at several stages in the metal-production process. Applying biotechnology to dumps and tailings would reduce pollution from acid drainage and metal-rich run-off waters, while generating extra metal alues. *In-situ* leaching would negate the need for open-pit mine development. And the bacterial leaching of concentrates would be a less hazardous alternative to smelting and refining, which emits harmful carbon dioxide, sulphur dioxide and nitrous oxide toxins into the atmosphere. The conservative mining community has been locked into the paradigm of extraction–smelting–refining and the development of pollution-saving modifications to existing technologies. Insufficient investigation has been undertaken by mining strategists and policy researcher alike into the development and diffusion of metal-processing alternatives of a completely different nature.

NOTES

[1] A. C. Warhurst, Biotechnology for Mining: The Potential of an Emerging Technology: The Andean Pact copper project and some policy implications, *Development and Change*, **16** (1985), pp. 93–121.

[2] J. A. Brierley, C. L. Brierley and G. M. Goyak, AMT-BIOCLAIM: A new wastewater treatment and metal recovery technology, in *Fundamental and Applied Biohydrometallurgy'*, R. W. Lawrence, R. M. R. Branion and H. G. Ebner (eds) (Amsterdam: Elsevier, 1986), p. 501.

[3] On genetic engineering and leaching microbiology, see R. Acharya and R. Spencer, Biotechnology and the Mining Industry: The Bacterial Connection, Gaia Institute for the Study of Natural Resources, Maastricht, August 1989. These technologies are also discussed in A. C. Warhurst, Employment and Environmental Implications of Metals Biotechnology, World Employment esearch Working Paper (WEP) 2–22/WP, International Labour Office, Geneva, 1990.

[4] This section relies on C. L. Brierley, The Commercial Development of Metals and Minerals Technology, in P. R. Norris and D. P. Kelly (eds), Biohydrometallurgy: Proceedings of the International Symposium, Warwick 1987, Science and Technology Letters, Surrey, 1988.

[5] BIOMINET publishes and disseminates project results, news and events, annotated bibliographies and contract reports. For example, a recent report looked at the state of the art of genetic-engineering research in metals biotechnology, while another report was a pre-feasibility study of *in-situ* bioleaching of an underground Canadian sulphide mine. BIOMINET also undertakes training workshops. It has 550 members, half of whom are from industry, 17% from universities, 20% from private research organisations and 13% from government and media.

[6] C. R. Cupp, Biohydrometallurgy – Biomythology or Big Business, in Lawrence et al. (eds), *Fundamental and Applied Biohydrometallurgy*.

[7] J. Gana and S. Santa Ana, *Análisis de Proyecto 'Desarrollo de Procesos biológicos y su aplicación industrial en la lixiviación bacteriana de cobre de minerales chilenos'*, Documento de Trabajo, no. 3, CESCO (Centre for Copper Studies), Santiago, Chile, 1989.

GUIDE TO FURTHER READING

R. W. Lawrence, R. M. R. Branion, H. G. Ebner (eds), *Fundamental and Applied Biohydrometallurgy* (Amsterdam; Elsevier, 1986), p. 501.

P. R. Norris and D. P. Kelly, Biohydrometallurgy: Proceedings of the International Symposium, Warwick 1987, Science and Technology Letters, Surrey, 1988, p. 578.

UNIDO, Why Chile leads the world in researching bacterial leaching of copper ores, (mimeo), IDO/F/146, Vienna, October 1987.

A. C. Warhurst, Biotechnology for Mining: The Potential of an Emerging Technology, The Andean Pact Copper Project and Some Policy Implications, *Development and Change*, **16** (1985), pp. 93–121.

A. C. Warhurst, Employment and Environmental Implications of Metals Biotechnology, World Employment Research Working Paper (WEP) 2–22/WP, International Labour Office, Geneva, 1990.

21

Biotechnology and the Environment

Joachim H. Spangenberg

Biotechniques used for environmental purposes cover a wide range, including: conventional breeding (esp. of microorganisms), modern techniques such as mutation induction and stress-breeding, and new methods such as cell hybridization and genetic engineering. The latter, which forms the major topic in the public debate on biotechnology, can be used either in contained systems (probably most indirect applications will fall under this category) or by deliberately releasing modified organisms (necessary for the most direct applications). Both are regulated by European legislation, which had to be incorporated into the national legislation before the end of 1991. As this is not the case for the deliberate release of organisms modified by techniques other than genetic engineering (and as the EC legislation introduced administration procedures which will cost time and money to ensure an environmentally sound use of genetically modified microorganisms), users will probably prefer organisms which have been altered in other ways than by genetic engineering whenever it is an issue of deliberate release. The method to be chosen for contained use depends on efficiency and cost considerations.

Environmental impacts of biotechnology include impacts of industrial applications (probably more biodegradable products, fewer by-products and more waste water), agricultural applications (production of renewable resources, pest and pesticide resistant crops, changed input patterns, e.g. for fertilizers and other agrochemicals) and, in the long run, others like mineral leaching of heavy metals or increased exhaustion of oil wells.

Environmental applications of biotechnology can be subdivided into:

- indirect applications: e.g. the substitution for toxic substances by appropriately designed, less harmful substances, or research on degradation processes;
- direct applications: the reduction of emissions by biofilters and waste-water treatment plants, as well as the degradation of xenobiotic substances in toxic-waste disposal clean-up operations or the degradation of ubiquitous xenobiotics (man-made chemicals not existing in nature).

Indirect Application: Research

The most important environmental application of modern and new biotechnologies in the near future will be laboratory research, ascertaining how xenobiotics are degraded by microorganisms; which are the appropriate organisms (e.g. unusual ones like archaebacteria from extreme biotopes like the deep sea, hot springs, etc.); what are the necessary environmental conditions (e.g. availability or absence of oxigenium, specific hydrocarbons or metals) in the soil, groundwater or fermentation plant; which enzymes take part in the degradation process; where bottlenecks are limiting the speed of degradation, etc. Unfortunately the necessary question today is 'How can we improve the degradation of TCDD, DDT, HCH . . . ?' instead of being able to ask 'Which characteristics are necessary to make chemical substances easily biodegradable without dangerous by-products so that persistence and accumulation in ecosystems cannot occur and chronic environmental and health problems are avoided?'

Today, the physiology of only a few organisms is well known (E. coli, Drosophila), and especially, newly discovered species need intense research to even determine their major characteristics; for example, archaebacteria which can live and replicate at temperatures of more than 100 degrees Celcius.[1] These anaerobic, methane-producing bacteria are under intense research. They could serve in waste-water treatment plants to produce less sludge than conventional aerobic organisms, also degrading stable substances with a relatively low energy content. Their specific importance lies in the field of in-situ cleaning of contaminated soils, where sufficient availability of oxygen can hardly be guaranteed.

If biotechnology could contribute to an explanation of the relationship between chemical structures and toxic, carcinogenic, mutagenous and teratogenous characteristics of substances, a major success for environmental and health protection would be reached. The predictability of environmental impacts is an important goal for any effort to develop a 'predictive ecology'. Rapid progress is made possible by new technologies, with a focus in the field of degrading halogenated hydrocarbons.

Indirect Application: Biosensors and Diagnostica

Optimalizing degradation processes in bioreactors needs on-time data to allow for automatic process regulation to ensure appropriate environmental conditions for maximum turn-over (oxygen, temperature, food). This data can be gathered using biosensors, a combination of biological detection units and electric data processing. Whereas traditional methods of monitoring bioprocesses rely either on removing samples, with subsequent analytical chemistries, or indirect monitoring by sampling the gas phase, in many cases new biosensors can deliver physiologically relevant data more easily and more rapidly than conventional measurement, detecting e.g. intermediate degradation products which indicate disturbances in the mixed populations used for degradation. The high degree of biochemical specificity, achieved e.g. by using monoclonal antibodies, purified enzymes etc., also provides the capability to directly monitor key substances that influence and control all behaviour.[2] In this manner modern biotechnology in control processes helps make biotechnology in degradation processes work.

Different organisms have worked as bio-indicators long before they were systematically investigated and used, such as lichens indicating air pollution in industrialized areas for more than 100 years. Today, microorganisms like bacteria and yeast are used to detect toxic substances in waste water. In this manner BAYER AG uses conventional microbes from waste-water treatment plants with a built-in gene from marine fluorescent bacteria which exprime that gene when the matabolism is working undisturbed. Whenever toxic substances hinder the degradation process, the light is dimmed (which can be measured in automatic photometers) and toxic waste can be stopped on its way to the treatment plant. Whereas the BAYER strain U70pDB101 is won by genetic engineering, this is also possible with conventionally bred microorganisms, as different exhibitors demonstrated at the BIOTECHNICA trade fair 1988/89, avoiding problems with the deliberate release of genetically modified organisms. With new biotechnologies, biosensors can be specifically designed for different detection purposes, making this application one of the fastest growing markets in biotechnology.

Indirect Application: Fixed Enzymes and Proteins

A promising attempt to degrade xenobiotics seems to be the use of fixed enzymes. They

(1) are re-usable several times;
(2) do not undergo mutation processes, so they continuously provide he same degradation abilities;
(3) only make one, well defined step in the degradation process with no unknown intermediate products.

Biotechnology can help to isolate and produce appropriate enzymes with high stability and effectiveness.

Protein-loaded membranes also offer the opportunity to purify water by filtering ut heavy metals. These can be isolated using membranes with adhesive proteins which 'collect' heavy metals until they are completely saturated. Afterwards they can be washed (like ion-exchange columns) and used again. Work is being conducted to regulate the adhesion and washing process in such a way that the different heavy metals can be won separately, making it possible to recycle them.[3] Membrane material and adhesive proteins can be won from algae or bacteria used in confined systems for production purposes, without using fertile organisms in semi-open systems bearing the risk of unavoidable releases of organisms into the environment. For these purposes, archaebacteria are also under intense research. They have been found to offer rare surface proteins with interesting filter characteristics.

Direct Applications

The ability of microorganisms to degrade naturally occurring substances is an essential precondition for the existing ecological balance. For all naturally produced substances there are degrading microorganisms that use the substance as food. Some xenobiotics are not, or – much more frequently – are very slowly degraded, due to the fact that their chemical structures only very rarely occur in nature. In these cases the degradation process can be accelerated by feeding the degrading organisms structurally analogous, easily digestible substances to initiate an adaptation process to the target chemicals. Additionally, desorption and solubilisation techniques have to be used to increase the availability of the hardly soluble substances in the liquid phase of the medium.[4]

Direct Application: Confined Systems

For nearly every organic substance there is a microorganism able to degrade it. However, in many cases the organisms do not mineralize the target substances, but do only the first step of the necessary degradation process, accumulating intermediates which may even be of higher toxic risk than the originating substance. Nonetheless they can be degraded too, by other specialized microorganisms. The task of allocating all necessary enzymes for full mineralization can be achieved by different means – e.g. by mixed bacteria populations in which every species (or strain) carries one or a few of the necessary enzymes. To avoid the accumulation of intermediates it is necessary to form a stable biosystem in which all species are present at the right time in the appropriate place with the necessary number of organisms in a specific status of activity. Systems such as these only form slowly, have to be in use continuously and need careful management, especially if the number of species involved exceeds three or four. A technical means to achieve a stable dynamic in such systems is the use of immobilized organisms, fixed on a carrier with a large surface. This makes the transport of intermediates from one organism to another easier and hinders the reassociation of degradation products which could easily occur in free-flowing media.

As the management of microbial ecosystems with different biological optimum conditions is very complex (and not always successful), it seems desirable to have the complete set of enzymes in one organism. This can be achieved by stress-breeding of one species, cross-breeding of different species or a combination of both. Stress-breeding denotes offering the organisms no other carbon sources than the target substance and its intermediates. If the degradation chain is short and the substances involved have no or only limited toxic characteristics, this process can then be successful, resulting in organisms with degradation capacities that do not exist naturally. This method of breeding employs the existing genetic information for partial degradation and the natural exchange of genetic material between organisms of the same or different species (conjugation, horizontal gene-transfer). So an organism with degrading capacities for highly toxic (to microorganisms) chlorinated nitrophenols could be selected. Stress-breeding combines the charateristics of an organism that degrades chlorinated nitrophenols (4C2NP) to 4-chlorine-brenzcatechin with the ability of another microorganism to mineralize this substance. Necessary preconditions are the selection of appropriate organisms and the definition of specific environmental conditions (in the first above-mentioned degradation step, it was the availability of additional carbon and energy sources in a nitrogen-free medium, making 4C2NP the only available source of nitrogen).[5]

Much attention is given today to the degradation capabilities of anaerobic microorganisms. They do not need an expensive oxygen supply, produce much less biomass (which must be stored in a dumping site), and instead transform 90% of the hydrocarbons to carbon dioxide and methane. The biogas can be used as an energy source. This in-vivo evolution of new degrading capabilities uses incidentally occurring natural processes such as mutation and gene transfer, which can be accelerated by technical means only to a limited degree. If the evolution process is too inefficient, genetic engineering may be used to transfer relevant genetic information to organisms for everyday use, such as the pseudomonas bacteria, provided appropriate and well-defined sets of genes are available. The resulting 'patchwork organisms' can either be used in confined systems or be released into the environment. In this case a careful risk assessment would be a necessary precondition.

Direct Application: Semi-Confined Systems

Whereas confined bioreactors only allow a minimum of release (an ideal confined system would have no emission of organisms at all), semi-confined systems such as waste-water treatment plants and bio-filters result in a stady flow of relevant amounts of (modified) organisms into the environment.

New characteristics can emerge in a waste-water treatment plant without direct action aimed at the target of optimizing the degrading organisms. As the plant is a large bioreactor in itself, all processes described may occur, but probably at a low speed. This can be enhanced by adding specialized organisms to the sludge, hoping that either the organisms will establish themselves in the aquatic ecosystem of the waste-water treatment plant or the relevant genes will be transferred and appropriately used by the established population of microorganisms. In both cases relevant environmental impacts are possible: a microorganism that makes heavy metal soluble would make it possible to produce sludge (which today absorbs more than 90% of the heavy metals of a plant) which could be used as agricultural fertilizer. But if the same organism were to mobilize all the heavy metals deposited in the major European river beds the ecosystem of the North Sea might collapse.

Risk scenarios like this will probably limit the application of genetically modified organisms in waste-water treatment for the foreseeable future, giving preference to a combination of confined (specialized organisms degrading xenobiotics) and semi-confined systems (conventionally bred organisms degrading all other biodegradables).

Another field of direct application of microorganisms to combat environmental pollution are biofilters and biowashers that reduce gaseous emissions. They are able to degrade inorganic gases such as $NH4$ and $H2S$ as well as organic substances, including nitrogen-containing hydrocarbons, without transferring the problematic substances into other environmental elements such as water or soil. Biowashers function as showers, dissolving gaseous substances in the liquid (mainly water) and thus cleaning the exhaust air. The washing water is subsequently treated in small biological waste-water treatment plants and then re-used. Using foam instead of a washing liquid enlarges the surface for solution processes and can make the process more efficient.

Biofilters consist of a carrier with many caverns, resulting in a large surface area that is covered with bacteria. Dissolving happens in the contact zone of the bacterial layer with the affluent air; the bacterial degradation occurs within the filter, while the microorganisms multiply, mineralizing a broad range of substances quickly and with a high degree of efficiency. Other technical processes that use permeable membranes for the biodegradation of poorly soluble or gaseous substances such as Toluol or Dichlorethane are under development. Using these kinds of biotechnical systems, it is possible to get an efficient clean-up of gaseous affluents at a price significantly lower than that of conventional chemical or physical treatment. The most decisive factor in the microorganisms' degradation capacity is appropriate environmental conditions; therefore technical optimalization of the equipment seems to be more important than the use of genetically altered degrading organisms, but special breeding programmes are already under way.

Direct Application: Deliberate Release

The main environmental application considered for the deliberate release of microorganisms is the cleaning-up of old dumping sites and other contaminated soil. Moving contaminated soil to disposal sites does not destroy the toxic substances, and although incineration does, it destroys all other organic matter as well, leaving only dead matter.

The organisms can be applied to the soil either in situ (without moving the soil) or to dug-out soil at an on-site treatment plant. In this case the soil is isolated from the ground, water, and food; and specialized microorganisms, adapted to the specific kind of contamination, are added. Detergents can help increase the solution of hydrophobic substances, and repeated mixing processes homogenize the substrate and provide oxygen supply, accelerating the decontamination process.

This method obiously cannot be used if the contaminated soil is covered by buildings, but also gaseous emissions after the removal of the covering earth layer during digging can cause problems. Even for underwater soil contaminations, as found in different lakes formerly used as toxic-waste disposals in the petrochemical area near Houston, Texas, in-situ treatment of the contaminated mud by application of oxygen and minerals was necessary to avoid emissions of toxic gases.

In-situ application includes spraying a foam layer to clean up the top 30 to 40 cm, tilling the soil with special machinery to a depth of one metre, and deep drilling (if necessary down to the groundwater aquifer). In every case microorganisms, oxygen and minerals are applied to the soil, and if the groundwater is contaminated too, it is removed, treated in a biological waste-water treatment plant and replaced.

In-situ clean-ups must be continued until a given threshold concentration is reached (at least). But even if limiting factors such as oxygen supply, lack of nutrients and minerals are maintained by artificial supply, minimum concentrations are not easy to achieve, because the metabolization process slows down as the concentration decreases. A zero contamination cannot be reached, as the degradation process stops at a given minimum quota. Its quantitative value depends not only on the kinetic parameters of growth and metabolism but also on the thermodynamics of the overall transformation reaction. This can be influenced by adding a secondary substrate which enables the mineralization process to continue, so that higher degradation levels can be achieved.[6] The halogenated organic compounds are likely to be prime targets for such biotehnological processes because of their widespread utilization and the biodegradability of many of the most commonly used compounds.

Technical developments under way include means to provide suitable environmental conditions to facilitate maximum biodegradation rates,[7] as well as the selection, breeding and genetic construction of more efficient strains of microorganisms. The use of new or altered microorganisms can cause problems if not all intermediate products and environmental interactions are fully known, if the modified organisms replace others inherent to the soil or if the multitude of toxic substances on old dumping sites causes mutations or other changes in the behaviour of the released organisms (e.g. blocking/deblocking physiological pathways). To limit environmental risks and to ensure efficient degradation a genetically optimalized organism would have to:

- tolerate a wide range of temperatures, salinity, acidity, and toxic substances;
- guarantee efficient degradation also of low concentrations (some ppbs in the case of dioxin), without producing any toxic or

otherwise dangerous by-products or endproducts (such as vinylchloride from PER degradation);

- have a high degradation capacity that is better than that of naturally occurring microorganisms;
- die after the degradation is completed;
- be unable to transfer parts of its genome to other organisms.

As no known organism is able to satisfy all these requirements (and no means to create one is visible), the use of microorganisms for soil and groundwater clean-up will probably be restricted to endogenous microorganisms in highly polluted, well defined landsites, where there is probable homogeneous contamination of limited areas. Genetically modified organisms will not be the appropriate tool for the degradation of ubiquitous substances such as DDT.

NOTES

[1] The temperature limit for life in liquid water has not yet been defined, but is likely to be somewhere between 110 and 200 degrees Celsius, since amino acids and nucleotides are destroyed at temperatures over 200 degrees Celsius. See T. D. Brock, Life at High Temperatures, *Science*, **230** (1985), pp. 132–138.

[2] J. S. Schultz and M. Meyerhoff, Status of Monitoring in Biotechnolgy, *Enzyme and Microbial Technology*, 9/11 (1987), pp. 697–699.

[3] B. Volesky, Removal and Recovery of Metals by Biosorbent Materials, in C. P. Hollenberg and H. Sahm (eds), *Biotec 2* (Stuttgart/New York: Fischer, 1988), pp. 135–149.

[4] Ramos Rojo et al., Laboratory Evolution of Novel Catabolic Pathways, in Hollenberg and Sahm (eds), *Biotec 2*, pp. 65–74.

[5] H. J. Knackmus, Mikrobieller Fremdstoffabbau, *BioEngineering*, **6** (1989), pp. 113–115.

[6] A. J. B. Zehnder and G. Schraa, Biologischer Abbau xenobiotischer Verbindungen bei niedrigen Konzentrationen und das Problem der Restkonzentration, *Das Gas- und Wasserfach*, 5 (1988), pp. 79–83.

[7] P. Morgan and R. J. Watkinson, Microbiological Methods for the Cleanup of Soil and Groundwater, *FEMS Microbiology Reviews*, 63/4 (1989), pp. 277–299.

GUIDE TO FURTHER READING

C. F. Forster and D. A. J. Wase, *Environmental Biotechnology* (Chichester: Ellis Horwood, 1987).

C. P. Hollenberg and H. Sahm (eds), *Biotec 2 – Biosensors and Environmental Biotechnology* (Stuttgart/New York: Fischer, 1988).

C. Knorr and J. Spangenberg, *Environmental Biotechnology, A Study for the Commission of the European Communities* (Brussels, 1990).

H.-J. Rehm and G. Reed (eds), *Biotechnology*, Vol. 6a, Biotransformations, and Vol. 8, Microbial Degradations (Weinheim: Verlag Chemie, 1986).

R. J. Scholze et al. (eds), *Biotechnology for Degradation of Toxic Chemcls in Hazardous Wastes* (Park Ridge, NJ: Noyes Data Corporation, 1988).

G. S. Omenn (ed.), *Environmental Biotechnology* (New York: Plenum Press, 1988).

22

Military Implications of Biotechnology

Jaap Jelsma

How to assess the implications of innovation
for weapon development: a dynamic approach

Biotechnology has established itself as a development with large potentials, of which, in the civil sector, quite a number have become true already. Will the future also bring us a major development in the military domain, maybe even a biological arms race?

The answer to this question can only be a speculative one. In fact, pinpointing the starting point of a military trajectory in biotechnology is only possible in retrospect, and even then it may be difficult. This is not to say, however, that an early assessment of the probability of such a development is impossible or senseless. On the contrary, such an assessment may contribute to the pursuit of timely measures aimed at avoiding a possible military exploitation of bio-technology.

Especially since World War II, the design and production of wea-pons has become a *technological* development, which can be assessed in different ways. More than often, technology assessments review a *state of affairs*, with a strong accent on the technological factor, i.e. on what is possible in a technological sense. However, in modern tech-nology this factor is subject to rapid change which limits the value of such assessments strongly. Moreover, the technological factor is not the only relevant parameter. As we will show below, social factors are at least as important. Therefore we start from the presumption that weapon development is a *social process in time,*[1] in which at least four elements have to interact in order to sustain an ongoing development:

- scientific or technological breakthroughs;
- a political environment that furnishes motives to exploit such breakthroughs for weapon development;
- the attitude of scientists;
- entrenchment of new weapons in military planning, i.e. in military operational practice, doctrines and bureaucracy.

Interaction of these elements can occur in different ways, generating different types of weapon development. Three such types are distinguished, each type being relevant to the assessment under discussion.

Different Types of Weapon Development

Type 1: Military R&D in wartime

The most salient political situation which stimulates weapon development is that of an armed conflict between states. In the case of war, weapon R&D obtains an unequivocal meaning in terms of a national emergency, for which the cooperation of scientists is easily mobilized.[2]

In this context the systematic search for biological weapons began. On the eve of World War II the first serious R&D programmes aimed at the creation of biological weapons for mass destruction were set in motion in the UK and Japan, soon followed by those in the US and Canada.[3]

Type 2: Military R&D in peacetime

After the Second World War military R&D was not abandoned. On the contrary, it became an institutionalized activity, especially in the US and the UK (and probably in the USSR as well). The Japanese expertise gained from the wartime research was appropriated by the US on behalf of their own programme on biological weapons.[4] The political legitimation for ongoing weapon development was based on the growing tension between East and West, coupled with the belief that, in the light of that tension, every conceivable technological lead should be explored on behalf of national security. However, as mentioned above, new weapons have to entrench not only into political, but also into military thinking. In this sense biological weapons did not function well because they did not meet the requirements of the military for controllable, robust and predictable weapons.[5] Mainly for this reason biological weapons could not easily be incorporated into military logistics, and therefore they failed to win a status comparable with nuclear and chemical weapons. At the end of the sixties major improvements of biological weapons were foreseen in circles of US defence, due to 'dramatical breakthroughs' in molecular biology. The expectation was that biological weapons could gain military significance as yet.[6] However, certain factors hampered the exploration of the scientific innovations. In those days, working on biological weapons was a controversial issue for scientists;[7] moreover, there was ample opportunity for jobs in the civil sector of molecular biology. In the second place, a political trend towards detente and disarmament started in the early seventies. As an important result of this trend the Biological Weapons Convention (BWC) was signed in 1972. Under this treaty the development, production and stockpiling of, but not research for, biological weapons became forbidden.[8] The fact that militarily attractive biological weapons did not yet exist undoubtly facilitated the conclusion of the treaty.[9]

Type 3: Military application of civil technological development

In 1974 the real breakthrough in biology occured when artificial recombination of DNA became possible. This tool not only created far-reaching possibilities for gaining new knowledge about living organisms on the molecular level; at the same time it held the promise of making biological knowledge technologically usable in an unprecedented way: it opened the perspective of 'genetic engineering'. Indeed these challenging new possibilities evoked a debate about their risks for human health and the environment, but hardly about their military potential. Most of the few scientists who did discuss the issue denied the military perspectives of biotechnology, or treated it as of little danger, referring to the BWC.[10] The closure of the debate on risks marked the start of a stormy technological development. The 'new biology' or 'biotechnology' began to bear fruit in the early eighties, bringing an increasing stream of new products to the market.

The coming true of the 'biological revolution' coincided with a turnaround of American domestic and foreign policy, leading to a hardening of international relations between the superpowers (the 'new Cold War'[11]). American military planners and politicians began to underscore the significance of biotechnology for national security.[12] A new intertwining of political and technological factors provided biological research for military purposes in the US with fresh momentum.

This momentum was reflected in rising budgets for military research in biology, in a growing list of Pentagon-sponsored biological projects at universities and in the revival of test-facilities for military biological work which had fallen into decay during the Nixon and

Carter days.[13] Scientists, the third factor in the process, were willing to cooperate. Not only do military projects in general yield attractive funding in the long run. The fundamental character of most of the military projects on biotechnology triggered the professional interest of the scientists, and offered them the social legitimation they needed: indeed, the outcome of many projects can be imagined to be useful for civil purposes as well.[14] Moreover, as in the fifties, they were under pressure to cooperate in an endeavour which became, in the changed political climate, a respectable activity and a yardstick for patriotic behaviour.[15] There were also scientists who opposed this trend but they could not stop it.[16]

Thus there is a third type of weapon development, in which civil technological development precedes the military development.[17] In our case the political climate, in which biotechnology took on a military strategic meaning, came into being when the technology had already shown its potentials in the civil sector. These civil successes made the political case for a military application of biotechnology only stronger, by raising expectations that the weak spots of biological weapons might be taken away in the future, while improved chemical weapons might also be within reach.

The most salient technological options on which such expectations are based are summarized in the next section, which reviews a number of sources, supplemented with assessments of my own.[18] If certain leads are already being explored this has been indicated. Most of the available information concerns the American state of affairs.

Options for Improvement of Biological and Chemical Weapons

Effect and controllability

Important in this respect is research on *toxins* of natural origin. Such toxins, which can be used as lethal weapons as well as incapacitantia, are in most cases far more effective than toxins of chemical origin. Moreover, they are considered as working faster and in a more controllable way than weapons based on bacteria or viruses. Therefore military interest in toxins has grown strongly, and also the number of toxins that are considered militarily relevant has grown steadily. The recombinant DNA method offers a number of options to improve toxins as well as their antidotes:

(1) The possibility to produce in quantity all kinds of toxins, which were difficult to obtain formerly. This possibility has greatly stimulated research on the structure and function of toxins. When

the structure has been elucidated, one can try to design variants of the toxin, using computer-aided modelling methods developed in civil research on drugs and toxins.[19] Such variants are meant to have a stronger biological effect or a more effective binding to the receptor. For military use more invasive organisms could be designed too. A more realistic option is the biotechnological development and production of antidotes against toxins. Such antibodies, which can also be of civil use, have already been used for protective purposes in the Gulf War.[20]

(2) Combination of specific properties and preferred dissemination characteristics in one and the same organism. For instance it has been suggested that one could build a toxin into a harmless virus that infects people through the respiratory tract. In the civil sector this type of strategy is followed successfully in agriculture by splicing genes coding for toxins that kill parasitic insects, into bacteria and viruses that live on the crops to be protected.[21] In a similar way, specific antibiotic-resistance factors or changed antigenic characteristics may be built into pathogens, in order to circumvent the enemy's natural defences (but also one's own, unless specific precautions are taken). Furthermore, the design and production of cell-specific cytotoxic agents (immunotoxins), which has been explored for civil purposes, may become of military relevance. Immunotoxins are hybrid molecules consisting of antibodies and (parts of) toxins, which kill specific cells recognized by the antibody.[22] Such hybrids are being tried out against tumour cells, but the approach is, in principle, useful for all kinds of other cells. The perspectives of the approach have recently broadened considerably now that antibodies, in their native state, can be produced in bacteria instead of in mammalian cells.[23]

(3) Accelerating the effect of toxic or incapacitating organisms. Seen from a military viewpoint the quickest possible effect is desirable to make the weapons work in a more predictable way. This problem too has its counterpart in the civil sector. In agriculture chemical pest control has remained dominant over the use of biological pesticides, mainly because of the slow working of the latter. An option to reduce this disadvantage of biological pests is to shift the expression of the toxic gene towards an earlier stage of the life cycle of the host organism.[24]

A related problem is the control of the *persistence* of toxic organisms used as a weapon. Here again we refer to agriculture as the civil sector where a similar problem exists related to the control of biological pesticides. The persistence of genetically modified biological pesticides is considered as a potential environmental problem, especially in those cases in which the host organism forms spores (bacterial spores are often very persistent). One of the options for solving this problem is to try, by genetic modification, to bereave the organism of its property to form such spores. Building in properties that bind organisms to environmental factors, for example low temperatures, is another option.

Biotechnology also opens perspectives on *improved chemical weapons*. Recombinant DNA techniques can give a new impulse to the understanding and manipulation of the transmission of nerve impulses on the molecular level. Such knowledge could lead to the design of more sophisticated nerve gases and their antidotes. Indeed, molecular neurology has become a flowering research field with military relevance. In the mid-eighties the US Army started the sponsoring of research projects concerning the structure and function of enzymes, transmitters and receptors that play a role in neural transmission.

Defence and detection

In the field of defence, biotechnology offers new methods for the preparation of antidotes against toxins (see above) and of *vaccines* that can be used for protection against bacteria, viruses and vector-bound diseases. Through cloning of DNA fragments it is possible, in the design, production and improvement of vaccines, to work with relevant (non-toxic) parts of viruses instead of with the organism as a whole. This means that viral antigenic molecules can now be produced on a large scale with relative ease, through methods which are generally considered as much safer than the traditional ones. An improved vaccine against anthrax has already been developed.[25] This approach may eventually lead to the production of wholly synthetic vaccines, which may be administered orally instead of through the bloodstream or even, as an aerosol, through inhalation. The latter option, though considered strongly speculative, opens the possibility of unnoticed immunization at a distance. If that becomes possible, protection of non-combatants or one's own troups against a biological weapon can be realized covertly, and thus the use of the weapon itself could be concealed. This perspective enhances the military interest in aerosol immunization, which has been intense over the years.

Research on vaccines has a meaning for the military not only as a defence against an attack with a viral weapon, but also as protection against exotic diseases which occur in regions of possible future military operations, especially Third World countries. In this field there is an evident overlap with civil research, for instance in the search for a vaccine against malaria.

Another important research field to be mentioned here is that of early *detection* of biological and chemical weapons on the battlefield. This research, using very sensitive biotechnological methods like the hybridoma technique, takes place under the Advanced Bio-Chemical Technology Program, set up by the Pentagon in the early eighties. As early as 1982 a Swedish institute for defence research advertised a field-detector for the ultra-sensitive and fast detection of aerosols of bacteria. The device had been developed for military use, but the institute considered it useful for civil purposes too, e.g. in environmental monitoring. However, such detection devices may be misled by an enemy using organisms with changed antigenic characteristics (see above).

Production and dissemination

The technological application of recombinant organisms achieved quick successes in the field of drug production. This seems to indicate that in most cases scaling up to industrial production will be possible in a relatively short time after (the much more time-consuming) research has yielded a useful prototype of a recombinant organism. From the viewpoint of military 'preparedness' this could be an attractive characteristic of biotechnology,[26] especially in a situation were research is a free activity but production is forbidden, as it is under the termes of the BWC. It also shows that civil and military process technology (bioreactor technology and down stream processing) overlap. This overlap will make the control of military applications of biotechnology more difficult, as the nuclear case has illustrated extensively.[27]

Another relevant aspect is the protection of microbes against damage by a-biotic factors. Rapid decay by such factors as mechanical damage during production and spraying, or radiation from sunlight, drying, varying temperatures and decay within the body, has always been a weak spot of previous biological weapons. However, owing to the rising commercial importance of microbes in biotechnology one may expect an ongoing improvement of techniques to protect microbes against such factors. Resistance to radiation is due to a relatively small number of genes which could readily be transplanted into microbes of commercial and military interest.[28] Coating techniques (microencapsulation) may contribute to stabilizing biological weapons in the future, making them more robust and reliable.[29] Such techniques also have relevance for the production of toxins, making them applicable in a wider scope of environments.[30]

New Materials

Biotechnology does not hold promise only for the production of weapons. It may also yield new materials and substances ('strategic biomolecules') which can be attractive for military use. The $16 million Biomolecular Engineering for Materials Applications (BEMA)-program of the US Navy, which started in 1983, gives an impression of this,[31] it specifies the following options:

- engineered enzymes and other catalysts;
- anticorrosion and antibiocorrosion systems;
- antibiofouling coatings or systems for ships;
- production of molecules used in adhesives, coatings and lubricants;
- novel sensing technology (sound and chemical sensors).

Additional suggestions for exploration of biomaterials were made at a joint conference of the US Navy and the National Research Council's 'Committee on Biotechnology Applied to Naval's Needs' in the fall of 1983:[32]

- prevention of microbial contamination of fuels;
- food manufacturing;
- production of growth factors and other ways to accelerate wound healing;
- production of blood substitutes, a protein microcapsule to transport oxygen, fibronectin, clotting factors and other plasma products;
- making an enzyme to convert blood type A to the universal type O;
- producing human–human and mouse–human hybridomas to make antibodies that passively protect people against bacterial and other infections.

A more recent example of a useful product resulting from a military biotechnological research project is silk made by a bacterium carrying the silk-producing gene of a spider. The silk produced is of such quality that it can be used for the manufacturing of bullet-proof vests and helmets.[33]

Future Developments

Several types of future developments are conceivable, depending on the way in which the necessary factors for weapon development will manifest themselves.

1. Technological options as mentioned above may fail to yield the expected breakthroughs in biological weapon development. In that case a crucial ingredient for the process of entrenchment of such weapons is lacking, and most probably the process will stagnate again. With respect to the production of toxins, such outcome is improbable, however, since biotechnology has already yielded too many positive results. In the field of vaccines too there has been considerable progress. It is true that vaccines are not weapons, but possession of effective vaccines may lower the threshold to the use of weapons against which such vaccines offer protection. The same applies to antibodies for use against toxins. Such antibodies, and also improved vaccines, have been administered to American soldiers during the Gulf War, which proves the entrenchment of these substances into military practice.[34]

2. If military biotechnology does result in attractive prototypes of real biological weapons (i.e. well controllable harmful microorganisms and viruses), then entrenchment of such weapons in military planning on a prototype scale would be probably unavoidable.

At first sight such development, in the present situation of decreased tension between the USA and the USSR, may seem incredible with respect to the superpowers.[35] On the other hand, it remains to be seen when and where precisely the recent detente will lead to a retreat from weapon development. In the past the giving up of viable weapon options has appeared to be very difficult, especially when these options have become embedded in military institutions. Moreover, military circles in the West have not given up seeing Russia as a threat, despite its present deplorable economic situation.[36] Though the fact that US Congress recently approved unanimously a bill that renders the provisions of the 1972 BWC into domestic law[37] may indicate that the political support for exploring the biological weapons option is declining again, this does not mean that research has stopped. On the contrary, recombinant DNA safety guidelines have been relaxed recently to enable the Pentagon to conduct studies on two highly virulent agents.[38] For the short term, the expectation is that the military research of the superpowers and some of their allies in the field of biotechnology will continue. This may lead the 'type 3' development to change into one of 'type 2', which bears the risk of further social entrenchment of a military biotechnology, and of proliferation of its products to other countries. From the recent past there are several indications – in the form of rumours as well as facts – that quite a number of smaller nations are interested in biological weapons, especially in the Middle East, and that some of them have been supplied with organisms and equipment by Western countries:

- In October 1988 Robert M. Gates, former deputy director of the CIA, said that the proliferation of chemical and biological weapons may constitute 'the most immediate threat to world peace today'. CIA director William H. Webster stated that 'at least 10 countries' are working on producing biological weapons. He did not name the nations, but US officials and analysts have mentioned Israel, Syria, Iran and Iraq as among those with active research pro- grammes.[39]
- ABC-television reported in January 1989 that Iraq had produced biological weapons in an underground facility in Salman Pak near Baghdad, and that Iran does experiments with such weapons. At about the same time a spokesman of the government of West Germany declared that his government had indications that Lybia is considering developing biological weapons at a planned institute for microbiology.[40]
- A few days later US senator McCain made charges that the US had provided Iraq with the highly virulent tularemia bacterium, a strain which was a standardized part of the former US arsenal of biologi- cal weapons. McCain added that 'we have every reason to assume that Iraq may weaponize two of the three most lethal biotoxins – anthrax and tularemia – and it may well be on the way to weaponiz- ing the third: equine encephalitis'.[41]
- In August 1989 it became known that the Iranian Research Or- ganization for Science and Technology had approached first a Canadian toxicology laboratory, and subsequently the Dutch scien- tific microbial archive, in order to obtain toxine-producing fungi (probably mycotoxines).[42] In response the Dutch Ministry of Foreign Affairs, in December 1989, published a list of toxic microor- ganisms that should not be exported abroad without previous con- sultation with the ministry. A few days later Dutch newspapers reported that a Dutch firm had supplied a bioreactor to Iran for (it said) the production of vaccines. This supply illustrates the overlap in civil and military biotechnological equipment mentioned above.
- On the eve of the Gulf War, and during the war itself, additional evidence appeared that Western (especially German) firms had supplied Iraq with toxins and biotechnological equipment.[43]
- A recent intelligence report of the US Navy states that, except NATO countries and the USSR, at least 14 other countries possess chemical weapons; biological weapons are said to have become increasingly popular among developing countries.[44]

The above evidence indicates a growing interest of smaller countries and local superpowers in acquiring biological weapons. This is not to say that these countries are already involved in genetic engineering of the organisms they are accused of working on, but this may change in the future. Moreover, the quantity production of a conventional biological weapon (i.e. based on a natural organism) will be facilitated by biotechnological equipment, obtained from the civil market.

3. Armed conflicts can easily trigger a 'type 1' development. In a situation of acute conflict the interest in biological weapons, and in defences against such weapons, will grow quickly, and their development may be initiated or speeded up. In the Gulf War, for instance, the biotechnological products (antibodies and vaccines) used by the American troops were still in the (unlicensed) development stage.[45]

NOTES AND REFERENCES

[1] Using this approach we follow recent developments in the field of technology dynamics, see W. E. Bijker, T. P. Hughes and T. Pinch (eds), *The Social Construction of Technological Systems* (Cambridge, Mass: MIT Press, 1987), esp. pp. 17–51.

[2] Scientists can have, in their own eyes, legitimate reasons to cooperate in weapon projects, see, for instance, J. B. Conant, *My Several Lives, Memoirs of a Social Inventor* (Harper and Row, 1970), chapter 5.

[3] See, for an extended history of biological weapon development, the SIPRI-study *The Problem of Chemical and Biological Warfare* (Uppsala: Almquist and Wiskell, 1971). See also R. Gomer, J. W. Powel and B. V. A. Roling, Japan's biological weapons: 1930–1935, *Bulletin of the Atomic Scientist*, October 1981, pp. 43–53, and B. J. Bernstein, Churchill's secret biological weapons, ibid., January/February 1987, pp. 46–50.

[4] P. Williams and D. Wallace, *Unit 731, The Japanese Army's Secret of Secrets* (Hodder and Stoughton, 1989).

[5] US Army BW Manual, quoted in *The Problem of Chemical and Biological Warfare*, vol. 2, p. 311.

[6] *The Problem of Chemical and Biological Warfare*, vol. 2, pp. 314–325.

[7] Negotiations between the Pentagon and the National Research Council (NRC) of the National Academy of Sciences (NAS) over the start of a military research programme in genetics came to nothing because of the 'reluctance to involve NAS–NRC in such a controversial endeavour', *The Problem of Chemical and Biological Warfare*, pp. 314–325.

[8] The complete text of the treaty can be found in E. Geissler, *Biological and Toxin Weapons Today* (Oxford University Press, 1986), pp. 135–137.

[9] *The Problem of Chemical and Biological Warfare*, vol. 2, p. 327.

[10] E. Milewski, RAC Discussion on Biological Warfare, *Recombinant DNA Technical Bulletin*, 5, no. 4 (1982), p. 188, and S. Krimsky, *Genetic Alchemy* (Cambridge, Mass.: MIT Press, 1982), p. 106.

[11] H. Davis, The New Cold War, in H. Davis (ed.), *Ethics and Defense* (Oxford: Basil Blackwell, 1986), pp. 173–187.

[12] 'Perhaps the most significant event in the history of biological weapons development has been the advent of biotechnology', the US Army reported to a House Committee in 1986. This and other examples are to be found in C. Piller and K. R. Yamamoto, *Gene Wars* (New York: Beech Tree Books, 1988), p. 115. See also, *Soviet Military Power* (Department of Defense, Washington DC, 1984).

[13] The US defence budget for CBW-research has grown from $16 million in 1980 to $75 million in 1988; in 1986 the 'program aimed at developing defenses against potential biological warfare agents' had 100 researchers at 50 universities under contract (*Science*, 20 May 1988, p. 981). For a survey of projects, see *McGraw-Hill's Biotechnology Newswatch* 4, no. 13 (1984), pp. 8–9, and *Genewatch* 2, no. 2 (1985), pp. 14–15. The former US military biological test-institute Dugway in Utah is being made operational again (*Science*, 8 April 1988, p. 135).

[14] See, for example, the discussion in the US Recombinant DNA Advisory Committee (RAC) about a Proposal to Clone Shiga-like Toxin Gene from E. Coli (Shiga toxins cause diarrhoea in third-world countries), *Recombinant DNA Technical Bulletin* 7, no. 3 (1984), p. 130–141.

[15] At a conference at Stanford University in 1985 Under-Secretary of Defense Perle mentioned 'a not easily dismissed obligation of all scientists to contribute . . . to the national security', see R. Perle, National Security and Scientific Inquiry, National Press Club, 3 May 1985, *Proceedings Media Outreach Program* (American Association for the Advancement of Science, Washington DC, 1985).

[16] S. Wright and R. Sinsheimer, Recombinant DNA and biological warfare, *Bulletin of the Atomic Scientist*, November 1983, pp. 20–26; S. Wright, The military and the new biology, ibid., May 1985, pp. 10–16; S. Wright, New designs for biological weapons, ibid., January/February 1987, pp. 43–46; B. Hatch Rosenberg, Updating the biological weapons ban, ibid., January/February 1987, pp. 40–43; *Science*, 15 June 1984, p. 1215.

[17] An earlier example of this type of development is that in the field of lasers. Lasers originated from civil R&D and were subsequently applied for military purposes.

[18] The major sources with respect to biological weapons on which this section is based are E. Geissler, *Biological and Toxin Weapons Today*, esp. chapters 2, 3 and 4; C. Piller and K. R. Yamamoto, *Gene Wars*; *McGraw-Hill's Biotechnology Newswatch* of 21 September 1981, 19 October 1981, 19 April 1982, 16 May 1982, 7 June 1982, 3 October 1983 and 2 July 1984. References to other sources are made in the text. Comparisons made between civil and military development are my own.

[19] See, for a recent example of the use of molecular model building in

structure analysis, the work on the three-dimensional structure of heat-labile enterotoxin from Escherischia coli, T. K. Sixma et al., *Nature*, 351 (1991), p. 371. An example of protein engineering can be found in R. E. Bird et al., *Science*, 242 (1988), p. 423.

[20] *New Scientist*, 12 January 1991, p. 23.

[21] J. B. Kirschbaum, Potential Implications of Genetic Engineering and Other Biotechnologies to Insect Control, *Annual Review of Entomology*, 30 (1985), pp. 51–70.

[22] E. S. Vitetta et al., *Science*, 238 (1987), p. 1098.

[23] A. Plückthun, *Nature*, 347 (1990), p. 497, and *Bio/Technology*, vol. 9 (1991), p. 545.

[24] It has been suggested, for instance, that it is quite possible to clone genes in baculoviruses (which are insect pathogens) under the control of promotor-genes that are expressed earlier during infection than the normally used promotor, expressing its gene at the end of the life cycle of the virus. See M. D. Summers and G. E. Smith, Genetic Engineering of the Genome of the Autographa californica Nuclear Polyhedrosis Virus, B. Fields, M. A. Martin and D. Kamely (eds), *Genetically Altered Viruses and the Environment* (Cold Spring Harbor Laboratories, 1986), pp. 319–339, esp. pp. 329–330.

[25] *New Scientist*, 12 January 1991, p. 23.

[26] In 1986 a US defence spokesman indeed made suggestions in this sense, see *Nature*, 323 (1986), p. 57.

[27] The overlap between civil and military biotechnology may be conceived also as an indication that an enemy who has a capability in civil biotechnology can develop biological weapons too; for that reason civil biotechnological knowledge and goods should be considered to be of strategic importance for national security. Indeed, this is the import of a report written by a 'US Government Interagency Working Group on Competitive and Transfer Aspects of Biotechnology'. The group, consisting of representatives of several ministries, big industries, science, NASA, and the CIA, reported in 1983 to the presidential science advisor: 'National security in both military and commercial sense are inextricably intertwined. There is a common knowledge base as we have pointed out. Technology and equipment used for commerce can also be applied to the development and manufacture of useful products.' See Biobusiness World Data Base (US Department of State, Washington DC, 27 May 1983). Since then, several countries have subjected biotechnological equipment and organisms to export controls.

[28] R. Novick and S. Shulman, in S. Wright (ed.), *Preventing a Biological Arms Race* (Cambridge Mass.: MIT Press, 1990), chapter 5.

[29] For instance, the biological insecticide Gypchek appears to have a considerably prolonged effect in sunlight when coated with a sunscreen called Orzan LS, see *Environment*, 30, no. 2 (1988), p. 23.

[30] By incapsulating the toxin of Bacillus Israelensis in an oil suspension it can be used against larvae of flies that live on the surface of pools; the oil

suspension prevents the toxin from sinking down (interview with researcher of ITAL, Wageningen, the Netherlands, 1986).

[31] *McGraw Hill's Biotechnology Newswatch*, 3, no. 10 (1983), p. 1.

[32] *McGraw Hill's Biotechnology Newswatch*, 3, no. 19 (1983), p. 1.

[33] Associated Press press release, February 1990.

[34] *New Scientist*, 12 January 1991, p. 23.

[35] Despite the fact that 'America has won the Cold War', the USSR must, in a military sense, continue to be viewed as a superpower, among other reasons because of its huge military R&D potential.

[36] Comments in recent issues of *Military Technology* are illustrative in this respect.

[37] *Nature*, 345 (1990), p. 192.

[38] The Pentagon studies concern investigations on yellow fever and Venezuelan equine encephalitis virus, conducted at the Plum Island Research Facility in New York; *Technology Review*, May/June 1990, p. 80.

[39] *Washington Post*, 19 January 1989.

[40] *Washington Post*, 20 January 1989.

[41] *Washington Post*, 24 and 29 January 1989.

[42] *New York Times*, 13 August 1989.

[43] *Der Spiegel*, 33 (1990), pp. 81/84, and 41 (1990), p. 148; H. Leyendecker and R. Rickelmann, *Exporteure des Todes, Deutscher Rüstungsskandal in Nahost* (Göttingen: Steidl Verlag, 1991).

[44] *New York Times*, 18 March 1991.

[45] *New Scientist*, 12 January 1991, p. 23.

GUIDE TO FURTHER READING

SIPRI, *The Problem of Chemical and Biological Warfare* (Uppsala: Almquist and Wiskell, 1971), and the SIPRI Yearbooks.

E. Geissler, *Biological and Toxin Weapons Today* (Oxford: Oxford University Press, 1986).

C. Piller and K. R. Yamamoto, *Gene Wars: Military Control over the New Genetic Technologies* (New York: Beech Tree Books, 1988).

N. A. Sims, *The Diplomacy of Biological Disarmament* (Basingstoke: Macmillan Press, 1988).

S. Wright, *Preventing a Biological Arms Race* (Cambridge Mass.: MIT Press, 1990).

Part IV

Diffusion of Biotechnology

Introduction to Part IV

From the point of view of economic and social impact, scientific breakthrough or the invention of new technologies per se are insignificant. Rather, it is through the *diffusion* of new technologies that economic and social effects occur.

The diffusion of biotechnology, however, is perhaps the least rigorously studied of all the themes in biotechnology. In part the reason for this is the complexity of the process of diffusion. The matter is not simply one of analysing the proportion of potential users who adopt the technology over time, although this is an important part of the diffusion process. To begin with, the process of adoption and use lead to further changes in the technology, with the result that the processes of innovation and diffusion merge. Furthermore, the process of adoption has complex determinants that differ in the case of different biotechnologies and over time. It is accordingly necessary to distinguish between economic determinants of adoption (such as the effect of the technology on profitability which, for example, was an important influence on the fate of single cell protein discussed elsewhere in this book), and social determinants such as the social acceptability (or lack of acceptability) of the technologies, such as social attitudes to hormones in food or the environmental release of genetically altered organisms. Rigorous studies of the diffusion of biotechnologies would take these kinds of factors into account.

The diffusion of biotechnology also has an international dimension with important implications both for international competitiveness and for the wellbeing of developing countries. Some of the implications are taken up again in part VII of this book, which examines policy implications. In the final chapter of this section the diffusion of biotechnology to Cuba is examined, showing how the use of biotechnology has been adapted to suit the particular circumstances of this country.

Diffusion is also uneven across different sectors of economic activity. In the case of pharmaceuticals, for example, diffusion of the technologies and results have been fairly rapid, while in the case of agricultural applications, although many of the techniques have been widely adopted, results are being achieved far more slowly.

While acknowledging that diffusion has been under-studied and that more research is needed in this area, the aim of this section is to provide a taste of some of the major issues that affect the diffusion of biotechnologies.

23

University–Industry Relations in Biotechnology

Martin Kenney

More than any other modern technology, biotechnology has its origins in university science that was funded by government agencies with the idea that research could lead to the development of new pharmaceuticals that would improve human health. The explosion of commercial interest in biotechnology has ruptured the traditional boundaries separating the universities and industry and created new mechanisms for interaction between these two institutions. This chapter briefly reviews the developments of university–industry relationships in biotechnology with special emphasis on the U.S. and highlights some of the difficulties in these relationships.

The biotechnology industry is international in scope, and all major chemial and paraceutical companies are actively involved in funding and conducting biotechnology research. Similarly, the scientific research which led to the techniques underlying biotechnology was pioneered in a number of countries. However, the first efforts to commercialize these scientific techniques occurred in the U.S. This was aided by the better development of the institutional framework in the U.S. for creating small biotechnology startup firms. Further, the barriers between U.S. universities and the industrial sector have always been more permeable in the U.S. than in Europe or Japan.

Three different genres of university–industry relations have developed to commercialize university biotechnology research. Any of these types of relations may exist at a single university simultaneously. Thus, different professors can be linked with different companies. The first type of linkage is that of a professor who has started a company to commercialize an aspect of research. The second relationship is an institutional long-term affiliation, involving multimillion dollar sums, between a single firm and a laboratory or university department. The final type of relationship involves the development of biotechnology centres or institutes formed by corporate consortia which often have a component of federal or state involvement. We examine each of these types in turn. In the conclusion we discuss the implications of these massive new connections between the university and industry for society at large.

Professors and Entrepreneurial Companies

It is difficult to pinpoint the first occurrence of the idea that biotechnology based on the new techniques of genetic engineering and molecular biology might become profitable. The watershed event, however, was the formation in 1976 of Genentech, a new startup company. The founders were Robert Swanson, a venture capitalist, and Professor Herbert Boyer, a Nobel Laureate molecular biologist at the University of California, San Francisco Medical Center. They secured the first investment of $100,000 from Swanson's former employer, the venture capital firm Kleiner Perkins, and began their research in Boyer's university laboratory. The key to this relationship is that both Swanson and Boyer got major equity positions in their new company. By 1980 when Genentech went public, both Swanson and Boyer's stakes in the company were worth tens of millions of dollars.

The Genentech experience pointed out graphically that the molecular biology research carried out in the university could yield impressive financial returns to the scientist. In the late 1970s and even more so in the 1980s university molecular biologists affiliated themselves with or even founded companies while retaining their university professorships. Thus, in this period at least 200 small firms in the U.S. were started by university professors and venture capitalists to take advantage of university-conducted research. This 'biotechnology goldrush' was largely a U.S. phenomenon, though some startups attempted to recruit European professors and one company, Biogen, went so far as to open a Geneva laboratory (since closed).

The professors who were intimately involved in startups received consulting fees, research grants and, most important, equity in the startup. When Genentech went public in 1980 Herbert Boyer was worth over $100 million. Many other university-based molecular biologists had similar experiences of becoming wealthy when the company they held stock in was sold to the public. Obviously, with such enormous capital gains at stake there was great pressure on other professors to use their expertise for personal gain as well.[1]

At the time many believed that this privatization activity by individual professors was a passing fad. However, the revolution that occurred in biology has continued to develop new commercial possibilities and thus companies intent on exploiting academic research have continued to be created. The current wisdom is that any top-notch molecular biologist is already connected with some startup firm. Though this is an exaggeration, it contains an important element of truth: this entire branch of science has been commercialized.

The enormous value that professorial equity stakes in companies can assume has created potential conflicts of interest between the professors' academic responsibility and their personal financial situation. For example, the student–teacher relationship can be violated as the professor becomes more concerned with his/her financial stake. Professors may become so involved in their company that they abdicate their teaching responsibilities. Or a professor can direct students into research topics useful to his/her firm. Another problem is that professors may appropriate the unpublished results of their own or a student's research for a company in which they have a stake. In each case the role of the professor as a teacher is subordinate to financial interests.

Cases have been reported recently of professors testing pharmaceuticals produced by companies in which they had an equity stake. The dangers in such cases are that the judgment of these professors may be clouded by their personal financial interests. This is especially problematic because invariably professors concerned have not acknowledged their pecuniary interests in the research results. The potentially huge capital gains to be had from the commercialization of research results creates enormous pressures for secrecy, and can even lead to the falsification of results. Thus, in this type of relationship the professor as an individual is the conduit for increasing corporate influence in the university.[2]

Large Long-Term Contracts between Companies and Universities

The rapid development of biotechnology was a surprise to the chemical and pharmaceutical companies. These large corporations had little or no in-house expertise in the area of molecular genetics and thus were unable to rapidly assimilate the commercial potentials opened up in the rapid scientific advances. By the late 1970s, though, it had become apparent that their future in their respective industries might be dependent on developing products based on the tools of molecular genetics. To gain access to this information it was necessary to gain access to the same pool of professors that the venture capitalists were recruiting to assist in the startup companies.

The mechanism developed was a relatively long-term (usually longer than three years) contract between the university and a particular company. These contracts provided that in return for funding the research the company would have a series of rights to the research results. The contracts were all unique and were negotiated differently, but in every instance the corporate aim was to secure access to this new technology. For the university unit involved, the advantage was that research money was available without the necessity of time-consuming preparation of grant applications and the concomitant insecurity regarding funding. For university administrators there was the possibility of charging overheads on the grant and thereby securing financial benefits.

Between 1980 and 1983 over $140 million was funnelled into long-term contracts and this sum continued to grow throughout the 1980s. These contracts can be funded in the tens of millions of dollars and extend as long as 10 to 12 years.

The largest of the early contracts, for $70 million and lasting 10 years, was between Hoechst, the German chemical giant, and Massachusetts General Hospital of the Harvard Medical School. In this case, Hoechst funded the creation of a molecular biology department at MGH. The privileges that Hoechst received for this funding were: (1) the right of first refusal for funding any university projects; (2) an ability to prevent department members from consulting with other companies; (3) the right to access to all postdoctoral and graduate researchers; (4) the right to have four company researchers in the laboratory at any one time; (5) symposia and research reports for Hoechst; (6) the right to review all manuscripts 30 days before submission to a journal; and, perhaps most importantly, (7) the rights to exclusive licences on all commercially exploitable discoveries. With this contract Hoechst was not only able to secure access to university researchers but was also able to set new standards of control.[3]

In contrast to faculty entrepreneurship these long-term contracts formalize relationships between companies and researchers. The advantage, of course, is that the university administration can monitor the contracts. Conversely, however, it provides the private sector with an important opportunity to shape the research agenda. The backing of a few elite professors with gigantic sums helps to ensure their increased influence and power in the institution, thus validating their research agenda. These professors are given the power to threaten to accept a position at another institution and thereby transfer huge sums of money. Thus, industrial funding can drastically influence the power balance in an institution, and provide a measure of agenda-setting power.

For the professor there are also risks. For example, when the contract nears completion the professor faces an impending funding shortfall. The industrial sponsor is then in a particularly strong position and can induce greater concessions from the professor and the university. This can easily lead to a shifting of the lab's agenda to more commercially relevant research. The funding agent has enormous power to affect research directions, either subtly or more bluntly.

The long-term relationship permits the dominance by single companies of laboratories and departments and almost invariably is accompanied by first access and exclusive licences for the sponsor companies. This can be conceptualized as having university administrators approve the rental of a certain part of the university to the private sector for a certain period of time. The long-term result of such trends will be a university with a research agenda increasingly determined by the needs of a single corporation.

The Research Centre Phenomenon

The research centre model is quite different from the previous two in that it has traditionally involved funds provided by the public sector (either federal or state funds) and contributions by more than one company. The objective of these centres is to facilitate the transfer of information and 'knowhow' from university researchers to corporate scientists in exchange for research funding. The usual justifications for government involvement in research centres have been that it will either further international competitiveness or assist in local economic development. These are almost invariably flawed or incorrect.

The research centre concept was pioneered in the 1970s by the National Science Foundation in a variety of fields.[4] By the 1980s, though, with the promise of enormous profits, a plethora of biotechnology centres were started using either federal or state funds. Two contrasting types of companies have been organized to take part in these centres. The first type aims to secure the participation of a number of similar companies. For example, a university research centre might consist of a number of competing pharmaceutical companies. The second genre would organize a centre around a group of non-competing companies (from different industries). For example, the Cornell Biotechnology Institute consists of Eastman Kodak, General Foods and Union Carbide. These three companies are all interested in biotechnology; however, they do not compete.[5]

The success of the research centre concept is predicated largely on the ability of the firms involved to cooperate and on the ability of university scientists to conduct work of commercial interest to the constituent companies. Nevertheless, the enormous number of centres being developed provides reason for scepticism regarding their success. Often the companies joining these centres see their membership more as a charitable contribution to the university than as a method of acquiring more knowledge. Conversely, the universities and the scientists often see them as little more than another method of garnering income to fund research.

Anecdotal evidence gathered thus far indicates that few of the centres have made much progress. However, most have been in existence for less than five years and thus it is too early to predict the impacts accurately. The success of many centres seems questionable because of the tremendous duplication of effort as each state and university attempts to develop its own research centre – all of which may ultimately result in the dilution of effort, a redirection of resources and opportunity for marginal researchers to garner large amounts of funds. The centre programme was conceived as a measure to increase the rapidity of information transfer, and yet their performance has been quite irregular.

Pitfalls in the University–Industry Connection

The common belief among most observers was that the development of university–industry relationships would be an unalloyed good. The reasons given in all of the OECD countries was that this would yield a more efficient transfer process and increase economic competitiveness. Further, the funds could be used for supporting research personnel and purchasing new equipment. Thus, both the university and industry were expected to be winners.

This commercialization has had some subtle effects on the way molecular biology is conducted. This is because of the tremendous value being placed on biological information. Before commercialization biological materials and information, the coin of academic biology, was traded freely. For example, materials formerly exchanged informally are now exchanged only after elaborate disclaimer forms are signed promising that discoveries made with the new materials will not be patented and the materials will not be exchanged with other researchers. Thus, the information exchange which was so crucial for academic progress has become truncated and weakened by the injection of commercial motives into traditional academic concerns.

The university's emphasis on patenting discoveries has steadily increased because of the concern among corporate sponsors about being able to reap the benefits of their investments. This increase in patenting injects important new concerns into the research process. First, secrecy is necessary to protect patentability. Second, care must be taken to establish who should be involved in the research, so as to exclude 'non-essential' members. Third, scholarly papers must be written in such a way as to not contradict the patent claims. Once again, the injection of these ulterior motives has impacts, at one level, on the environment and ethos of the university and, at another, on the very heart of the scholarly enterprise.

Probably the most serious consequences are those related to the collegiality between faculty members. There have been reports of professors refusing to serve on graduate students' dissertation committees together because of their allegiance to different companies. Similar problems can be seen in the lack of willingness of faculty members to discuss their most recent results at departmental functions due to a fear that competitors will find commercial value in the advances. In such an atmosphere, distrust and secrecy replace the characteristic scholarly dialogue so necessary for the life of the university.[6]

The university's adoption of the norms of industry is fast occurring in biotechnology. In the short run this might lead to the faster adoption and diffusion of new biotechniques, but in the long run the consequence might be an erosion of the university as a source of independent research. Institutions are being created which value commercialization of research over free information flow and publication. The ultimate cost may be the university as an independent institution.

Europe and Japan

The biotechnology industry is international, and, as in the Hoechst–MGH case, many multinational corporations have invested in U.S. universities. On a lesser scale, U.S. corporations have invested in non-U.S. universities (e.g., Monsanto's investment in protein engineering at Oxford University).[7] The role of the university in the commercialization of biotechnology is different not only between Japan and Europe, but also among European countries. It is difficult to make any generalization regarding these very diverse countries except to say that venture capital-financed startups have played a far smaller role. Concomitantly, few professors in other countries have been able to capitalize on their specialized knowledge. Thus, the commercialization of university biotechnology expertise has been far more orderly.

In Europe, the university's role in the commercialization of biotechnology varies cross-nationally. For example, in the U.K. there have been both university spinouts and two companies that were formed by the British government to commercialize certain patents developed in British universities. However, with the exception of a few corporations, British industry has shown little interest in research in British universities. The French government actively developed some small companies to commercialize biotechnology. In Germany, however, the universities have proved somewhat more resistant to entrepreneurial involvement with the corporate sector and few professors have launched startup firms.

Japan is the most anomalous, in that until recently professors in its prestigious national universities were forbidden even to consult with industry. The weakness of the venture capital markets also discouraged professors from starting small companies. Thus, the major routes the U.S. used to commercialize biotechnology were not open to Japanese firms. The tactics Japanese firms adopted were unique; they hired young molecular biology graduates and dispatched their best researchers to laboratories in Japan and overseas to acquire hands-on capabilities. The Japanese government has also organized research consortia of companies and of university professors which will also assist in technology transfers.[8]

Conclusion

Biotechnology was the first major technology that was developed entirely in university research laboratories. As a result the commercialization process led to the creation of new links between university and industry. As we have seen in the U.S. context, this process has had important impacts on the ethos and goals of university research. It is still too early to evaluate the full extent of changes the injection of the profit motive into the university environment has had. Evidence is accruing, though, that serious breaches of the public trust have already occurred and it seems certain that more will occur.

The involvement of both national and local officials also bears watching. The allure of high technology, large sums of money, and economic development on the basis of research have resulted in new participants in guiding research. These officials have no intrinsic interest in science outside of its potential economic benefits. Their short-term goals provide an unstable base of support and industry offers little more. The industrial support boom so hailed by some will most probably yield far fewer long-term benefits than first believed, but the barriers between the universities and industry may have been permanently broken.

NOTES

[1] For the early history of the biotechnology industry, see Martin Kenney, *Biotechnology: the University–Industrial complex* (New Haven: Yale University Press, 1986).

[2] For further discussion of these problems, see Kenney, *Biotechnology*. Also see Nicholas Wade, *The Science Business* (New York: Priority, 1984).

[3] U.S. Congress, Committee on Science and Technology, *Commercialization of Academic Biomedical Research* (Washington, D.C.: United States Government Printing Office, 1981).

[4] National Science Foundation, *Development of University–Industry Cooperative Research Centers: Historical Profiles* (Washington, D.C.: National Science Foundation), 1982.

[5] Kenney, *Biotechnology*, pp. 42–47.

[6] Kenney, *Biotechnology*, chapter 6.

[7] Kenney, *Biotechnology*, pp. 204–205.

[8] Herman Lewis, Biotechnology in Japan. *Scientific Bulletin*, Department of the Navy, Office of Naval Research Far East, Vol. 10, No. 2 (April–June, 1985).

GUIDE TO FURTHER READING

Kenney, Martin (1986). *Biotechnology: The University–Industrial Complex*. New Haven: Yale University Press.

Lewis, Herman (1985). Biotechnology in Japan. In *Scientific Bulletin*, Department of the Navy, Office of Naval Research Far East, **10**(2) (April–June 1985).

National Science Foundation (1982). *Development of University–Industry Cooperative Research Centers: Historical Profiles*. Washington, D.C.: National Science Foundation.

Reams, Bernard (1986). *University–Industry Research Partnerships*. Westport, CT: Quorum.

Twentieth Century Fund (1984). *The Science Business*. New York: Priority.

U.S. Congress, Committee on Science and Technology (1981). *Commercialization of Academic Biomedical Research*. Washington, D.C.: U.S. Government Printing Office.

U.S. House of Representatives (various). Various hearings by the Subcommittee on Investigation and Oversight and the Subcommittee on Science and Technology of the Committee on Science and Technology.

24

Popular attitudes, information, trust and the public interest

Mark F. Cantley

Introduction

The need for public understanding, in a high-technology-based society

In any society, the advance of science and its technological applications are under some degree of social control.

Where science and technology demand significant resources, or their applications threaten established interests, their further development will depend upon social consent. This consent is readily given if the science and technology are seen as important to competitive capability, as in military or economic contexts – or when they offer obvious benefits, such as a cure for cancer. Where the relevance is less obvious and their impacts unforeseeable (e.g. because long-term or indirect), the consent (and resources) will be less readily obtained; and where they arouse significant apprehension or threaten established interests – economic, social, intellectual or other – the necessary consent and resources may be replaced by outright opposition.

In a developed and densely populated world, all societies depend increasingly on advanced science and technology. The effective functioning of democracy in these circumstances therefore depends upon a sufficiently large proportion of the electorate having an adequate understanding of the nature, achievements and possibilities of science and technology. An apt analogy is the link between universal literacy and the extension of the democratic franchise; the development of either without the other causes political instability.

Key characteristics of biotechnology: the 'package deal'

These general observations apply with particular force to biotechnology, because it is pervasive in its effects across many economic sectors, and builds on recent discoveries as subversive as those of Galileo or Darwin.

On the one hand, there are positive popular perceptions of traditional fermented foods and drinks, and of the more recent triumphs of sanitation and medicine over disease. On the other, there is concern about unfamiliar and little understood technologies such as genetic engineering, and still more so about the juxtaposition of such novelties with the familiar processes of birth and procreation, and the ethical, religious and cultural values and feelings associated with identity, privacy, human (and animal) rights, and the nature of man.

Life has been defined as 'licking honey off a thorn' – a package deal. Political autocracies, of left or right, have often wished to promote creativity and innovation to modernise their military and economic power, but creativity and innovation spill over into political demands that challenge autocracy. Similarly with biotechnology, which 'packages' four closely inter-related characteristics:

- It has inherent scope for many further scientific discoveries, and hence vast but unpredictable technological potential.
- It is an indispensable element in the unremitting global economic competition, agricultural and industrial, between countries, between blocs and between companies, which will cause the decline of some and the rise of others – Schumpeter's 'gale of creative destruction'.
- It offers well-founded hopes of further reducing hunger and controlling disease.
- It will continue to provoke widespread popular apprehension about non-understood science and socially or culturally unacceptable innovation.

Public information: a strategic factor for biotechnology

These four strongly interacting and continuing factors lead inexorably to one conclusion: public understanding, attitudes and acceptance will be of increasing strategic significance for the progress of biotechnology.

This view was highlighted in the January 1986 report by the US National Academy of Sciences, 'Biotechnology: An Industry Comes of

Age'.[1] Contrasting with earlier US reports stressing R&D, entrepreneurial spirit, academic–industrial collaboration and similar factors, the report quotes (with some emphasis) T. O. McGarity of Texas School of Law: 'In the end, whether or not these new biotechnologies really get off the ground in this country is going to depend upon whether we can erect a regulatory regime that can secure public trust.'

What are the determinants of public understanding, of public trust? The formal education system? The popular media? The deep-rooted values which define and reflect a national culture? However mysterious, these determinants of trust and understanding are of strategic importance; arguably, the most important single factor in a society's self-management, learning and survival, whether the 'society' is global, regional or national. A striking example over the past century has been Japan's swift integration of science and technology and its adaptation to the world trading system, while retaining a traditional cultural identity and weathering the shocks of war and transformation to democracy.

The following section elaborates the above issues, with particular emphasis on the management of biotechnology at European Community level.

Science, Society, Regulation and Public Attitudes

Freedom of scientific enquiry and societal control

Galileo, Milton and recombinant DNA were brought together in 1978 when Professor Freeman Dyson was invited to testify at hearings of the Subcommittee on Science, Research and Technology of the US House of Representatives.[2] The subcommittee sought to educate itself concerning the broader issues raised by recombinant DNA. F. Dyson suggested 'an analogy between the seventeenth-century fear of moral contagion by soul-corrupting books and the twentieth-century fear of physical contagion by pathogenic microbes', and quoted Milton's arguments in *Areopagitica* for accepting some risks rather than treating all uncertainty by restraint. Milton had been arguing for a free press. Milton's observations on the silencing of Galileo and the consequent general decline of intellectual life in seventeenth-century Italy were seen by F. Dyson as pertinent to the rDNA debate:

I could recount what I have seen and heard in other countries, where this kind of inquisition tyrannises; when I have sat among their learned men, for that honor I had, and been counted happy to be born in such a place of philosophic freedom, as they supposed England was, while themselves did nothing but bemoan the servile condition into which learning amongst them was brought; that this was it which had damped the glory of Italian wits; that nothing had been there written now these many years but flattery and fustian. There it was that I found and visited the famous Galileo, grown old, a prisoner to the Inquisition, for thinking in astronomy otherwise than the Franciscan and Dominican licencers thought.

Scientists frequently cite Galileo, in defence of their freedom of enquiry. James Burke, Moderator of the September 1985 European Parliamentary Colloquium on *Europe and the Challenge of New Technologies*, has also recounted the Galileo story, but goes on to link the issues to democratic control: It is these structure-generated limitations on the freedom of action in science which set the boundaries beyond which it is unsafe to go. Within those boundaries the structure also dictates what research is to be considered socially or philosophically desirable.'[3]

But who controls the structure? J. Burke argues for 'a type of "balanced anarchy" ' in which all interests could be represented in a continuous reappraisal of the social requirements for knowledge, and the value judgements to be applied in directing the search for that knowledge. The view that this would endanger the position of the expert by imposing on his work the judgement of the layman ignores the fact that science has always been the product of social needs, consciously expressed or not. Science may well be a vital part of human endeavour, but for it to retain the privilege which it has gained over the centuries of being in some measure unaccountable, would be (argues, J. Burke) to render both sience itelf and society a disservice: 'It is time that knowledge became more accessible to those to whom it properly belongs.'[3]

Democratic Control, Popular Understanding and Confidence in the Regulator

For such reasons, increasing attention is now being devoted to the link between democratic control and popular understanding. If there is 'ignorant democracy', control without understanding, there is danger not only to science and technology, but ultimately to society itself. Such was the motivation behind the UK Royal Society report, 'The public understanding of science', produced by Sir Walter Bodmer and his committee.[4] Perceptions of risk are particularly relevant to the debate on biotechnology regulation for the safety of health and environment; as the Bodmer report puts it, in the absence of widespread understanding, we 'shy at kittens, and cuddle tigers'. There is more damage to human DNA from tobacco smoke than is ever likely to result from rDNA laboratory work: 'Tobacco causes more death and suffering among adults than any other toxic material in the environment . . . involuntary exposure to cigarette smoke causes more cancer deaths than any other pollutant.'[12]

Issues are not resolved by statistical analysis, however, but by public confidence: as Otway expresses it: 'If the public cannot evaluate the risk, they will evaluate the regulator.'[5]

The high public esteem of the US Food and Drug Administration, the low standing of the US nuclear power industry, are equally illustrative of the working of 'Otway's law'.

A significant example: the steroid hormone ban

It was therefore a significant blow to Europe's veterinary scientific community, to the various industrial interests involved, and to the Commission services concerned when they were confronted in 1985 with resolute public (and consequent Parliamentary) opposition to a draft Directive which proposed to permit the use of certain substances in animal rearing. The scientific advice was clearcut, based on many man-years of careful work, and the prudent advice of three scientific advisory committees (Veterinary, Animal Nutrition, and Food). The substances (steroid hormones) were nature-identical. But the issue was no longer scientific. It was political: the politics of public confidence, public opinion and democracy; and the same parliamentarians who a few weeks previously had applauded at the Strasbourg conference on the challenge of new technologies, now voted 117 to 10 for the hormone ban. The Commission had to give way, as Vice-President Frans Andriessen explained: 'In public opinion, this is a very delicate issue that has to be dealt with in political terms. Scientific advice is important, but it's not decisive.' The scientists were bitter: 'If you legislate in haste, you repent at leisure.' The industrialists, cynical and threatening: 'Sure we can survive without hormones. But we are a science-based company, and if things are going to be banned in Europe on non-factual grounds there's no future for us here.'[6]

But these scientific and industrial reactions quoted overlook an important aspect, on which the judgement of the politicians may well be more sound. Did the reactions reflect absence of confidence in *science* – or in one or more of the farmer, the industrialist and the regulator? Brian Wynne[7] expressed it clearly:

The Lamming Report is pretty clearcut in its conclusions about the three nature-identical hormones, and one would not be able to challenge them unless one was a specialist in the relevant research fields. Accepting their conclusions however, there are some crucial qualifications to their acceptance which are, I think, directly relevant to the question of whether, as the authors of your documents say, public opposition to the 'scientifically well-prepared' Commission proposal in favour of the hormones is irrational. Notice that (in about three separate places) the Lamming Report says that the use of the three nature-identicals is acceptable *so long as* they are used under specified conditions (point of injection, dose, waiting period afterwards, etc. etc.). From a scientific perspective these qualifications are secondary – to do with enforcement of regulation, which is not their concern, and fair enough. But to *regulators*, i.e. to the Commission, they are central. If they cannot be enforced, then as the scientists say, the use of the hormones is not acceptable. I submit, as I have said on parallel issues like the 2, 4, 5-T issue, that it is on this aspect of enforceability that the public is making its tacit judgement. It doesn't believe that the necessary conditions for their safety will be socially enforced – and who is to contradict them?! This is an empirical social judgement, on which the public is as fit to judge as anyone. (The fact that we are then blackmailed by threats that a ban isn't enforceable either, does not help to overcome this point.) So I think those who lament the irrationality, of the public as against the 'rationality' of the Commission and its experts are failing in their responsibility. The scientific experts may be correct, but it is up to the decision makers (and advisers) to pick up on the conditions which the scientists give, and examine those properly. They are too ready to reiterate what the scientists say, dropping the key conditions on the way. This is a general problem, not confined to hormones, and it has led me to argue that we ought to place a good social scientist with an empirical nose on every scientific advisory committee, to highlight the sociological questions in the risk assessment process. They do this in Canada by the way.

Thus what might at first glance appear to be a public rejection of science is more plausibly related to an issue of confidence in the adequacy of regulatory control. Nor is the example an isolated or exceptional one. As biological science continues its swift, well-funded advance, the breakthroughs will lead to further innovations. There will be more hormones, more precisely targeted molecules with highly productive effects, perfect substitutes, capable – if properly applied – of meeting every objective test of safety; but increasingly remote from popular understanding, and productive of disquiet about potential misuse.

The link between regulatory policy and public information

This close relationship between regulatory policies and public information and trust was underlined at the workshop, 'Regulating Industrial Risks', held at the Commission's Joint Research Centre, Ispra, in October 1984. The basic questions addressed were:[8] 'How can an appropriate balance be maintained between industrial progress based on technological innovation and the potential risk from these new developments to health and environmental well being? Can risk regulation processes be sufficiently effective and responsive to reconcile accelerating technological advances with growing pressures to avoid ill-effects from that progress?'

Amongst the key observations recommended by this workshop for consideration when developing specific policies, the following are particularly relevant to Community initiatives in biotechnology regulation and public information:

(3) Successful regulations cannot be based solely on scientific information, because scientific consensus appears not to be achievable and the regulatory process has to resolve social and political conflicts that extend beyond scientific considerations. Furthermore, the scientific community may contain divergent viewpoints and sometimes experts appear as advocates of a specific viewpoint.
(4) Effective regulations must not only be scientifically sound, but must be practically implementable and command the respect of the organizations, groups, and individuals affected.
(5) The most effective 'style' of regulation, in a particular context, takes account of regulatory experiences elsewhere and is adapted to the deep-rooted political and cultural conventions, as well as to the general administrative procedures in a particular country or region.
(6) Communications media (TV, radio, press) are integral elements in political processes and, therefore, inevitably play a significant role in shaping regulations, in the allocation of resources to regulatory institutions and risk research, and in influencing public support for or against regulations. Policymakers must, therefore, give adequate attention to the media.

Measuring public attitudes: in America, latent concern

The gap between 'scientific facts' and public perceptions is not easily measured, but there have been serious attempts. Jon D. Miller of Northern Illinois University in a 1984/85 study, 'The Attitudes of

Religious, Environmental and Science Policy Leaders Towards Bio-technology'[9] employed again the stratification into four categories which had been developed in other public policy contexts, a stratification of use in considering target groups and appropriate communication channels for effective public information:

1. The *decision-makers* themselves – few in number, and with the power to make binding decisions on any given set of policy matters.
2. *Policy leaders*, outside the decision-making circle, but with detailed knowledge and special interests in the field, representing constituent groups either formally or intellectually, and influencing the other groups.
3. Less expert but generally interested and aware laymen, typically leading or influencing opinion in their communities: the *attentive public*. J. D. Miller and colleagues had in earlier work estimated that approximately one in five American adults was attentive to science and technology matters in 1979, and that this proportion had doubled since the late 1950s. J. D. Miller (1984) found that by 1983 approximately a quarter of American adults were attentive to science and technology policy. (Although, in the OTA study cited below, the survey 'found that almost half of American adults – 47% – describe themselves as either very interested, very concerned, or very knowledgeable about science and technology'.)
4. Most numerous, the remainder of the population, the *non-attentive public*; generally not following all issues, but capable if interested and aroused by a particular topic of effectively exercising a veto power (e.g. on US involvement in Vietnam).

Miller's study focused on the attitudes of religious, environmental and science policy leaders, i.e. a group of type 2, and careful efforts were made to define the target and sample populations. Response rates were high: around 80%. The results were broadly reassuring for the biotechnology industry in the USA.

1. 'For most of the religious, environmental and science policy leaders included in this study, rDNA is not a major science policy issue at the present time, but there is evidence of an emerging awareness.'
2. '(They) do not feel informed about rDNA as a process or technology or about its possible benefits or risks to our society.'
3. 'All three of the leadership groups in this study displayed positive attitudes towards rDNA research and a wide array of applications. ... There was broad agreement that the current level of governmental regulation of rDNA was "about right", and there

was support for allowing the field testing of genetically-engineered materials.'

The more recent survey of US public opinion, conducted for the Congressional Office of Technology Assessment, is similarly reassuring.[13] Two-thirds of the American public claim that they understand the meaning of genetic engineering. The report summary gives highlights:

Americans say they approve of a variety of specific applications of this emerging technology. A majority believes the risks of genetic engineering have been greatly exaggerated, and 58 per cent of Americans feel that unjustified fears of genetic engineering have seriously impeded the development of valuable new drugs and therapies.

and

Public support for research in genetic engineering and biotechnology is widespread. Eighty-two per cent of Americans say such research should be continued. Forty per cent say that government funding for biological research should be increased and 43 per cent say it should remain the same. Ten per cent of Americans say they favour cutting research funds.

Less reassuring, however, are the responses to the OTA survey's question about public trust in the regulatory authorities: there is 'a potential credibility problem in government's role in biotechnology. Americans feel that Federal agencies are less believable than university scientists about statements of potential risks associated with biotechnology. Moreover, in disputes involving statements concerning potential risks, they say they are inclined to believe environmental groups over Federal agencies by a margin of 63 to 26 percent.' Consider this result in conjunction with Otway's Law: if the public, unable to evaluate the risk, evaluate the regulator, and find him wanting – what then? The lack of trust may imply a latent or potential opposition, which can be set alight by pressure group activity focused on any specific and concrete challenge, such as a heavily-publicised field release.

Public attitudes in Europe: harmonising diversity

One aim of the European Community institutions is to create – preferably by 1992 – a harmonised regulatory environment throughout the Member States. This should provide, simultaneously:

- a high standard of protection for health and environment;
- the maximum of international common standards and practices, (e.g. in line with the OECD recommendations [18]) to facilitate trade and avoid competing on safety-related aspects.

In practice, it is clear that different regulatory environments, and the different attitudes prevailing in public or government circles, do feature significantly in influencing the location of investment and of product launches. The countries benefitting from these differences will not readily accept restrictions which are not seen to be rationally based. In promoting harmonised regulatory attitudes, the Commission of the European Communities has in biotechnology sought to promote 'expert concertation' at Community level. The expert groups involved apply only the 'quality, safety, efficacy' criteria; but decisions at national or Community level are ultimately determined by democracy: 'Vox populi, vox Dei', as St Augustine remarked in another context.

To assess public attitudes in Europe, the diversity of political structures, languages and cultures demands separate measurement by country, and is reflected in the diversity of the results. Several years ago (in 1977 and 1979), the Commission of the European Communities financed systematic survey of public attitudes towards science and technology. [10, 11.]

Of the various questions posed, two have some relevance to biotechnology, and the responses indicate the wide divergence between national attitudes in the different Member States at that time; differences which are still reflected in current national attitudes to the development of regulations on biotechnology, as the extracts from the survey analysis suggest (table 24.1). It is not difficult to conceive of relationships between the attitudes measured, for example the maximum values ringed in the table, and political, legislative/regulatory, and consumer attitudes towards some aspects of biotechnology in the countries concerned.

It is well-known that *Denmark* has been the first country to legislate specifically and strictly for rDNA legislation.[16] At the other extreme, *Italy* has not instituted any national legislation or registration of rDNA work, equally reflecting the democratic judgement. It may be that public awareness of the consequences of thalassaemia and other haemoglobinopathies, relatively more prevalent in Mediterranean countries, has influenced attitudes towards favouring genetic research. But the controversy over single-cell protein in Italy has turned opinion against synthetic food; while *France* is noted for its historic, some would say, archaic, laws on food constituents.

The fact that these old statistics seem to have predictive value 10 years later, suggests deep-rooted and slow-changing attitudes; *German* unease about genetic research is second only to Denmark's, and is reflected in a critical stance towards gene technology, by no means limited to the 'Greens'.

The *Dutch* position is interesting. From a position of relative disquiet at the time of the first rDNA debate (1978–82), when the Netherlands limited research activity severely (e.g. by 10 litre containment rules), a high quality, extensive and systematic provision of public information, in schools and through the various media, has led today to a situation summarised as follows by the government's Committee on Foreign Investment:[17] 'As for recombinant DNA activities, a DNA commission reported in August 1983 to the Dutch parliament that genetic engineering is controllable and acceptable. The parliament has used this advice to decide that there will be no special law or regulation promulgated for genetic engineering. The existing regulation will be maintained, but centralized "advice" pertaining to the risks of genetic engineering will take place. The Netherlands adheres to all safety regulations. The Dutch biotechnology industry considers Dutch regulations rather strict, but acceptable. Internationally Holland plays a leading role in formulating proposals to regulate the safest application of genetically manipulated organisms.'

The Dutch experience encourages the hope that trustworthy provision of clearly presented information can diminish fears about conjectural risks. It provides a hopeful pointer for the types of action considered in the final section.

Towards Action: the Public Authorities and the Public Media

Corporate efforts, or public authorities?

All analyses of the problems discussed above lead to recommendations for more effective public information; but few have gone beyond these recommendations into specifics. One interesting insight offered by Karl Heusler is to recognise the risk – and the hostility which may be engendered – by 'information overload'; like Otway, he concludes [19] that *trust*, rather than additional information, should be the goal. Hence the manner and content of presentation matter more than the quantity.

Measurement of public attitudes is expensive, and has many pitfalls. Few of the various public information initiatives in biotechnology have been linked with systematic experimental measurements of attitudes 'before' and 'after', and such measurements are always at risk of gross perturbation by the coincidence with some newsworthy event elsewhere which is connected in public perception with the same topic. Monsanto, for example, undertook careful attitude measurement and informed the local community scrupulously in advance of their proposed field trials in Charles County, Missouri (of a *Pseudomonas* bacterium into which they had cloned the gene for BT toxin, to defend seeds against corn root worm). But the timing coincided with news of alleged unauthorised field release in California, where the injection of trees on the premises of AGS led to widespread public criticism, news coverage, and ultimately the imposition of sanctions by the Environmental Protection Agency. News coverage and political opinion in St. Louis, Missouri swung rapidly against the Monsanto trial.

Table 24.1 Biotechnology-relevant questions and responses, from the survey of attitudes to eight research areas (all figures represent % of respondents), 1979

Country / Question	EC	B	DK	D	F	IRL	I	L	N	UK
Genetic Research										
Worthwhile	33	38	13	22	29	41	49	37	36	32
Of no particular interest	19	20	10	16	22	20	19	31	17	21
Unacceptable risks	35	22	61	45	37	22	22	18	41	36
Don't know	13	20	16	17	12	17	10	14	6	11
	100	100	100	100	100	100	100	100	100	100
Development of Research on Synthetic Food										
Worthwhile	23	16	13	34	10	23	11	25	23	34
Of no particular interest	21	26	21	16	20	29	20	39	30	25
Unacceptable risks	49	44	50	36	66	38	65	25	42	36
Don't know	7	14	16	14	4	10	4	11	5	5
	100	100	100	100	100	100	100	100	100	100

The scale of the potential problem is thus almost beyond the scope of any one company, even the largest; and the credibility of company efforts is always diminished by obvious self-interest. Yet there is a clear public and general interest in ensuring that there is a sufficiently widespread understanding and trust to avoid or to reverse the partial

breakdown of confidence in science and technology. Such understanding, in public and political circles, is equally necessary to ensure
a rational approach to regulation.

It is therefore incumbent upon the public authorities, at national and
Community level, to promote greater understanding and awareness.
Public trust cannot be bought, but as K. Heusler emphasises, may
gradually be *earned* through trustworthy behaviour, in a climate
where general awareness has been systematically built and maintained by intelligently conceived provision of information, as in the
Dutch example.

The need for a systematic communication plan

The US needs were clearly enunciated at the National Academy 1985
workshop:[1]

The best way to go about educating the public, according to William
Ruckelshaus, former head of the Environmental Protection Agency,
would be through an 'elaborate, comprehensive, and sophisticated
communication plan'. Such a plan should recognize that different
audiences require different messages. It could focus first on those who
will most directly affect biotechnology – Congress, the regulatory
agencies, industry, the press, environmentalists, academics. It could
also take advantage of specific events – an important technical development, a particular experiment, regulatory approval of a product –
to further public understanding of the field. Taken together, such
efforts could begin to close the gap that has traditionally existed in the
United States between scientific developments and public understanding. Concludes Ruckelshaus, 'We need to do a much better job,
not just in this area but across the board, as we try to grapple with the
complexity involved in public participation in decisions of enormous
scientific uncertainty.'

Such a structured approach is necessary, but needs to be complemented by a more direct approach to the public; which for effective
impact has to be visual. One has only to consider the impact of cartoons
– unpleasant messages, critical of biotechnology (or of the behaviour of
scientists), and inaccurate; but what is the effective response in a free
society? Who, moreover, is to respond 'in the public interest'?

Publications and conferences are numerous and readily organised,
but they seem limited in influence as compared with the scale and
impact of radio and television, with their audiences of millions. The
quality of impact of television may be transitory, but it can serve to
stimulate interest in further study, by videotape and/or book. Such

'multi-media' approaches are used increasingly by both educators and broadcasters. The Open University in Britain has produced an excellent 6-programme series of videotapes on biotechnology, supported by documentary materials. Teleac in the Netherlands have also broadcast a balanced biotechnology series, accompanied by a well-written popular book, and ZDF in Germany made a 13-part series. But many of these reach less than 5% of the peak audiences; the views – and the viewing – of the other 95% also matter, if one is to go beyond 'preaching to the converted'.

The impact of television, and the need to reach a wider audience

The impact of television on public perceptions of science and scientists was systematically investigated by G. Gerbner.[14] Leaving aside the science programmes, he focused upon the portrayal of scientists within general entertainment programmes, and has shown the preponderance of a fairly negative stereotype, as compared with other professionals. He then through survey measurements and careful stratification of his samples, by weight of television viewing and by attitude to science and technology, established that there is a significant negative correlation between hours of television viewed and favourable attitude to science; and this correlation occurs both within groups generally favourable to science and within groups unfavourable. To overcome such stereotyping will not be the work of any single initiative, but such research confirms both the significance of the broadcast media, and the need for science to break out of the confines of the 'popular science' slot and trespass into general entertainment. This is not an easy thing to do, given the fixed allocation of broadcasting time to each category of programme, and the intense competition for time between the various categories. Biotechnology knows already the problems of a subject which spreads horizontally across the interests of different university departments, and of several Ministries in any government. Access to the public broadcast media is also jealously (and rightly) controlled by various public authorities. While often subject to some accusations of political bias, the authorities are particularly suspicious of industrial sponsorship.

If the public authorities are to accept any co-finance in sponsoring the production of materials for broadcasting, there needs therefore to be a strong guarantor of the objectivity and public interest in the editorial control of such productions. The Commission of the European Communities might be able to fulfil such a role, and has already offered support to one such initiative (see Box 24.7). Its resources, however, are extremely limited. Its role can only be a catalytic one, mobilising resources found elsewhere, be it from programme producers, broadcast authorities or the biotechnology industry; but the right catalyst can greatly amplify the effectiveness of these other resources. The Commission services have for several years discussed with educators and television producers, from various Member States, the possibility of a European series on biotechnology. In spite of some promising discussions, all productions to date have remained essentilly national in scope or character.

Box 24.7: The 'Biot 2000' multi-media series, and the Bio-Zon story

> The Commission was approached in 1984 by Aegis SA, a Luxembourg-based firm, with a proposal for 'Biot 2000': a multi-media, 'docu-drama' series in biotechnology, using actors and a fictional story to 'package' the essentially documentary material in a manner likely to make it of interest to a larger audience, and of greater realism and impact to the viewer.
>
> Sonny, whose family are seed merchants near Leiden, goes to Cambridge for his doctoral research in molecular biology; also studying there is Niamh, an Irish girl, daughter of 'Frank the Fixer', the famous sugar baron. From her studies of carbohydrate chemistry, she dreams of a sugar-based biodegradable plastic, to bring employment to her people in the West of Ireland. They meet at the student debate on *in vitro* fertilisation, in November 1985:[15] fall in love, start a company, 'Bio-Zon', with venture capital, and will proceed to encounter and overcome a series of obstacles at once dramatic and romantic, regulatory and technical, economic and personal . . .
>
> Co-finance for preparing the series has been provided by the Commission, industry and a Member State government.

The forthcoming huge expansion of satellite broadcasting will multiply existing capacity for TV-broadcast-hours by a factor of five. This may lead to the growth of 'narrowcasting', and could offer a significant opportunity for the expansion of Community-backed programmes – including those relating to science and technology policy. We are open to suggestions.

REFERENCES

1 Olson, Steve: Biotechnology: An Industry Comes of Age (National Academy Press, 1986, based on an academy conference on genetic engineering, 27–28 February 1985).

2 Dyson, Freeman: *Disturbing the Universe* (Harper and Row, 1979; Pan Books, 1981).

3 Burke, James: *The Day the Universe Changed* (BBC Publications, 1985).

4 Royal Society working group, under the chairmanship of Bodmer, Walter: The public understanding of science (The Royal Society, London, 1985).

5 Draft paper, The social dimensions of biotechnology, July 1982; by FAST unit (Forecasting and Assessment in Science and Technology), Commission of the European Communities.

6 Quotations are from a report in the *Financial Times*, 24.01.86, by A. Gowers and I. Dawnay: Another shot of politics for the beef farmer.

7 Private communication to M. F. Cantley, 10.2.1986.

8 Maini, J. S., Peltu, M., and Otway, H.: Regulating Industrial Risks; Executive Summary of a workshop (IIASA (International Institute for Applied Systems Analysis), Laxenburg, Austria, November 1985).

9 Miller, Jon D.: The Attitudes of Religious, Environmental and Science Policy Leaders Towards Biotechnology (Public Opinion Laboratory, Northern Illinois University, April 1985).

10 Commission of the European Communities: Science and European Public Opinion; XII/922/77, 1977.

11 Commission of the European Communities: The European Public's Attitudes to Scientific and Technological Development; XII/201/79, 1979.

12 Brown, Lester R., et al.: State of the World, 1986 (A Worldwatch Institute report, W. W. Norton & Co., 1986).

13 US Congressional Office of Technology Assessment, May 1987: New Developments in Biotechnology. Background Paper 2: Public Perceptions of Biotechnology.

14 Gerbner, George: Science on Television: How It Affects Public Conceptions, *Issues in Science and Technology*, Spring 1987.

15 The edited film of the debate, 'The Risks of Ignorance', presented by James Burke, is distributed by Aegis S. A., Grande Rue, 62, L-1160 Luxembourg.

16 Danish Government: The Environment and Gene Technology Act; Act No. 288 of 4 June 1986 (Authorised English translation).

17 Biotechnology in the Netherlands: A Strong Foundation – A Shining Future (issued by Ministry of Economic Affairs, Commission for Foreign Investment in the Netherlands, 1986).

18 *Recombinant DNA Safety Considerations* (OECD, Paris, 1986).

19 Heusler, K. (1987): The Commercialisation of Government and University Research: Issues in the Public Acceptance of Biotechnology, paper given at Canada–OECD joint workshop, Toronto, 7–10 April.

20 Reported in Zilinskas, R., and Zimmerman, B. (1987): *The Gene-Splicing Wars* (Macmillan). Reviewed by Eric Ashby in *New Scientist*, 25 June 1987.

25

The Genetics and Politics of Frost Control

Sheldon Krimsky

Most people understand that when the temperature drops below freezing their tomato plants will die unless some action is taken to protect them. But who can imagine spraying a microbe on the plants to prevent the onset of ice crystals? What does the formation of ice have to do with bacteria? What might have been a plausible story line for a science fiction novel has proved to be a serious research programme in plant genetics.

Scientists developed a microbe that they believed could reduce the temperature at which ice crystals would form on vegetation. It was the first genetically modified bacteria sanctioned by a federal agency for release into the environment and it sounded off a national debate in the United States. The development of this microbes, called ice minus, illustrates the interplay of science and human imagination. First, it was the scientific imagination that established the causal connections between bacteria and frost. And second, it was the public imagination that brought world attention to the consequences of tampering with the earth's microflora.

The discovery of ice minus dramatizes to us that the natural world operates in unexpected and often counter-intuitive ways. Nature scoffs at a Cartesian mechanism. Physical, chemical and biological states relate through complex webs of interdependent and interpenetrating processes. Boundaries between disciplines are what we impose. They are not nature's design. Perhaps because of its mystery and open-endedness, science has become a vast reservoir for the literary imagination. If ice minus were not a product of the laboratory, it might well have appeared in the new genre of biofiction.

Ice Minus Bacteria

The story of ice minus begins around 1974 when scientists discovered that certain soil bacteria, such as *Pseudomonas syringae* and *Pseudomonas fluorescens*, efficiently catalysed ice formation at temperatures several degrees below zero.[1] Water does not always freeze when it is cooled below zero. It can remain in a liquid (supercooled) state when ice nucleating particles are not present. However, when ice nuclei are present in the form of inert particles of matter or bacteria, then freezing takes place at or very near zero degrees.

Plants are damaged by extracellular ice that forms within their tissues. Ice crystals mechanically disrupt biochemical processes. If ice nuclei are not present when the plant is exposed to moisture, water will supercool. The supercooled water is capable of remaining in a liquid state at temperatures that approach $-40°C$. Thus, frost injury to plants requires low temperatures, moisture, and ice nuclei.

It has been estimated that about 11 per cent of the land mass of the continental United States is allocated to the cultivation of frost sensitive plants. Losses to crops from frost injury is reputed to be about $1 billion annually.[2]

The agricultural sector has experimented with various techniques to minimize the impact of frost on important crops, particularly in those regions where freezing temperatures, within the subzero range of ten degrees, occur for short periods during critical stages of the growing season.

Traditional methods of protecting crops from frost damage are premised on keeping the air temperature above freezing in the vicinity of the harvest. Farmers have resorted to burning smudge pots, watering the soil, utilizing wind machines to keep warmer air circulating, and applying foam-like insulation around the plants. Reductions in frost injury under field conditions have also been achieved through the application of bactericides that are used to destroy ice nucleation microbes.[3]

The discovery of gene splicing provided a novel approach to protecting plants from frost damage. It involved replacing the indigenous ice-nucleation bacteria with strains that have had the ice nucleation genes excised.

Pseudomonas syringae is a ubiquitous organism that resides on the leaf surfaces of plants. The role of this bacteria in the agricultural ecosystem has been studied since 1902 when the organism was first isolated.[4] But it wasn't until the early 1970s that scientists learned this organism was unusually efficient at seeding ice crystals from subfreezing water. Most organic and inorganic substances that are ice-nucleating agents (INA) are active below temperatures of −10°C. *P. syringae*, however, catalyses the formation of ice crystals at temperatures between −1.5°C and −5°C. This characteristic of the microbe attracted special interest from plant pathologists who study frost damage to agricultural crops.[5]

The mechanism of ice nucleation for *P. syringae* has been traced to the bacterial synthesis of a protein. The release of this protein provides a site for crystallization of moisture as the temperature drops below zero.[6] Strains of *P. syringae* that lack the protein, fail to exhibit ice nucleating activity at the higher subzero temperatures.

About half the strains of *P. syringae* that have been tested are ice-nucleating agents. Also, through chemical mutagenesis, scientists have been able to transform an INA *P. syringae* into a non-INA type. One method devised to reduce the crop damage during subfreezing temperatures is to colonize the seedlings of a plant with non-INA bacteria. If the antagonistic bacteria take over the niche of the INA type, frost formation could be delayed, at least until the temperature drops below −10°C. This process requires displacing about 0.1% to 10% of the bacteria on leaf surfaces since that represents the percentage of ice nucleating strains.[7]

Steven Lindow of UC-Berkeley did some of the path-breaking work in studying the relationships between INA bacteria and frost formation on plants. He discovered how bacteria can limit the damaging effects of moisture on frost-sensitive plants at temperatures above −5°C. He also demonstrated that by reducing the numbers or ice-nucleation activity of the INA bacteria, frost injury to field-grown plants declined.[8]

Although the number of INA species is relatively small, the prevalence of the bacteria in the environment is quite high, particularly on certain food crops. *P. syringae* are widely distributed in nature and believed to be one of the largest sources of bacterial ice nuclei that are active at high temperatures (near zero degrees). By reducing the population density of INA bacteria, scientists believed they had an effective and environmentally safe method of protecting crops from sudden frosts. With an estimated billion dollar annual loss to agriculture from frost damage, there was a strong financial incentive to develop the new technology.

Commercializing a Deletion Mutant

The Oakland, California biotechnology firm Advanced Genetic Sciences (AGS) initiated an R&D programme to commercialize a microbial frost inhibitor. Of the several possible approaches for developing antagonistic bacteria to displace INA *P. syringae*, the chosen path was to modify the microbe by removing the ice nucleating gene, a so-called deletion mutant (denoted as INA⁻). Initially, AGS accomplished this by exposing INA⁺ (the strain with the ice-nucleating gene) to chemical mutagens. The chemical mutant was produced in the laboratory and field-tested. Special permits were not required since the product was not a genetically engineered microbe. AGS reported that the INA⁻ produced by mutagenesis slowed down the effects of frost damage. But the company believed it could improve the effectiveness of the product.

AGS then deleted the ice-nucleation gene from *P. syringae* using recombinant DNA techniques and this came under greater public scrutiny. In the spring of 1984, the company submitted a proposal to the National Institutes of Health's (NIH) Recombinant DNA Advisory Committee for field-testing the genetically engineered *Pseudomonas* strain. The proposal enunciated that the rDNA deletion mutants of *P. syringae* (INA⁺) are more stable, better defined genetically, and probably more fit than INA⁻ strain resulting from chemical mutagenesis. In its submission to the NIH committee, the company disclosed that it had removed about 1000 base pairs of DNA sequences and that the changes were non-revertable.[9]

AGS maintained that the rDNA mutant strains were, from the viewpoint of ecological risk, no less safe than the strains isolated in nature or produced by chemical mutagenesis. Furthermore, the company contended that the rDNA-derived strain was superior to its chemically-derived counterpart for inhibiting frost damage. Thus, the natural and genetically-modified strains were considered alike in all respects with regard to risk but unlike in some respects that pertain to ecological issues. Meanwhile, industry spokesmen and some government officials pointed to the hypocrisy of regulating a process and not a product. If the product is potentially hazardous, they argued, it should not matter whether it was derived by chemical mutagenesis, gene splicing, or cultured from nature.

Advanced Genetic Sciences submitted its field-test proposal to the RAC in March 1984. By October, NIH notified the company that its field experiment would be approved pending further data and a monitoring plan. However, since the U.S. Environmental Protection Agency was advancing a new regulatory regime for biotechnology, in November AGS inactivated its application to NIH and submitted a notification of a field test to EPA's Office for Pesticides and Toxic Substances. After a review period of 13 months, AGS met EPA's requirements for its Experimental Use Permit (EUP).

The permit issued to AGS was for field-testing ice minus at a site undisclosed to the public in Monterey County, California. A Christmas tree farmer who read about the unprecedented test gathered a few signatures and registered a complaint to the county Board of Supervisors. They cited the secrecy of the site and the unknown risks associated with the organism. Within several weeks after the letter was received, the Monterey Board of Supervisors declared a moratorium on the field tests and held hearings that attracted national media coverage.

The action taken by the supervisors brought international attention. A telegram to the Board, which was highly critical of the test, came from the Green Party members of the German Parliament who warned of 'grave dangers to the environment'.

Monterey citizenry became more incensed over the field experiment when it was disclosed that the site selected was adjacent to local residences. To make matters worse, a disgruntled employee of AGS blew the whistle on the company by disclosing that ice minus had been injected into trees located on the roof of the facility without EPA approval. A subsequent EPA investigation charged that AGS had falsified information on the EUP application (a charge latter dropped) and failed to perform tests according to agency requirements.

The accumulation of allegations and the company's poor record in communicating with local authorities, severely tarnished AGS's credibility within Monterey County. EPA withdrew the EUP for the field test. And nearly a year and a half after the first test was planned, the Monterey Board of Supervisors passed an ordinance regulating the release of GEOs based upon the county's authority over land use.

Following a parallel track, UC-Berkeley scientists Lindow and Panopoulos received EPA approval to field-test their ice minus strain at the state agricultural field station in the northern California community of Tulelake. Two women from the town began an organizing effort to stop the test after they had learned about ice minus from a Monterey citizen activist who addressed the Tulelake community. The women built a small grass-roots committee on the model of the Monterey activists, calling themselves the Concerned Citizens of Tulelake (CCT). They notified local governments in contiguous communities, drew up a petition, lobbied various groups, sent out a stream of irate and informed letters to federal, state, and local authorities, and they organized their own local forum on the release of 'mutant bacteria'.[10] The CCT also played a significant role in shaping local news coverage about ice minus.

By June 1986, the CCT was successful in convincing both the Modoc County and the Siskiyou County Boards of Supervisors to pass resolutions opposing the ice minus experiment. Undaunted, UC-Berkeley continued its plans to test ice minus, threatening legal action against local or county opposition.

The law suit came instead from the CCT with the help of Jeremy Rifkin and the Californians for Responsible Toxics Management (CRTM). The court issued a restraining order blocking UC-Berkeley from undertaking the field test. Within several weeks, the parties to the suit reached a negotiated settlement under which the University of California would prepare an environmental impact review (EIR) and the citizens' groups would withdraw their petition.

While UC was scoping out its EIR, AGS decided not to pursue a field test in Monterey County but proceeded to enter negotiations with other California communities. This time the company took a proactive role in communicating directly with farmers and civic organizations. It declared that it would not pursue a site in secrecy or under conditions where community approval was not forthcoming.

AGS finally met EPA requirements for the field test in the fall of 1986. By February of the following year, AGS had endorsements for its tests from two county boards of supervisors and approval from the EPA of three sites. An environmental coalition including Earth First, the Berkeley Greens and FET failed in a final effort to gain a court injunction against the test planned for Brentwood, California. The first federally sanctioned deliberate release of a genetically modified bacteria took place amidst throngs of reporters on 24 April 1987. Five days later the UC-Berkeley test was carried out without the attention of the national media.

From the time the two Berkeley scientists submitted the field-test proposal to NIH, it took nearly five years for the legal, political, and regulatory requirements to be satisfied. Because it was first, ice minus carried a lot of weight. It symbolized a new era for biotechnology, the transition from the laboratory to the field.

Despite the attention it received, there was broad consensus among scientists that ice minus was probably among the safest agricultural products one could create with the new technology. The organism had genes deleted and not added and there was no evidence that the product could outcompete its parental strain. As a public controversy, ice minus was history just as soon as the field was sprayed by the moon-suited scientist. The precedent had been set. The first major barrier to releasing engineered organisms was overcome.

Two issues remained. Would ice minus become a viable product? And, can the scientific community reach some consensus on evaluating the risks of other deliberate releases? Ice minus was the catalyst that fostered a dialogue between two communities in science, which, until this time had been well insulated from one another. The future of environmental release in the U.S. will depend upon whether ecologists and molecular geneticists can reach a consensus on risks and whether they can communicate that consensus to the general public.

NOTES

[1] L. R. Maki, E. L. Galyon, M. Chang-Chien, and D. R. Caldwell (1974). Ice nucleation induced by *Pseudomonas syringae*, *Appl. Microbiology*, **28**, 456.

[2] S. E. Lindow (1985). Ecology of *Pseudomonas syringae* relevant to the field use of ice-deletion mutants constructed in vitro for plant frost control. In: *Engineered Organism in the Environment: Scientific Issues.* H. O. Halvorson, D. Pramer, and M. Rogul (eds) (Washington, D.C.: American Society for Microbiology, 1985), pp. 23–35.

[3] S. E. Lindow (1983). Methods of Preventing Frost Injury Caused by Epiphytic Ice-nucleation-active Bacteria. *Plant Disease*, **67**(3), 327–333 (March 1983).

[4] Susan S. Hirano and Christen D. Upper (1985). Ecology and Physiology of *Pseudomonas syringae*. *Bio/Technology*, **3**, 1073–1078.

[5] S. E. Lindow (1982). Population Dynamics of Epiphytic Ice Nucleation Active Bacteria on Frost Sensitive Plants and Frost Control by Means of Antagonistic Bacteria. *Plant Cold Hardness and Freezing Stress.* A. Sakai and P. H. Li (eds) (New York: Academic Press, 1982), pp. 395–416.

[6] Hirano and Upper, 1985.

[7] Lindow, 1983.

[8] Lindow, 1982.

[9] Advanced Genetic Sciences, proposal to field test genetically engineered *Pseudomonas* strains containing artificially introduced deletions in ice-nucleation genes, 22 March 1984, submitted to the National Institutes of Health Recombinant DNA Advisory Committee.

[10] Sheldon Krimsky and Alonzo Plough, *Environmental Hazards: Communicating Risks as a Social Process.* See chapter 3, The Release of Genetically Engineered Organisms into the Environment: The Case of Ice Minus, (Dover, MA: Auburn House Press, 1988). pp. 75–129. Also, Sheldon Krimsky, *UDSAETNING AF GENSPLEJSEDE ORGANISMER I MILJØET* 'IS MINUS-SAGEN' GPN DEBAT 3. NOAH. Tema: Sagen om is-minus Dakteriet.

26

Cuba's Entry into New Biotechnology: A Case Study on Diffusion

Martin Fransman

The Cuban case illustrates dramatically what can be achieved when a firm commitment is made to the development of biotechnology capabilities and their application to a wide range of areas in accordance with the country's economic priorities. In this section the Cuban case is examined in greater detail, paying particular attention to the way in which this country entered the field of new biotechnology, the areas in which new biotechnology has been applied, and the institutional changes that have been brought about in order to facilitate the development of biotechnology. Finally, based on this case study, conclusions will be drawn regarding the lessons for other developing countries.

1. The General Approach

In terms of Cuba's scientific and technological development the crucial watershed occurred after the Cuban Revolution in 1959. Until this time Cuba depended primarily on its agricultural activities, which excluded sophisticated processing and research and development, and on tourism. In this way the foreign exchange was earned which financed imports of manufactured products, largely from the United States. In the period following the Cuban Revolution a new set of priorities was established. Most important from the point of view of the development of the biological sciences in general, and biotechnology in particular, was the emphasis that was given to the role of science and to the development of the national health service. Frequent reference is made by Cuban scientists to the conviction prevailing at that time that the future development of Cuba was inextricably bound up with the future development of science in the country. It was this conviction that inspired a rapid growth in the school and higher education system. At the same time, an important result of the revolution was the expansion and extended delivery of medical services to all sections of the population. This meant that within a short period of time Cuba was able to develop a relatively sophisticated medical system which included training and research facilities in universities and other national institutions. It was this medical system which was later responsible for Cuba's rapid and successful entry into new biotechnology.

However, new areas of science and research do not emerge automatically; their emergence depends on new groups of scientists and researchers, committed to the new fields of study and devoted to the institutional changes that are required to realise the new scientific research. From this point of view it is significant to observe that the new institutions which evolved in Cuba to develop the biological sciences and biotechnology emerged in a pluralistic rather than a linear way.

At the apex of Cuba's scientific planning establishment is the Cuban Academy of Sciences, which was originally established in 1861 but which was substantially restructured after the revolution. The Academy contains the Superior Scientific Council which consists of about 77 distinguished scientists elected from the Academy's various institutes, from the Ministry of Higher Education, and from industry. The Academy also contains a number of other smaller but influential advisory groups. However, it is significant that the Academy does not totally dominate or control the scientific establishment. For example, only about 10 per cent of the total number of Cuban scientists and engineers work in Academy institutes.

The Ministry of Higher Education, with some degree of autonomy from the Academy, has also played an important role in the establishment of scientific institutions. From the point of view of the development of Cuban biotechnology, an important example is the establishment of the National Centre for Scientific Research (CENIC), which was the major biomedical and chemical research centre and was set up in 1965 in order to stimulate research in new areas. CENIC has a staff of approximately 1,000 and is divided into four main divisions: biomedicine, chemistry, bioengineering, and electronics. CENIC has played a significant role in research and in training scientists who subsequently have become involved in other spin-off institutes.

An important example is the Centre for Biological Research (CIB) which was established in January 1982. The establishment of CIB is of particular interest as a result of its innovative and unbureaucratic origins. In 1981 a 'Biological Front' was established essentially outside the existing bureaucratic framework. The Front consists of scientists and policy makers with an interest in extending and developing biological research in various directions. It served to co-ordinate and articulate the interests of those in the different ministries and institutes who wished to strengthen Cuban involvement in biotechnology. While the leaders of the existing scientific establishment were closely involved with the activities of the Biological Front, the Front was set up as a high-level policy-making body with relative autonomy from the Academy and the various Ministries involved in the biological sciences and their areas of application. From this position the Front supervised the establishment of CIB and later the Centre for Genetic Engineering and Biotechnology, CIGB. By helping to give birth to CIB and CIGB the Biological Front served to increase pluralism in the Cuban scientific system. While biotechnology could be developed in existing institutions, such as those under the control of the Academy of Sciences and in CENIC, this new set of technologies could also be advanced through new institutions such as CIB and CIGB.

CIB began with a staff of six researchers in a small laboratory. Its major initial mission was the production of interferon for use as an anti-viral agent. In part the interest in interferon resulted from the outbreak in late 1980 of dengue hemorrhagic fever, which affected approximately 300,000 people and resulted in 158 deaths. However, in addition to this pragmatic goal, CIB also aimed to use interferon as a 'model' for the development of a wider range of capabilities and assets. In other words, interferon would be used as a springboard for the development of a Biotechnology-Creating System with expertise in the areas of genetic engineering and bioprocessing. CIB grew rapidly and by 1986 was divided into four laboratories: genetic engineering, immunology, chemistry, and fermentation. In addition to the production of interferon, CIB produces its own restriction enzymes and its research also involves the synthesis of oligonucleotides, the cloning and expression of a number of other genes, and the production of monoclonal antibodies for diagnostic purposes. Although recombinant DNA research was also done in a number of other institutes, notably CENIC and to a lesser extent the Cuban Institute for Research on Sugarcane Derivatives (ICIDCA) which was established in 1963, CIB became in the early 1980s the major location in Cuba for the development of capabilities in new biotechnology.

When CIB opened in January 1982 it began to produce human leukocyte alpha interferon using a method (which did not involve genetic engineering) developed by Kari Cantell of the Central Public

Health Laboratory in Helsinki. Cantell gave assistance by transferring his method to CIB and was surprised at the speed with which the Cubans mastered the method. Having mastered this conventional method for producing interferon, CIB embarked on recombinant DNA-based techniques for producing various kinds of interferon. In this latter task a central role was played by scientists such as Dr Luis Herrera, who was Vice-Director of CIB. Herrera's background is particularly interesting because it illustrates personally the way in which Cuba was able to enter the field of new biotechnology. In 1969 Herrera studied molecular genetics (working on yeast) at Orsay University in Paris. The following year he took up a post as researcher at CENIC where he started a laboratory dealing with the genetics of yeast. Yeast was of interest in Cuba because it was used in order to convert sugarcane derivatives into single cell proteins which were used as animal feed, thus substituting for impoted soya feeds, the Cuban climate not being suitable for the growing of soya. Research on yeast was partly aimed at improving yeast strains in order to increase the nutritional value of the single cell proteins by eliminating some of the undesirable nucleic acids. Under the auspices of ICIDCA there were in total 10 plants, each producing 12,000 tons per annum of single cell protein for animal consumption. In developing their work researchers in this laboratory became interested in new biotechnology. In 1979 Herrera returned to France to study molecular biology and genetic engineering. With the formation of the Biological Front and the establishment of CIB in 1982 he joined this institute as its Vice-Director. In 1983 he once again went to France where he spent time at the Pasteur Institute. Representing a new breed of young, post-revolution scientists who were quickly able to master the latest international research techniques, he has since established an international reputation for his research in new biotechnology. Although, in the case of Dr Herrera, entry into new biotechnology involved access to European institutes, Cuban biotechnology and CIB in particular have also benefited from Soviet science. A notable example is the group of chemists working in CIB and mostly trained in the USSR. With a strong background in organic chemistry some of these scientists moved on to the synthesis of oligonucleotides and the synthesis of DNA. Other groups in CIB are involved in immunology, including immunochemistry and protein purification, and fermentation.

There is widespread agreement that the Cuban mastery of new biotechnology has been impressive. One example is the conclusion arrived at by a team of UNIDO experts appointed to find a Third World location for the new International Centre for Genetic Engineering and Biotechnology. This team visited the major Third World countries involved in biotechnology and concluded that the Cuban biotechnology programme was one of the best they had seen. Other examples are the assessments made by distinguished foreign visitors to Cuba. While acknowledging that the Cubans are not attempting to do world frontier basic research, many of these visitors have been impressed with the level of achievement of Cuban biotechnologists.

2. Interferon as a 'Model'

Some further comments are in order on the use by the CIB of interferon as a 'model' for the development of new biotechnology capabilities.

The first point to be made is that the development of core scientific capabilities in the area of new biotechnology in CIB drew on the *already well developed science base* that existed in Cuba by the time the CIB was set up in 1982. Mention was made in the last section, for example, of the earlier research done in CENIC on the molecular genetics of yeast. In entering new biotechnology, therefore, Cuba was not starting *ab initio*. In this way, Cuban entry into new biotechnology was facilitated by a pre-existing stock of substantial scientific capabilities. Clearly, many developing countries are not in as fortunate a position.

The second point is that interferon was an appropriate choice for Cuba largely as a result of the country's well-developed health sector. This meant that the development of interferon with the use of genetic engineering techniques was not simply a 'pure' research activity, but was an example of scientific work being closely linked to the production of useful output, namely the delivery of medical services, a high priority in post-revolutionary Cuba. This link established a unity between 'science push' and 'demand pull' determinants of technical change, which in turn ensured that this part of the science system was not 'alienated' from the needs of the rest of the socioeconomy. Interestingly, interferon has also been used as a 'model' by many Japanese companies entering the field of new biotechnology. In their case, however, the need determined from the corporation's point of view was for a way of acquiring new biotechnology capabilities at the same time as producing a commercialisable product. Interferon, it was believed, provided an example of one of the first new biotechnology-based commercial products. For other developing countries, however, a more appropriate 'road' for the development of new biotechnology may exist, depending on the circumstances and priorities of the country. For Brazil, for example, the ethanol from sugar project may have provided an appropriate road. In other Latin American countries the development of mineral-leaching bacteria for the purposes of mineral extraction may provide an appropriate way of entering new biotechnology.

Thirdly, the possibility of using interferon as a 'model' for the development of other applications and products illustrates the pervasiveness of new biotechnology. This point is further supported in the Cuban case by the history of the Centre for Genetic Engineering and Biotechnology (CIGB).

3. Realising Economics of Scope: the CIGB and the Pervasive Applicability of New Biotechnology

Encouraged by the success of CIB in developing new biotechnology capabilities and impressed with the potential of this set of technologies, the Biological Front recommended the establishment of a new and larger institute which would carry on and extend the work of CIB. Accordingly, on 1 June 1986 the Centre for Genetic Engineering and Biotechnology (CIGB) was established on a new site near CIB.

CIGB was structured in terms of the following five groups, each dealing with a specific problem area:

- Proteins and hormones. The aim of this group is the production of proteins using recombinant DNA techniques for applications in the areas of human medicine and veterinary science. This group continues the work done in CIB on the chemical synthesis of oligonucleotides and DNA.

- Vaccines and medical diagnosis. The aim of this group is to develop vaccines against diseases prevalent in Cuba and other tropical and subtropical areas through the cloning of the surface proteins of viruses, parasites, or bacteria. The group also works on developing monoclonal and polyclonal antibodies and DNA probes for the purpose of detecting and diagnosing various illnesses.

- Energy and biomass. The research of this group involves the transformation of various kinds of biomass via the use of chemical methods and enzymes. For example, research is done on yeasts and fungi which transform the sugar by-products molasses and bagasse into proteins for animal consumption. A new strain of the yeast candida has been developed which increases the production of an amino acid important for both human and animal nutrition. In this way CIGB will extend research in this area done in ICIDCA and CENIC.

- Plant breeding and engineering. This group does research on improved plant varieties using genetic engineering and other biotechnologies such as cell culture. Nitrogen fixation is one area singled out for study.

- The genetics of mammalian eukaryotic cells. This group uses the cells of higher organisms for the cloning of genes and the production of proteins.

Thus, by using interferon as a 'model', first CIB and then CIGB have been able to develop core scientific capabilities in the area of new biotechnology and apply them to a wide range of areas consistent with Cuban development priorities. However, the research of CIGB has also been defined to include an emphasis on complementary capabilities, namely downstream bioprocessing. This has been done by making provision for a pilot bioprocessing plant in CIGB.

4. The Importance of Downstream Bioprocessing

However, the development of an effective biotechnology-creating system involves more than the mastery of the core scientific capabilities. One feature of such a system is the possession of the necessary downstream bioprocessing capabilities. In order to develop the latter kinds of capabilities CIGB has established a pilot plant.

Two groups work with this plant, the one specialising in the fermentation process and doing research on the optimisation of productivity, and the other working on questions of purification. Both of these groups are involved with the difficulties that are confronted in scaling up the bioprocessing with the use of larger bioreactors. A major problem confronted by both groups is that there is little experience in Cuba regarding bioprocessing and scale-up. Furthermore, unlike in the case of many of the core scientific capabilities, where research is done in universities and where the results are usually made public, a good deal of research on bioprocessing is done in private companies with the findings kept commercially secret. Bioprocessing, requiring sophisticated engineering skills and specialised inputs, frequently constitutes more of a constraint in developing countries than the mastery of the core scientific capabilities. The same point was made to the present author by senior officials involved in the planning of biotechnology in the People's Republic of China during a visit in 1987.

In the Chinese case, in strong contrast to the Cuban example, the core scientific capabilities were rapidly acquired largely as a result of scientific interchanges with the United States. However, major constraints exist in China on the downstream bioprocessing side which depends on the capabilities of Chinese industrial and engineering enterprises.

5. Conclusion

The Cuban example is particularly illuminating regarding the way in which the core scientific capabilities needed for biotechnology-based industry may be developed. In Cuba this involved an evolutionary process whereby specialistic scientific institutions were created and adapted to the development and application of the *science base* on which biotechnology rests.

As soon became apparent, however, the creation of scientific institutions and the strengthening of the science base were the easiest parts of the job. The reason was that policy-makers in biotechnology had ready control over the mechanisms that were needed for the creation of these institutions and for fortifying the science base. Far more complicted, unfortunately, was the task of generating the *industrial and technological base* essential for applying and commercialising biotechnology. Activities such as the scale-up of bioprocesses first researched under laboratory conditions, and the turning of these processes into commercially viable biotechnology-based products, require a different set of competencies.

While these competencies complement those found in the science base, they are fundamentally different from the latter. Furthermore, they are less subject to influence by the policy-makers. It takes much longer for these competencies to be accumulated in sufficient quantity and quality. Moreover, in countries such as Cuba and China, one of the unintended consequences of the planned economy regime was an industrial structure that was not particularly responsive to the demands emanating from new industries based on new technologies such as biotechnology. For example, the engineering industries required to provide the capital goods and equipment needed for bioprocessing proved themselves to be poorly suited for the task.

While other developing countries will not necessarily confront the problems of centrally planned industrial structures, underdevelopment, almost by definition, poses industrial and skill difficulties that impede attempts to diffuse biotechnology-based competencies. These kinds of problems will obviously have to be carefully tackled (as is discussed in more detail in the references mentioned in connection with this chapter).

Nevertheless, on the optimistic side, the Cuban case illustrates just how much progress can be made in the diffusion particularly of the necessary bioscience that must accompany biotechnology. The Cuban experience also shows how important is the commitment of the political authorities who are needed to create the required institutions and allocate the funds. As shown in this chapter, the Cubans were particularly lucky to have leaders (which included Fidel Castro) who had made a strong commitment to biotechnology, seeing it (rightly or wrongly) as a major tool for dealing with some of the country's most pressing problems. It was this commitment, together with the flexibility inherent in the workings of international political economy, that was sufficiently great so as to overcome many of the constraints imposed by Cuba's problematical relationship with the U.S. This relationship made it more difficult for the Cubans to access Western science and technology than most of their developing country counterparts.

It is, therefore, possible to conclude on an optimistic note. This is possible precisely because of the progress the Cubans have made in the face of significant constraints on the diffusion of bioknowledge to their country. Other developing countries will accordingly want to take account of the Cuban experience in making their own plans for biotechnology.

ACKNOWLEDGEMENT

The study on which this chapter is based was undertaken by the author as part of the feasibility examination for the United Nations University's Institute for New Technologies (INTECH), Maastricht, The Netherlands. The author would like to thank INTECH's Director, Professor Charles Cooper, for permission to refer to the study here. Gratitude must also be expressed to Dr Luis Herrera and all his colleagues in Cuban biotechnology for their hospitality and insight into their country's biotechnology programme.

REFERENCES

Fransman, M. 1994. Biotechnology: Generation, Diffusion, and Policy In Cooper, C. (ed.), *Technology and Innovation in the International Economy*. Cheltenham: Edward Elgar.
Fransman, M. and Tanaka, S. 1994. Government, Globalisation and Universities in Japanese Biotechnology', Research Policy (forthcoming).

Part V

Impact of Biotechnology

Introduction to Part V

Worldwide interest in biotechnology is strong because of the technology's expected impact on economy and society. Will our life be much different as the result of applications of biotechnology?

A number of biotechnology-based drugs has come to the market, and a large number of new drugs is in the pipeline. Most of these drugs replace other drugs. Some offer a cure where none has been known before. But the new drugs as such will probably not fundamentally alter our health systems. The impact will vary from one national health system to another.

The biggest change will probably be caused by new diagnostics. Many diseases can be identified at a much earlier stage. Much more will be known of the predisposition for specific diseases (with all the difficult ethical questions which this raises). In addition, improved possibilities for self-application of diagnostical kits may alter the relationship between doctors and patients. Some of these problems have already been mentioned in the section on applications.

Much more far-reaching than in the health system will be the application of biotechnology to agriculture. It may fundamentally change the position of agriculture in the economy. The close link between agriculture and food production will be loosening, as agriculture will produce more raw materials for industry.[1] At the same time, the potential to produce food from non-agricultural sources will increase. The links between agriculture and industry will become much closer. In this process, agriculture may come under competing pressures from the agrochemical inputs industry on the one hand and the processing industry on the other.

With a rapidly increasing world population, hopes run high that biotechnology will make a substantial contribution to feeding billions of additional inhabitants. The technological potential, indeed, is there. But it is actually applied in a social setting which makes it highly questionable whether the potential is really going to be used for this purpose.

The poorest developing countries are those countries which also have the highest share of agricultural products in their exports. Instead of expanding food production, many developing countries will mainly increase cash-crop production for export. Since applications of biotechnology in highly developed countries will increase production there as well, more supplies may chase far fewer markets. As a result, export income, especially for the poorest developing countries, may come under additional pressure.

The employment impact of biotechnology is very much related to the structural changes in agriculture. Industrial applications of biotechnology will not make much of a difference for employment. The biotechnological route of production will create as little employment as alternative forms of highly automated chemical processing. Additional employment can only be found in research and (temporarily) in the machine-building sector which provides the means of production for every industrial restructuring process.

In the long run, our present petrochemical-based industrial base will probably have to give way to a new industrial system in which bio(techno)logical production processes will play a crucial role. Since these processes take place at normal temperature and pressure, they will be much less energy-intensive than the high temperature/high pressure processes which actually dominate the chemical-processing industry. The present concern with global warming processes may speed up such a change. But it is still too speculative to try to describe such a change in detail.

NOTE

[1] Cf. Guido Ruivenkamp, *De invoering van Biotechnologie in de Agro-Industriële Produktieketen* (Utrecht: Uitgeverij Jan van Arkel, 1989).

27

Biotechnology: feeding the World?

John Hodgson

Biotechnology will never feed the world. That is down to farmers. But biotechnology will not even help very much unless it can overcome a very significant hurdle. The biggest obstacle to biotechnology's bene-fiting the needy in the developing world is not the lack of appropriate technology, even though most of the research and development in agricultural and animal biotechnology is not directed towards Third World needs. Nor is ownership of intellectual property rights (IPRs) the main issue, even though, without clarity in IPR, it will be very difficult for the owners of the majority of plant and animal biotechnol-ogy – companies in the developed world – to feel comfortable let-ting their technology into developing countries. No, by far the biggest barrier in delivering the products of agricultural biotechnol-ogy to the needy in developing countries is the virtually complete absence of market mechanisms which normally drive the R&D process.

Without them, the success of the venture depends wholly on that most valuable but fickle of commodities – individual commitment, reliability and drive.

There are markets in Third World agriculture which operate like any other – those for cash crops, usually export crops. Seed companies like Pioneer HiBred (Des Moines, Iowa) and Sandoz Seeds (Basle, Switzer-land) through acquisition now have extensive seed interests in de-veloping countries as part of their strategies for global technology exploitation. Smaller companies, too, want in. According to Peter Barfoot, Business Planning Manager at the Agricultural Genetics Company (AGC, Cambridge, UK), the company is considering a num-ber of joint ventures or other arrangements in countries like India, Brazil, Thailand and Singapore based on technology-for-equity deals.

There is certainly business to be done in the developing world, but it will bring little benefit to the needy.

Tripartite Arrangements

Several companies and high-technology laboratories are involved in developing biotechnologies specifically for the underprivileged in the Third World. The strategy has been 'technology tinkering' – adapting the principles of commercial plant biotechnology to the crops used and problems faced by small farmers in developing countries. Thus AGC is collaborating with the International Potato Institute (CIP, Lima, Peru) in inserting its proprietary cowpea trypsin inhibitor (CpTI) gene into sweet potato and potato: with support from the UK government's Overseas Development Administration (ODA, London), AGC will perform the molecular biology while CIP will take care of the breeding and field-testing. Under the agreement with AGC, CIP can distribute any resulting products without restrictions in developing countries. According to Peter Barfoot, AGC is willing to extend this kind of arrangement to other international institutes and other proprietary genes.

The tripartite agreement – between technology owner, international breeding institute, and donor – is a pattern repeated elsewhere. Monsanto (St Louis, MO) has granted free licences on its virus coat protein disease-protection technology for crops and territories in which it has no immediate commercial interests. One consequence of that move has been the formation of the International Laboratory for Tropical Agricultural Biotechnology (ILTAB) at the Scripps Research Institute (SRI, San Diego, CA). ILTAB is headed by Roger Beachy, who developed the coat protein technology, and supported by the *Office de la Recherche Scientifique et Technique Outremer* (ORSTOM, Paris) and SRI. According to Beachy's colleague, Claude Fauquet, ILTAB is focusing on transferring resistance to African, South American and Asian viral diseases of cassava, rice, tomato and sweet potato. Donors like the Rockefeller Foundation (New York) and USAID (United States Agency for International Development) provide additional support for specific projects at ILTAB.

In a separate project, organised by the International Service for the Acquisition of Agri-biotech Applications (ISAAA, Lisboa, Mexico), Rockefeller funded the development of virus-resistant potatoes by scientists in Mexico using Monsanto's technology. ISAAA aims to act as 'an honest broker . . . matching needs and appropriate proprietary technologies'. Clive James, ISAAA's leading light and formerly a Deputy Director of *Centro Internactional de Mejoramiento de Maíz y Trigo*

(CIMMYT, Mexico), has spent the last two years persuading technology owners to grant free licences and donors to support the development of that technology for developing countries. As a result, the McKnight Foundation awarded ISAAA a $1 million grant for core services, and the Resources Development Foundation (Washington, DC) committed $0.5 million. USAID will give specific support to develop a DNA probe system invented at Washington State University (Pullman, WA) for diagnosing *Xanthamonas campestris* infections of crucifers: the technology will be made available through the Asian Vegetable R&D Centre (AVRDC) to Third World clients.

ISAAA's steering committee boasts representatives from Monsanto and the Sandoz Seeds subsidiary, Zaadunie (Enkhuizen, Netherlands), and from leading academic centres of plant biotechnology such as the University of Wisconsin (Madison, WI), The John Innes Institute (Norwich, UK) and the Max–Planck Institute (Cologne, Germany). ISAAA also has the support of companies, like ICI Seeds (Haslemere, UK), which are not yet formally involved: 'Clive James can cherry-pick through the technology that we have.' says ICI's Keith Pike. 'We are offering training facilities or pieces of technology. We think [ISAAA] has a lot of merit.' The World Bank (Washington, DC) may also be entering the brokerage game with the formation of its Biotechnology Acquisition Unit under Gabrielle Persley. That Unit would have the great advantage of having the World Bank's money and experience right behind it.

Perspectives

By lining up these projects and plans end to end, it is possible to create the impression that something substantial is happening in biotechnology aimed at developing countries. But that would be an illusion.

Reality checks are in order. In order to establish a truer perspective, consider the scale of current activities, the technical status of projects and, most importantly, the developing country contexts in which the products must eventually function. Biotechnology's efforts on behalf of developing countries seem puny beside commercial programmes in agricultural biotechnology. According to analysts County NatWest WoodMac (Edinburgh, UK), for instance, the top twenty or so agrochemical and animal health companies between them spent around $300 million on biotechnology R&D in 1990: the top agricultural biotechnology companies spent around $70 million (see table 27.1). The spending for the developing world is at least an order of magnitude lower.

Table 27.1 R&D spending in agricultural biotechnology (data from County NatWest WoodMac and company sources)

Organisation	Crop	Animal Health (in $ million1990)	Total
Abbott (Chicago, IL)	6	1	7
American Cyanamid (Princeton, NJ)	4	4	8
BASF (Ludwigshafen, Germany)	5	1	6
Bayer (Wuppertal, Germany)	8	2	10
Calliope (Béziers, France)	5	0	5
Ciba-Geigy (Basle, Switzerland)	15	3	18
DowElanco (Midland, MI)	4	0	4
Du Pont (Wilmington, DE)	20	0	20
Eli Lilly (Indianapolis, IN)	0	10	10
Hoechst (Frankfurt, Germany)	4	6	10
ICI (London, UK)	18	0	18
Kemira OY (Espoo, Finland)	3	0	3
Merck& Co. (Rahway, NJ)	0	6	6
Monsanto (St Louis, MO)	18	43	61
Novo Biokontrol (Bagsvaerd, Denmark)	8	0	8
Pitman-Moore (Harefield, UK)	0	10	10
Rhone-Poulenc (Lyon, France)	13	5	18
Sandoz (Basle, Switzerland)	17	1	18
Sanofi (Paris, France)	11	6	17
Shell (London, UK)	3	0	3
SmithKline Beecham (Exton, PA)	0	13	13
Solvay (Brussels, Belgium)	5	10	15
Upjohn (Kalamazoo, MI)	4	14	18
Virbac (Carros, France)	0	3	3
Total	171	138	309
Biotechnology companies			
Agracetus (Madison,WI)	6	1	7
AGC (Cambridge,UK)	7	0	7
Bio Technica Intl (Overland Park, KS)	8	0	8
Calgene (Davis, CA)	11	0	11
Cambridge Vet. Sci. (Ely, UK)	0	2	2
Crop Genetics (Hanover, MD)	6	0	6
DNA Plant Technology (Cinnaminson, NJ)	13	0	13
Ecogen (Langhorne, PA)	6	0	6
Mogen (Leiden, Netherlands)	1.5	0	1.5
Mycogen (San Diego, CA)	10	0	10
Plant Genetics Systems (Ghent, Belgium)	8	0	8
Syntro (Lenexa, KS)	0	2	2
Total	76.5	5	81.5
Grand Total	247.5	143	390.5

The work is also much further behind. 'We're now at the beginning of development,' says ILTAB's Fauquet. 'Everything is in its initial phase. We're investigating possibilities – what genes, what plants, plant transformation, regeneration – there's still a lot to do.' He expects that development of usable technologies will probably take much longer than for the commercial crops despite advances in molecular biology and the understanding of plant physiology: 'Most of the plants are recalcitrant – we know nothing about transformation and regeneration. . . . We can regenerate a limited number of cassava varieties, for instance: one of them is of limited agronomic interest – but then only in South America; it's no good for Africa.' The development phase will also be slowed by manpower constraints. Fauquet estimates that there are, perhaps, twenty people worldwide involved in cassava transformation and regeneration. Compare that with around 1500–2000 tackling the equivalent problem in maize. That intensity of activity in Western crops creates competition which helps speed progress, he believes.

The relative paucity of investment in biotechnology for developing countries makes identifying developments that will actually help solve the problems of the underprivileged much more difficult. While Western agriculture can cover most of the numbers on the agbio roulette table, those acting on behalf of the Third World have to place their chips more strategically. They need an infallible system for picking winners – because there are not the resources to back losers and still come out on top.

Technology to Market

Picking the right technological solutions is only a start, however. To even start to do anything as grand as 'feeding the world', those solutions need to reach the end-users – the small-scale farmers. Most Western activities, including those involving ISAAA, ILTAB, AGC, World Bank and their backers, have collaborators in the developing countries. These are, almost exclusively, scientific organisations. Some have agreements with the international plants breeding institutes International Rice Research Institute (IRRI, The Philippines), International Institute of Tropical Agriculture (Ibadan, Nigeria), Centro Internacional de la Papa (CIP, potato), or CIMMYT (maize and wheat). Others are tied in with national plant breeding institutes or with university groups. It is entirely possible, of course, that those organisations will not serve as black holes into which technology, when it eventually emerges, disappears. But, unless their structure and outlook changes markedly in the next few years, that may well happen.

The case for international and national research and breeding programmes being good channels for technology transfer is not strong. In a 1990 briefing paper, Peter Oram of IFPRI (Washington, DC) pointed out a number of weaknesses in the agricultural research sector of developing countries. Among them were deficiencies in training, poor links between the agricultural research services and universities, poor management at senior level, weak links from the research service upward (to national policymakers) and downward (to the extension services which take the technology to the farmers).

Another analysis (Ewell, P., 1989: Linkages between on-farm research and extension in nine countries; OFCOR Comparative Study no. 4, International Service for National Agricultural Research, The Hague) expands on the weakness of the link between research and extension workers who, after all, are the distribution mechanism for the products that biotechnology will produce. Ewell concludes that extension workers are obliged to promote whatever technology comes down to them even if it is not adapted to local conditions and demands. He believes also that there is a sociological problem: extension workers and researchers are separated widely in educational status, salary and social class. They get the blame from both farmers and researchers when technology fails.

The upshot of this situation is that research and plant breeding institutes in developing countries are frequently isolated from their clients, the small, subsistence farmers. On the other hand, their links with cash-crop farmers are much stronger: their ability to pay for research notwithstanding, the cash-crop farmers are much more likely to be able to articulate their problems to the researchers and to use the products that are developed.

That would go counter to the ethos of the technology developers. 'The rice [ILTAB is developing] in Asia and Africa is for self-consumption', says Claude Fauquet. 'The tomato in Egypt for local consumption; sweet potato is not exported anywhere; the potatoes in South America are not of the type that is exported anywhere. We won't work for this kind of target [export crops].'

ICI's Keith Pike spent considerable time in Africa and Asia: 'We don't want to see technology go into university laboratories which are just as remote from the "market" as anywhere else. There has to be trickle-down to the farmer and trickle-up of information from the farmer.' Diversion of biotechnological resources intended for small-scale farmers has already occurred, some critics believe. 'There is a lot of air around Rhizobia,' says one. 'There is not enough scientific proof that it works well enough to be of benefit. In Zimbabwe, for instance, no small farmers use it. . . . In effect, this kind of research done in the name of the small farmer ends up just subsidising the big farmer.' The same critic thinks drought tolerance is another red herring: 'drought tolerance traits may enable plants to survive in the Sahel, but they don't increase yields of the crops. So in effect, they are no use to the farmer.'

Commercial Directions

One of the ways that the Consultative Group on International Agricultural Research (CGIAR), the umbrella for the international institutes, is considering tackling the problem, is to alter the incentive structures for researchers in developing countries. At present, job security and scientific progress are the main drivers: problem solving is secondary. Keith Pike is among those who would welcome a more goal-oriented approach from the research and breeding institutes: 'As a commercial business, we [ICI] measure success in pounds, shillings and pence. Perhaps we should count the added value in kilograms of additional yield or in the number of farmers helped. We must have some measure of success. Some institutes don't have objectives – they just do research.'

Without the kind of indices used by commercial companies – sales or profits, R&D expenditure per researcher – there are no measures of success or failure. Therefore, there are no mechanisms for deciding what is worth pursuing and what is a waste of time and resources.

With that in mind, six potential donors (the overseas development bodies in Norway, Switzerland, the Netherlands, and the UK, plus USAID and the World Bank) and three representatives from developing countries (Zimbabwe, Costa Rica and Thailand) are in the advanced planning stages of an intermediate centre (possibly to start in 1992) which would bridge the gap between the developing world and the developed in biotechnology.

Hans Wessels of the Dutch Ministry of Development Cooperation (The Hague), of the donors involved, explains that the donor organisations felt they should 'look for specific needs. . . . The problem is that developing countries are confused. . . . We want to establish a clearing house for impartial, technical advice.' The centre would be within the International Service for National Agricultural Research (ISNAR, The Hague) with around $2 million in initial funds and 3–4 permanent staff. In essence, the centre would undertake not 'technology assessment and transfer' but 'problem assessment', advising developing countries not just on what biotechnology had to offer, if anything, within the specific context prevailing in each country or region.

There are also advanced plans for what would be an entirely new kind of institute for developing countries – it would be modelled less along the lines of a research institute, more along the lines of a commercial biotechnology company. The *Institute International pour la Recherche Scientifique et Développement à Adiopoudoumé* (IIRSDA) already has support from the governments of the Ivory Coast (the host nation), France and Canada. It is intended that IIRSDA will undertake some basic research, but most will be applied. The research programme will be attuned to new techniques, especially in biotechnology, but will be directed, to a large extent, by its 'marketing' activities – contact with small-scale farmers, either direct, or through those much maligned go-betweens, the extension workers.

Sounds good – in theory. But, as many other start-ups in biotechnology have discovered, combining focused R&D with the need to reach diverse and demanding markets is a hard task. For at least one reason, IIRSDA – whether it succeeds or fails – is important: it acknowledges the need for a market-driven approach in the developing world. Without understanding that market, or at least recognising it – even given the best of intentions – taking biotechnology to the developing world would be just so much public relations.

28

The Impact of Biotechnology on International Trade

Gerd Junne

1. Introduction

New technologies have always affected the international division of labour. Biotechnology is no exception. It is expected to have a considerable impact on world trade, because it increases the scope for the substitution of one commodity for another. Such substitution processes have occurred again and again in history. The present situation, however, may differ significantly from historical experience in that (a) switches to a new raw material base may actually take place much quicker than in the past, (b) a large number of commodities will undergo major changes in supply and demand simultaneously, and (c) alternative sources for foreign exchange earnings may be more limited now than in the past (Junne, 1987a; Junne, Komen and Tomeï, 1989).

This chapter discusses different types of substitution processes in which biotechnology plays a role. These processes will contribute to the overall decline of commodity prices which is expected for the 1990s.

2. Different Substitution Processes

The present discussion deals mainly with trade in agricultural commodities, since it is in this field that the trade impact of biotechnology will be felt most. Four major types of substitution processes will be described below. In brief, the four processes concern:

(i) shifts in trade resulting from the *introduction of additional characteristics into existing plants*;
(ii) shifts resulting from *changes in food processing*;

(iii) shifts due to the *industrial production of plant components or substitutes*; and,

(iv) shifts due to the *unequal distribution of new production processes*.

2.1 Shifts as a result of new plant characteristics

Genetical engineering of plants has proved much more difficult than that of microorganisms. But more traditional biotechnology has already had a considerable impact on plant breeding especially by applying tissue and plant cell culture. This has helped to speed up traditional plant breeding and to reduce the lead time to develop new plant varieties.

The ease with which new characteristics can be added to plants (or existing ones deleted) contributes to a 'separation of the plant from its original environment' (Ruivenkamp, 1989). A better resistance to stress factors has made it possible to shift the geoclimatic limits concerning the growth of specific crops. As a result, some plants, exclusively produced in a subtropical or moderate climate, are now reared more and more in the north. An example is the development of forage grass to grow actively even in cold weather. This development would make it possible to shift some cattle production from South America and other southern countries to North America and northern Europe.

2.2 Trade shifts as the result of changes in food processing

Important early shifts in international commodity trade will not result so much from advances in the genetical manipulation of higher organisms such as plants, but from applications of new and modern biotechnology to microorganisms, a field in which much more experience has been gained. Advances in food processing, especially fractioning plant products into different components and 'reassembling' these components into final food products, have led to *separation of plants from their specific characteristics*. Many crops have become interchangeable. This has tremendously increased direct competition between producers of crops that hitherto used to cater to different markets (Ruivenkamp, 1989). The most outstanding example is that of the increasing competition between sugar and starch producers.

One of the results of cultivating maize in more and more moderate zones has been an increasing overproduction of corn in North America. This has stimulated producers' interest in alternative uses of their produce. As a result, research intensified in the 1970s on the enzymatic transformation of starch into High Fructose Corn Syrup (HFCS). The tremendous expansion of HFCS production between

1975 and 1985 and the resulting decline of sugar imports by the United States is the largest trade impact that biotechnology has had hitherto.

HFCS is not an entirely new product. Since the early 19th century it had been known that starch could be transformed into a sweetener. However, advances in the application of immobilized enzyme technology reduced production costs for corn sweeteners to such an extent that a switch from sugar to HFCS became a profitable option – at the high domestic price levels for sugar in the United States and Japan.

Total replacement of sugar in the United States by HFCS has reached around 6 million short tons of sugar, with one half of that amount being produced domestically and the other being imported. US sugar imports dropped from 5.3 million short tons in 1970 to about 2.2 million in 1987. This substitution process has levelled off, because penetration of those areas where technical and economic conditions allow it, has reached almost 100 per cent. Large-scale introduction in countries outside the United States and Japan is not to be expected.

While its liquid form has been one of the advantages of HFCS for industrial users, this very advantage, on the other side, has prevented the penetration of the consumer market for household use. Only a very small percentage of HFCS reaches the market in a crystallized form, because the energy costs of drying it make it uncompetitive in comparison with refined sugar. However, several less expensive procedures have recently been developed which may bring about another round of substitution. This could imply the end of US sugar imports.

Another important example for such substitution processes is the development of improved cocoa butter substitutes (Svarstad, 1988). While such substitutes have been on the market since the beginning of the twentieth century, most of them either did not sufficiently meet the specific desired properties or were too costly for commercial production. With the help of biotechnology, researchers hope to modify enzymes in such a way that they become fit for the production of cocoa butter substitutes. Such a development would have an obvious impact on international cocoa trade, if the substitutes were commercially viable and not banned by legislation.

2.3 Shifts due to the industrial production of plant components or substitutes

Biotechnology has not only increased the substitution of one agricultural product for another, but has also increased direct competition between agricultural and industrial products. Again, sweeteners provide a good example. Cane and beet sugar lost ground not only to HFCS, but also to industrially produced low-calorie sweeteners. Biotechnology has made it possible to produce different new low-calorie sweeteners. A very important role has been achieved by aspartame, made from two amino acids coupled by natural enzymes, now used in many low-calorie soft drinks. Its sweetening power is about 150 to 250 times as potent as sugar, depending on the formulation. Global aspartame demand, therefore, for more than a decade could be met by a few factories of one company. This is an obvious example of the ongoing process of 'dematerialization of production', which implies that the same 'use value' can be created with an ever smaller amount of material input. This trend works very much to the disadvantage of raw material exporting countries.

A number of low-volume, high-value compounds which are traded internationally, and which could not (or only with exorbitant costs) be produced by chemical synthesis, can now be obtained through either microbial fermentation or cell suspension culture. This is the expression of a third separation tendency, the trend towards the separation of the production of plant components from the land (Ruivenkamp, 1989) and an increasing *industrial* production of such components. Industrial production has a number of advantages: (a) it is less tied to specific seasons but can continue all year long; (b) it can ensure better quality than production influenced by the vagaries of nature; (c) it normally takes less time on account of the optimization of growth conditions; (d) it is less labour-intensive, because steps like planting, nursing the young plants, harvesting etc. can be avoided; and (e) it allows for easier production than natural processes as in the case of perennial crops, the production of which cannot easily be adapted to market conditions.

The most suitable candidates for industrial production are such high-value, low-volume substances as compounds of pharmaceutical value, fragrances, flavours, pigments and insecticides (UNIDO, 1988). A good example is the plan to produce the peptide sweetener thaumatin with *E. coli*, which could lead to a *substitute for the substitute* thalin, the thaumatin-based sweetener marketed by Tate & Lyle (RAFI, 1987b).

2.4 Shifts as a result of the unequal distribution of new production processes

Biotechnology can in many ways be used to make agriculture more productive. The diffusion of these new technologies (as with all other new technologies), however, does not take place at the same speed in all countries. Countries that are able to introduce the new technologies at a faster pace will consequently be able to increase their own market share, and in turn will displace other countries as exporters. Such shifts are actually taking place in the field of palm oil and cocoa.

The diffusion of biotechnological advances differs not only from country to country, but also from crop to crop. Whereas some crops (especially from industrialized countries) have received much attention in international biotechnology research, others have received much less. Where different cash crops compete with each other, breakthroughs with the help of biotechnology to increase production for one crop may be at the expense of another. Since the production of different crops is distributed unevenly over different countries, even the equal geographical dissemination of new technologies would have uneven effects on the trading position of different countries.

3. Changing Trade Patterns

Before looking at concrete changes in international trade flows, some general observations are necessary:

1. Biotechnology increases productivity in agriculture. Where productivity rises faster than demand, self-sufficiency is enhanced or, where already attained, surpluses are generated. Countries that have, hitherto, been importers of agricultural products, can become more self-reliant with the help of biotechnology. This means that increasing surpluses are chasing after fewer markets. As a result, prices decline. When exporting becomes less attractive, countries look to alternative uses for their agricultural produce and may use it as a substitute for products hitherto imported. In this way, a *chain reaction* may be triggered off which can affect a number of commodities from different countries.

2. If world demand for a specific commodity declines as a result of substitution, different exporting countries will not all be hit in the same way. Production cuts will be distributed *unequally*. Where a free market prevails, marginal producers drop out first. Trade in many commodities, however, is governed by long-term contracts. Most of Cuban sugar exports, for example, are sold to the Soviet Union and other (formerly) socialist countries at prices agreed in advance (Perez-Lopez, 1988). Cuba, therefore, is hardly hit by the substitution of HFCS for cane sugar.

3. If we want to get an idea of the trade impact of biotechnology, it is worthwhile to have a look at those trends that influenced development *before* biotechnology made itself felt. These trends are determined by the dominant forces and interests in international trade. These very same forces will have considerable impact on applications of biotechnology. As a consequence, biotechnology will probably contribute to developments that go in the same direction. It is analytically very difficult to separate shifts 'caused' by biotechnology from other ongoing changes in international trade flows.

4. It is difficult to be precise about the degree, timing, and pace of the expected substitution processes and about the extent of compensating alternative demand (e.g. from increasing South–South trade) for the commodities in question. The reason for this lies not only in the fact that it is difficult to forecast the pace of techno-economic development (e.g. the price of energy). *An additional element of discontinuity is introduced by political decisions*, on which the development and application of biotechnology depend to a large extent. These are not only decisions on the regulatory environment. Important decisions are also taken in the sphere of funding specific directions of biotechnology research and in creating a general economic environment which is supportive for the application of the results. These decisions can put the 'normal' sequence on its head with, first, specific (bio)technological developments taking place, which then have some implications for international trade. Instead, it often works the other way round. Governments restrict international trade in order to create an incentive for domestic suppliers to spend money on research for the development of substitutes. An example is the artificially high price for oils seeds in the EC, which the European Commission introduced to encourage the development of protein-rich oil seeds in the Community. Trade flows are affected not only by substitution processes that result from *past* technological developments, but also by political decisions that *anticipate future* technological possibilities and influence development in a desired direction.

3.1 *Political rather than technical determination of substitution processes*

It is the interaction between *technical* developments, *economic* considerations, and *political* pressure which gives shape to real changes in trade flows. This is well demonstrated by the largest trade displacement that biotechnology has hitherto 'caused': The substitution of domestic HFCS production in the United States for imports of cane sugar.

The substitution of HFCS for sugar has by and large been a United States (and to some extent Japanese) phenomenon exclusively. It is estimated that in 1986 HFCS consumption in the US was around 5.5 million short tons, and about a million tons more in the rest of the world, principally Japan. There are some good economic reasons why HFCS production is concentrated in the US: the US is by far the world's largest producer of corn. This special advantage, however, is insufficient to explain why the substitution of HFCS for sugar has only taken place in the US and not in Canada, where almost identical conditions prevail. Canadian HFCS production has grown only mod-estly since the late seventies. The reason is that Canada is one of the very few industrialized countries which allows its refineries to buy sugar on the free international market, where prices are normally below the world average cost of production. There is therefore no reason for domestic industrial sweetener users to switch from sugar, and virtually all Canadian HFCS production is *exported to the United States* (ICCSASW 1987).

Prices for raw sugar in the US have remained well above world market level since 1981 (Maskus, 1989). The American quota system had been abolished in the early 1970s when high world market prices for sugar guaranteed sufficient income for American producers. After seven years following more or less world market prices, the US gov-ernment enacted a new support scheme in 1981, when world market prices plummeted. Since then, the average price differential between world and domestic sugar prices has been 353%, reaching a maximum of 776% in June 1985 (Maskus, 1989). It is only with the support of the quota system that HFCS production has been profitable in the current period of low world market prices for sugar.

US sugar imports fell from 4.54 million tons in 1981 to 2.63 million tons in 1982. Import quota were subsequently further cut down rapid-ly, from 2.6 million tons in 1982/83 to 0.9 million tons in 1986/87. Three countries account for almost half the total of all quotas (Domini-can Republic, Brazil and the Philippines). The rest is distributed over 38 other countries.

The quota system has come under increasing pressure at the na-tional as well as the international level (and so has, as a consequence, the price advantage of HFCS over imported sugar). At the national level, all those interests that would profit from a general liberalization of international agricultural trade point to this obvious case of protec-tionism as a major obstacle for fruitful international negotiations. Internationally, the quota system has been challenged as being incom-patible with GATT rules. At the request of Australia, a GATT panel was established in September 1988 which recommended demanding that the US should either terminate these restrictions or bring them into conformity with the General Agreement. The US agreed to the adoption of the report even though it may have significant implications for the US sugar industry (GATT 1988b, 1989b).

The present protection of US sugar interests may be reduced when the actual provisions have to be revised which were extended through to the 1991/92 crop year by the Food and Security Act of 1985. The flexibility to shift back to increased imports of cane sugar, however, is limited, since closures have reduced annual cane-refining capacity in the US by an estimated 2.5 million short tons in the past few years (*Financial Times*, 27 September 1988).

The only other major sugar-importing country where HFCS has made important inroads is Japan. Japanese HFCS production grew from 84,000 tons in 1977/78 to an estimated 650,000 tons in 1987/88 (*International Sugar Journal*, 1989, 213). Japan imports both sugar and corn. Crott (1986) assumes that 'the inroads of HFCS might have been favoured by a difference of the taxation of sugar and isoglucose in import duties and subcharges'. The import of corn to produce HFCS instead of importing sugar has the political advantage of reducing the large balance of payments surplus that Japan has with the US, which continuously leads to political conflict. There is no comparable political pressure to reduce Japan's trade surplus with sugar-exporting developing countries.

The substitution of HFCS for imported sugar, however, expanded only until 1982 when a surcharge was placed on HFCS. The proceeds were used to help finance the domestic sugar support programme which resulted in a doubling of domestic beet and cane-sugar output to 950,000 tons. The surcharge has made sugar more competitive with HFCS. Since the surcharge was imposed, there has been a recovery of sugar consumption and an end to the rapid growth of HFCS use.

HFCS contributed about two-thirds to the substitution of sugar imports of about 700,000 tons between 1978 and 1984. Since sugar imports from some countries (e.g. Australia) were tied by long-term contracts, decreases of imports concentrated on Cuba, Thailand and the Philippines, while imports from other countries remained more or less steady. This is another example of the unequal geographical distribution of import cuts which result from substitution.

Beside the replacement of sugar imports, the OECD mentions *single-cell proteins (SCP)* 'from industrial substrates and potential competition with agricultural protein animal feed' as a second example of present or currently predictable trade impacts in agriculture (OECD, 1989). Again, political and economic factors are of prime importance, since the introduction of new single-cell products is actually 'limited more by economic, market and regulatory considerations than by technological constraints' (Litchfield, 1989). Hardly any substitution has taken place up to now (except in the Soviet Union) in spite of the fact that it would have been technologically feasible.

3.2 Expected impact on the trade of individual countries

On the basis of the observations made above, some estimates can be presented with regard to the impact of biotechnology on the trade of individual countries. The following remarks focus on developing countries.

In a crop-wise analysis, Panchamukhi and Kumar (1988) have presented a crude estimate of the likely annual loss of export earnings as a result of advances in biotechnology, which could amount to about US$ 10 billion by 1995 compared with export incomes in 1980. The percentages of exports which may be replaced that underly the estimate may be far too high. But given the fact that already a much smaller decline of effective demand can have a far-reaching impact on the price of the commodities in question, the estimated decline of export income may be, nevertheless, not too unrealistic.

Impact on Trade of Latin American Countries

An obvious negative impact will be the substitution of sugar imports, which will be especially problematic for the *Caribbean* sugar exporters, but much less for *Brazil*, which actually has some difficulty producing enough sugar for its ethanol programme and to meet its export obligations at the same time. Especially hard hit is the *Dominican Republic*, for which income from sugar exports sometimes reached up to 50% of total foreign-trade earnings. Sugar exports to the US declined by more than 75% in only six years, from 780,000 short tons in 1981 to 161,000 short tons in 1987 (ICCSASW, 1987).

Exports negatively affected probably include *Argentina's* meat export, which had declined already by about half during the 1980s (IMF, 1988). Biotechnology will help to increase overproduction in industrialized countries of products that can be used as animal feed. Overproduction in agriculture in general will lead to a more extensive use of soil in large areas, which probably will increase livestock production. It will lead to a better growth of fodder grasses in colder climates. Besides, applications of biotechnology to livestock (re)production will reduce the feed/output relation and thus make cattle-raising less expensive in highly industrial countries.

An area where exports from some Latin American countries could profit from biotechnology is the application of mineral leaching to copper production, which may be especially suitable for *Chile* and *Peru* (Warhurst, 1984, 1987).

Impact on Trade of Asian Countries

For the densely populated and population-rich countries of Asia, biotechnology may help to increase food self-sufficiency. This will be at the expense of the large food exporters, especially rice exporters like *Thailand, Burma*, and *Pakistan*. The increase of vegetable oil production in *India* will probably lead to declining palm oil imports from *Malaysia*.

The *Philippines* will have to endure the strongest impact on its exports, because two major export commodities are being hit at the same time, sugar and coconut products. Coconut oil may more and more be replaced by cheaper oils and fats on the world market. The decline of income from sugar exports has been even more dramatic: while sugar exports had been responsible for about one quarter of total export income in the 1970s, they declined from a share of 10% to about 1% between 1980 and 1987 with an absolute decline from SDR 480 million to 46 million in the same period (IMF, 1988).

A good example of the chain effects of substitution processes is provided by the impact that the reduction of tapioca imports by the EC from *Thailand* (unrelated to biotechnology) had on sugar exports: Thailand had to look for new markets for its starch-rich cassava and increased exports to *South Korea* (Berkum, 1988), where starch-based sweeteners already in the mid-1980s accounted for 18% of total sweetener consumption (Crott, 1986).

Some Asian countries have a tradition in running large-scale plantations and have built up the capacity to introduce technological advances quickly into production. They will probably be able to increase their share in the total of developing countries' agricultural exports. A case in point is Malaysia, which has not only increased palm oil production, but has also expanded the large-scale production of cocoa (which in Africa is mostly produced by small farmers). This was not due to biotechnology, but the resulting large-scale production makes it easier to diffuse any productivity-increasing technology.

Impact on trade of African Countries

The victims of the resulting 'South–South substitution' will be first of all African countries. Only the Ivory Coast seems to be able to keep or even increase its market share in world cocoa exports, because the country has built up considerable expertise in diffusing improved technologies in this field to the farmers. *Ghana's* cocoa exports have profited from the special quality of Ghana cocoa, but this advantage may be lost as a result of biotechnology applications which might be used to 'upgrade' less valuable cocoa from elsewhere so that it matches the quality of cocoa from Ghana.

Applications of biotechnology to oils and fats will probably also affect the export potential for groundnut oil. Export incomes from groundnut products have shown a tremendous decline in *Sudan* and *Mali* (IMF, 1988) and may also effect *Senegal*. The same negative effect may make itself felt in the case of palm kernel exports from countries like *Sierra Leone*. The production of low-volume, high-value substances (like flavours, fragrances, pigments, insecticides) may have negative effects on countries like *Madagascar* (cloves, vanilla), the *Comoros* (vanilla), and *Kenya* (pyrethrum). *Sudan* has already suffered from the decline of gum exports.

While the impact of biotechnology on Latin American and Asian trade will have some negative, but also some positive aspects, African countries will for a long time almost exclusively feel the negative impacts. It will take much more time to use biotechnology to increase local self-sufficiency in food-importing African countries because of the many bottlenecks in the diffusion of research results from international research institutions to agricultural extension services and African farmers (Junne 1987b).

4. Conclusion

The impact of biotechnology on trade flows is hard to measure. Uncertainty regarding the speed of technological development is not the main reason for the necessarily speculative character of most analyses. The main reason is that the impact depends to a very large extent not only on advances in technology, but also on economic factors and on political decisions. Biotechnology has no *direct* impact on commodity trade. The influence is always mediated by economic and political variables, such as the strategies of large companies that organize the international division of labour, and political decisions of governments which set the parameters for world trade. Political decisions can push as well as delay substitution processes. They are also decisive for the regional distribution of the effects of substitution. Biotechnology will probably strengthen trends that already existed before the introduction of biotechnology, because applications of biotechnology have a higher chance of being realized if they serve the strategies that major economic and political actors pursue.

The application of biotechnology will first of all affect trade in agricultural products. It will make many importing countries more self-sufficient and increase trade conflicts among overproducing countries. While overall agricultural exports from developing countries will probably stagnate, biotechnology will help to substitute products from specific (more developed) countries for commodities from other (less advanced) developing countries, contributing to a stronger concentration of agricultural production for the world market on fewer developing countries. While countries known as 'Newly industrializing countries' (NICs), given their technological capabilities, will also be able to boost their *agricultural* production, the less advanced countries (especially in Africa and the Caribbean area) will bear the brunt of the adjustment of trade flows. Divergent effects on commodity trade of different developing countries as a consequence will make it even more difficult for them than in the past to coordinate their position in international trade negotiations.

NOTE

A more elaborated version of the present text has been included in the volume *Microbial Technology: Economic and Social Aspects*, ed. E. J. DaSilva, C. Ratledge, and A. Sasson (Cambridge: Cambridge University Press, 1991).

REFERENCES

Berkum, S. van (1988). *Internationale Aspecten van het EG-Landbouwbeleid. De Relatie met vier ontwikkelingslanden.* The Hague: Landbouw-Economisch Instituut.
Buttel, F. H. (1989). How Epoch Making Are High Technologies? The Case of Biotechnology. *Sociological Forum*, vol. 4, 247–261.
Crott, R. (1986). The impact of isoglucose on the international sugar market. In *The biotechnological challenge*, ed. S. Jacobsson, A. Jamison, H. Rothman, pp. 96–123. Cambridge: Cambridge University Press.
Financial Express (1989a). Tissue culture breakthrough for Cardamom, *Financial Express* (Bombay, India), 24 October.
Financial Express (1989b). Tissue-cultured oil palms to help reduce import bill, *Financial Express* (Bombay, India), 25 October.
GATT (1988a). *International Trade 87–88*, vols I and II, Geneva.
GATT (1988b). United States sugar-trade policy under fire. *FOCUS (GATT Newsletter)*, 57, September/October, pp. 1–2.
GATT (1989a). *Review of Developments in the Trading System. September 1988–February 1989.* L/6530, Geneva.

GATT (1989b). US accepts ruling on sugar quotas. *FOCUS (GATT Newsletter)*, 63, July, p. 2.

Gist-Brocades (1990). New Products based on recDNA Technology. *Biotechnologie in Nederland*, 1990/1, p. 9.

Halos, S. C. (1989). *Biotechnology Trends: A Threat to Philippine Agriculture?*, World Employment Programme Research, Technology and Employment Programme, Working Paper WEP 2-22/WP 193, Geneva: International Labour Office.

ICCSASW (1987). *HFCS and Sugar: New Equation in the Sugar Market*. Special Publication of *Sugar World*, Toronto: International Commission for the Co-ordination of Solidarity Among Sugar Workers.

IMF (International Monetary Fund) (1988). *Balance of Payments Statistics*, vol. 39, Yearbook, Part 1 and 2, Washington.

Junne, G. (1984). Der strukturpolitische Wettlauf zwischen den kapitalistischen Industrieländern. *Politische Vierteljahresschrift*, 25, 2, 64–83.

Junne, G. (1985). Biotechnology and Consequences for Changing Relations between EC–USA, EC–Japan and USA–Japan? *Biotechnology Hearing*, European Parliament, Committee on Energy, Research and Technology (PE 98.227/rev.), 30 October.

Junne, G. (1986a). Die Verschiebung der Kräfteverhältnisse zwischen den USA, Westeuropa und Japan unter dem Einfluß der Biotechnologie. In *Technologie und internationale Politik*, ed. B. Kohler-Koch. Baden-Baden: Nomos, 139–167.

Junne, G. (1986b). Nuevas Tecnologías: Una Amenaza para las exportaciones de los países en desarrollo. *Efectos sobre la División Internacional del Trabajo*, Mexico: Secretaría del Trabajo y Previsión Social, 41–66.

Junne, G. (1987a). Automation in the North: Consequences for Developing Countries' Exports. In *A Changing International Division of Labor*, International Political Economy Yearbook, vol. 2, ed. J. S. Caporaso, pp. 71–90. Boulder: Lynne Rienner Publishers.

Junne, G. (1987b). Bottlenecks in the Diffusion of Biotechnology from the Research System into Developing Countries' Agriculture. *Proceedings of the 4th European Congress on Biotechnology*, Amsterdam: Elsevier, 449–458.

Junne, G., Komen, J. and Tomeï, F. (1989). 'Dematerialization of Production': Impact on Raw Material Exports of Developing Countries. *Third World Quarterly*, 11, 128–142.

Kelly, M., Kirmani, N., Xafa, M., Boonekamp, C., Winglee, P. (1988), *Issues and Developments in International Trade Policy*, Occasional Paper No. 63. Washington: International Monetary Fund.

Kramer, M. (1988). *Effektive Protektion von Rohstoffproduktion und -verarbeitung*. Stuttgart: Steiner Verlag Wiesbaden.

Litchfield, J. H. (1989). Single-cell proteins. In *A Revolution in Biotechnology*, ed. J. L. Marx, Cambridge: Cambridge University Press, 71–81.

Maskus, K. E. (1989). Large Costs and Small Benefits of the American Sugar Programme. *The World Economy*, 12, 1, 85–104.

OECD (1989). *Biotechnology. Economic and Wider Impacts*. Paris.

Ofreno, R. E. (1987). *Capitalism in Philippine Agriculture*. Quezon City, Philippines.

Panchamukhi, V. R. and Kumar, N. (1988). Impact on Commodity Exports. In *Biotechnology Revolution and the Third World. Challenges and Policy Options*, pp. 207–224. New Delhi: Research and Information System for the Non-Aligned and Other Developing Countries.

Perez-Lopez, J. F. (1988). Cuban–Soviet Sugar Trade: Price and Subsidy Issues. *Bulletin of Latin American Research*, 7, 1, pp. 123–147.

RAFI (1987a). *Vanilla and Biotechnology*. RAFI Communique, Brandon (Manitoba): Rural Advancement Fund International.

RAFI (1987b). *Biotechnology and Natural Sweeteners; THAUMATIN*, RAFI Communique, Brandon (Manitoba): Rural Advancement Fund International.

RAFI (1989). *Coffee and Biotechnology*. RAFI Communique, Brandon (Manitoba): Rural Advancement Fund International.

Rosario, L. de (1989). China's sugar output dips as demand rises. *Far Eastern Economic Review*, 2 March 1989, p. 83.

Rousseau, P. (1986). Biotechnologies for development – reflections on the protein chain. In *The biotechnological challenge*, ed. S. Jacobsson, A. Jamison, H. Rothman, pp. 124–147. Cambridge: Cambridge University Press.

Ruivenkamp, G. (1989). *De invoering van biotechnologie in de agro-industriële produktieketen. De overgang naar een nieuwe arbeidsorganisatie*. Utrecht: Jan van Arkel.

Senez, J. C. (1987). Single-cell protein: past and present developments. In *Microbial Technology in the Developing World*, ed. E. J. DaSilva, Y. R. Dommergues, E. J. Nyns and C. Ratledge. Oxford: Oxford University Press, 238–259.

Svarstad, H. (1988). *Biotechnology and the International Division of Labour*, Oslo: University of Oslo (Institute for Sociology).

UNIDO (1988). Growing Compounds from Plants. *Genetic Engineering and Biotechnology Monitor*, No. 25, pp. 58–60.

Wallerstein, I. (1989). *The Modern World-System III. The Second Era of Great Expansion of the Capitalist World-Economy, 1730–1840s*. San Diego: Academic Press.

Warhurst, A. (1984). *The Application of Biotechnology in Developing Countries: The Case of Mineral Leaching with Particular Reference to the Andean Pact Copper Project*, Vienna: UNIDO (IS.450).

Warhurst, A. (1987), New directions for policy research: biotechnology and natural resources, *Development*, 1987: 4, 68–70.

29

The Employment Impact of Biotechnology

Iftikhar Ahmed

Introduction

Empirical assessment of the employment impact of the new biotechnologies is handicapped by severe data constraints. Quantitative data is not available to adequately reflect the labour intensity of biotechnologies not yet released or disseminated. Moreover, data that exists is not made available for policy analysis, simply because it is in the hands of the private sector, which is reluctant to supply the commercially sensitive statistics, including those on labour use. This poses a major challenge to social scientists who are usually fed with *ex post* survey data on which to base their analysis of labour use.

To overcome the above problem, this chapter harnesses the fragmented and scattered data generated by geneticists and microbiologists for *ex ante* assessment of the employment implications of new biotechnologies using innovative methods. Essentially, a mixed methodological approach is followed. Analyses are based on hard data generated by country case studies which permit the verification of specific hypotheses. The other methodological approach is to apply deductive reasoning relating specific biotechnology breakthroughs to the magnitude and pattern of labour-utilisation problems encountered by individual countries.

Purpose

The purpose of this chapter is to deal with a wide range of employment-related issues in the context of the newly emerging biotechnologies. These include labour absorption in crop production, seasonality of employment, structural composition of employment and rural labour markets. The chapter also examines the relationship between employment generation and multiple cropping and between prevailing agrarian structures. The chapter explores the scope for designing and deploying biotechnologies to promote employment and examines the employment potentials of both genetic engineering and micropropagation technologies.

The chapter also deals with the differential impact on employment of a given biotechnology and, in this context, examines the implications of induced innovation theory and structural adjustment measures on the use of factor proportions.

Apart from micro-level assessments, the chapter deals with two essential aspects of the global impact of biotechnologies. These include the inter-industry repercussions on aggregate employment levels and the international division of labour.

Impact on Rural Employment

Based on a set of empirical case studies (Ahmed, 1991), this section highlights the findings on direct labour utilisation in agriculture, indirect employment created through the backward and forward linkages to crop production, the structure and stability of employment and the impact of the new biotechnologies on the rural labour market and on multiple cropping.

Labour utilisation in agriculture

Evidence from several ILO country case studies clearly suggests that with the application of advanced biotechnologies there is a saving in labour use for chemical means of plant protection. For instance the case study for Mexico (Eastmond and Robert, 1991) shows that the application of micropropagation techniques need not lead to labour displacement in citrus cultivation as this would be compensated by more intensive labour use in weeding, pruning, irrigation and harvesting from a reduction in crop losses (labour accounted for 78–82 per cent of total costs of citrus production). In fact, in Malawi and Kenya there was a substantial increase in labour use per unit of land following the application of the new biotechnologies (through the introduction of new practices) which also increased yields. In Kenya the doubling of labour intensity per unit of land was due to more labour needed for ridging before cultivating potato, and in Malawi for nursery and planting operations (see tables 29.1, 2 and 3).

Table 29.1 Biotechnology and farm size: potato and tea in Kenya 1987[a]

Key Indicators	Potato farms (N = 33)			Tea farms/ estates (N = 39)
	Biotechnology (BT)	Traditional technology (TT)	Relationship with farm size	Relationship with farm size (biotechnology only)
Labour productivity (gross output/ha in shillings)	33,210	16,382	Inverse for BT and TT	Inverse
Labour intensity (work-days/ha)	301	144	BT: unclear TT: inverse	Unclear
Labour productivity (kg/work-day)	124	100	Positive for both BT and TT	Positive (sh/work-day)
Labour's factor share (wages as % of value added)	27	23	Positive for both BT and TT	Positive
Capital use				
sh/w-day	3	3	Inverse for both	Positive
sh/ha	867	426	BT and TT	Positive
Intermediate inputs (sh/ha)	3,553	3,008	Inverse for both BT and TT	Inverse
Value added as a proportion of gross output (%)	89	82	BT: positive TT: inverse	Positive
Profitability (gross output minus operating costs in sh/ha)	20,816	9,916	Inverse for both BT and TT	Positive
Income ratio[b]	8	4	–	3 (small farms[c]) 2 (large estates[d])

[a] *Source*: Calculated from data in Mureithi and Makan, 1991.

[b] Ratio of income of the 30 per cent of richer farmers to income of the 70 per cent of poorer farmers.

[c] Up to 3 ha.

[d] Over 20 ha.

Another interesting feature noted for both the Philippines and Mexico (Eastmond, 1989) is the domination of women in the micropropagation laboratories. For instance, women constitute 80 per cent, 74 per cent and 85 per cent of the Philippine Society for Microbiology, Cell/Molecular Biology and Biotechnology Societies respectively (Halos, 1991). In both these countries these were considered as lowly paid jobs concerned with basic science, with previously limited linkage to industry. Moreover, the work in the tissue culture laboratories is tedious, requiring patience and perseverance.

Table 29.2 Change in labour (man-days) input per hectare in different operations by level of technology

Operations	Between T_1 and T_2	Between T_2 and T_3	Between T_1 and T_3
1. Nursery and new planting	+18.3	+6.8	+25.1
2. Pruning	+15.2	−15.4	+0.1
3. Weeding	+6.8	+2.8	−4.5
4. Plant protection	−1.9	0.0	−1.9
5. Plucking	−9.6	+7.1	−2.5
Total	+15.2	+0.8	+16.3

Source: C. Chipeta and M. W. Mhango 1991.

Table 29.3 Percentage distribution of labour input (man-days) per hectare by operations and level of technology

Operations	Level of technology		
	T_1	T_2	T_3
1. Nursery/new planting	0.0	21.2	28.3
2. Pruning	51.7	60.3	42.1
3. Weeding	17.4	6.4	9.0
4. Plant protection	4.2	1.3	1.8
5. Plucking	26.7	10.8	18.8
Total	100.0	100.0	100.0

Source: C. Chipeta and M. W. Mhango, 1991.

Structure and stability of employment

As a result of the application of Advanced Plant Biotechnologies (APB) to a range of crops in several case study areas, a high potential has been noted for an indirect and steady source of employment through a strengthening of the forward linkages to the juice processing plant (Mexico), poultry production (Nigeria), coffee, henequen, tequila and dairy industries (Mexico), and the tea industry (Kenya). Underemployment (e.g. 75–90 per cent) of the agricultural workers in the southeastern region of Mexico caused by seasonality of agricultural production can be reduced by applying APB to create and widen crop varieties which prolong the growing season and supply of ripe oranges. It not only cuts down agricultural underemployment but also reduces excess capacity in juice processing (e.g. greater utilisation of the orange-juice processing plant, which is idle for 6 months).

There is a structural change in rural and agricultural employment associated with the use of APB. Application of APB in China releases labour from agriculture which is absorbed in new and sideline activities in specialised occupations, with a change in social organisation of the delivery of these services. Increased labour use in the agriculture of Malawi and Kenya has been brought about through structural adjustments resulting from new farm practices associated with APB.

Indirect employment

The case-study evidence from Kenya shows that the backward linkages to input suppliers are enhanced by the use of the new technology, as shown by a larger flow of intermediate inputs per hectare for the cloned potatoes than for traditional farming (see table 29.1).

In general, it can be concluded that the chemicals used in rural areas are not produced there. Over 40 per cent of the chemical fertilisers are imported by the developing countries from the industrialised countries. The production process for both nitrogen and phosphate requires extremely large-scale capital-intensive plants. In fact, chemical fertilisers are among the most capital-intensive products made by man (Johnston and Kilby, 1975). A large chemical plant can cost anywhere between US$300 and 700 million (Doyle, 1985). Reducing dependence on chemical fertiliser through genetic applications will not lead to any serious labour displacement in Third World countries. This is particularly true of Africa, where a negligible proportion (1.2 per cent) of the world's total nitrogen fertiliser is produced (Keya et al., 1986). On the other hand, it will relieve the acute balance-of-payments problems faced by most developing countries, and even in those with their own fertiliser industry the elimination of dependence on chemicals offers the opportunity to reallocate vast amounts of scarce resources to more labour-intensive sectors.

In addition to the implications for the above traditional types of backward linkages to agriculture, a new type of backward linkage is emerging through the blending of two distinct types of workers' skills. Case-study evidence clearly demonstrates the blending of workers with 'low-tech' skills to engage in traditional agricultural work with 'highly skilled' technicians directly involved with the advanced biotechnologies. This was seen for the micropropagation of crops in Mexico – technicians and scientists in the laboratory and the greenhouse, who supplied plantlets for cultivation by traditional agricultural labour. In addition to the employment created for the traditional workforce in agriculture, 933 people, mainly scientific personnel, were employed by the Tea Research

Foundation of Kenya in 1986, generating an income of K704,371. About 500 plant scientists are engaged in cellular engineering in China. Similarly, scientific personnel can produce 8,000 to 10,000 potato plantlets per day through the application of micropropagation techniques, in Nepal, which can then be easily handled by semi-skilled workers for rooting them in sandbeds (Rajbhandari, 1988).

As regards employment created (cutting down seasonal underemployment) through the forward linkages it is clear from the ILO case studies that many more jobs could be created in orange-juice processing, poultry production and the coffee, tea, henequen, tequila and dairy industries in a range of Third World countries (Ahmed, 1991).

Rural labour market

Application of APB has increased the demand for hired labour (e.g. in Mexico for citrus and in Kenyan tea and potato production), boosted wages, improved labour's factor share and reduced rural–urban wage differentials. Gross earnings from APB in Kenya compare favourably with wage incomes in a modern sector job, important for dampening the pace of rural–urban flow of income seekers.

Despite the availability of chemical herbicides, virtually the entire Green Revolution (GR) area relied heavily on manual labour for weeding. Several characteristics of employment are important for an analysis of the social consequences of increased herbicide applications: (i) weeding is one of the most labour-intensive of all agricultural operations for GR crops; (ii) the GR led to a significant increase in the demand for hired labour in weeding; for example, in Sri Lanka, hired-labour use doubled (Hameed et al., 1977); (iii) overall labour use in weeding doubled or tripled over that of the pre-GR crops, for example, in Bangladesh (Ahmed, 1981) and the Philippines (Bartsch, 1977); (iv) small farmers adopting GR technology recorded much higher labour intensity in weeding than larger farmers (Ahmed, 1981); and (v) women constituted between 72 and 82 per cent of the labour used for weeding (Unnevehr and Stanford, 1985). In the above context, it is clear that the introduction of genetically engineered plant varieties will lead to a substitution of chemical herbicides for manual weeding, leading to a massive displacement of women's labour. The trends indicate that not only will the use of genetically engineered biotechnology plant varieties introduce a new fixed cost for farmers by forcing them to purchase the herbicide genetically tied to the seed supplied by the same company, it will also strike a colossal blow at the poor.

Multiple cropping

The single most important factor which contributed to greater labour use per hectare in the Green Revolution areas was the practice of multiple cropping facilitated by the early maturing varieties of GR cereals. The application of micropropagation techniques to potato could similarly help improve cropping intensity. Since potatoes in most Third World climates take only 40–90 days to grow (compared with 150 days in the temperate climates), they can easily be incorporated into the cropping patterns currently practised for cereals like wheat, rice and corn.

Thirty poor countries already have the capacity to micropropagate potatoes. It is a major source of food for the poor families in Africa, and some Asian countries like India, Sri Lanka and the Philippines. Indeed, micropropagation techniques have made potato the second biggest crop (by weight) after rice in Vietnam, and quadrupled the production in China over the past 30 years (*The Economist*, 13 October 1990.) In Vietnam as well micropropagation techniques have increased potato yields from 200 tonnes to 8,000 tonnes per year on 450 hectares of land within a period of four years (1980 to 1984) (Uyen and Zaag, 1985). These techniques have already brought about yield increases from 8 tonnes to 18 tonnes per hectare in Nepal (Rajbhandari, 1988). Micropropagation techniques for potato are attractive for employment creation and poverty alleviation for the following reasons: (a) year-round production of plantlets is possible; (b) saving on costs and difficulties of physical transportation of potato tubers to the fields for planting; (c) by generating plantlets directly from tissues of the plant, a substantial (in the aggregate) volume of tubers spared from planting can now be eaten by the hungry; (d) disease-free planting material could significantly reduce production fluctuations from diseases, potato being particularly vulnerable to as many as 268 diseases and pests, and indeed the late blight could wipe out more than 50 per cent of the total crop (Manandhar et al., 1988); and (e) while increases in yield benefit land owners, increases in cropping intensity would also benefit the landless by increasing the demand for hired labour.

There is good promise for genetic engineering breakthroughs for potato compared with most other plants because of the relatively easier possibility of incorporating genes for disease and insect resistance.[1]

Agrarian Structure

Resource-poor small farmers in the Third World tend to economise in the use of capital and make relatively greater use of the abundant supply of labour, particularly unpaid family workers. The evidence from Kenya (Mureithi and Mokau, 1991) suggests that channelling biotechnologies to the small farmers will lead to gains in output without sacrificing on-farm employment. Evidence from Kenya and Malawi shows that with the application of biotechnology labour's factor share increases. It is clear that the agrarian structure also has an influence on labour absorption in agriculture.

Genetic Engineering and Micropropagation for Job Creation

The following observations can be made of the possible employment impact of the genetic engineering breakthroughs, evidence on which has recently been compiled (Ahmed, 1991). The employment potential of the *transgenic* (containing a foreign gene) plants and microbes are discribed below.

(a) Pest and disease resistance and drought tolerance will reduce fluctuation in agricultural output and increase the demand for labour.
(b) Prolonging the shelf life of freely harvested agricultural produce will certainly enhance the employment of the poor producers confronted with inadequate marketing infrastructure.
(c) Genetically engineered microbes may stimulate labour use on small farms if these spill over from the rich neighbours' farms and fix nitrogen on the small plots of the poor neighbours or protect the small producers' limited crops from pests and diseases.

In addition to the indirect employment-enhancing potential of *transgenic* plants and microbes, employment could be directly increased if the Third World programmes of micropropagation techniques could be directed to labour-intensive food crops in order more cheaply to offer disease-free plantlets in large quantities and on time for planting by the vast majority of the farmers who are small cultivators.

Table 29.4 Commercial and government micropropagation laboratories: the Philippines[a]

Classification of laboratory	No	Grouping according to crops cultured			
		Orchids only	Orchids + ornamentals	Orchids + food/fibre	Food/fibre only
Government-run	7	3	1	1	2
Private	21	19	0	2	0
Total	28	22	1	3	2

[a] *Source*: Zamora A. B. and R. C. Barba, p. 51, table 1.

Table 29.5 Commercial and government micropropagation activities: Mexico

Firm (private/public)	Species	Total annual production of micro-propagated plants	Investment in tissue-culture laboratory (US dollars)	Number of employees in laboratory	Market % national	% international
Biogenetica Mexicana S.A. de C.V. (private)	Gerbera, Gypsophyla Dieffenbachia, Caladium, Spatiphyllum	100,000	90,000 (1984)	4	100	–
El Rancho, La Joya (Nursery) (private)	Orchids	100,000	100,000 (1985)	2	20	80
Invernamex (Nursery) (private)	Gerbera, Gypsophyla, Strawberry, Raspberry Blackberry	600,000	400,000 (1989)	8	100	–
Viveros 'El Morro' (Nursery) (private)	Spatiphyllum, Singonium	300,000	400,000 (1988)	7	–	–
FIRA (Bank of Mexico) (public)	African violet, Gerbera, Chrysanthemum Strawberry	100,000	45,000 (1988)	7	100	–

Source: Eastmond, et al., table 3.

Micropropagation techniques can be immediately deployed to enhance rural employment. The capacity to generate micropropagated disease-free planting material in Mexico (tequilina) and Nepal (potato) has already been demonstrated to be cheaper than the imported planting material. Indeed, this technique is already applied to potato in 30 poor countries, as explained in the section on multiple cropping. Singapore has the capacity (Plantek International) to provide disease-free coffee plantlets for large-scale plantings throughout South East Asia. Similarly, the government agency in Brazil (EMBRAPA) is able to produce coffee plantlets (*Biotechnology and Development Monitor*, 1990). We have already noted the capacities existing in Malawi, Nepal, Vietnam and Kenya.

This scientific capacity that developing countries possess is mainly being channelled to cover non-food crops and to meet the needs of the commercial and large-farm sector. This is clear from the developments in the Philippines, Mexico and India, although these may create employment indirectly for the hired rural workers.

The Biotechnology Department of the Indian Ministry of Science and Technology has supported (University of Delhi) micropropagation techniques for bamboo (40,000 plantlets by 1989–90), oil palm (5,000 plants by January 1990), coconut and natural rubber. Some commercialisation has been initiated or achieved for cardamom and bamboo (Mani, 1990). Similarly in the Philippines, out of the 28 commercial and semi-commercial tissue-culture laboratories owned privately or by the government, 22 are devoted solely to orchid propagation (see table 29.4). Only five laboratories propagate food/fibre crops and two laboratories which are solely devoted to food crops are both government owned.

Similarly in Mexico (see table 29.5), the micropropagation enterprises are geared to market plantlets for *nonfood* crops. Certainly, crops (maize, beans and other cereals) which constitute the basic Mexican diet receive little attention. Maize is grown primarily by the peasants and the application of micropropagation techniques to maize is not profitable at current prices.

A special technique developed in Japan could enable roughly 3 billion rice seedlings to be grown from a single seed in about 6 months. This opens up vast prospects for the hungry masses of the densely populated major rice-growing areas of the Third World. This technique saves on seeds (releasing more grains to feed the hungry) and plenty more seedlings to plant for the countless unemployed hands. In labour-scarce Japan, *robots* are being sought to meet the intensive labour demands for root separation of seedlings grown by the new technique (UNIDO, 1989).

As discussed in an earlier section, the above creates employment for the scientific personnel in tissue-culture laboratories and nurseries to blend skills with the traditional agricultural workers.

Differential Impact on Employment

The specific biotechnologies are reviewed for their employment implications in the industrialised and developing countries. These are the Bovine Somatotropin (BST) and Single Cell Protein (SCP) technologies in the dairy and animal protein feed industries respectively.

BST could boost overall employment and incomes in Mexico and Pakistan (Ahmed, 1991). In Mexico the BST technology could reduce the daily deficit of 12.5 million litres of milk and make it more accessible to the population (37 per cent of the population currently consume only 14.5 per cent of the available milk supply). It increases milk production in dairy cows by 10–25 per cent. This is like having extra milk without extra feed. Purchasing power could be increased by stimulating employment in the production and processing of milk and the feed industry, all of which are concentrated in a few hands. BST also holds prospects for Pakistan. Despite having $3\frac{1}{2}$ times as much pasture as Wisconsin and $1\frac{1}{2}$ times as many dairy cows, Pakistan produces only a quarter as much milk. Pakistan's cows are only 15 per cent as efficient as Wisconsin's. As a consequence, Pakistan has to spend about $30 million importing (mostly dried) milk each year (*The Economist*, 13 January 1990).

In contrast it is feared that the BST technology could lead to the demise of the family farm in the US. As regards the bias in favour of larger dairy farms, it is argued by four major manufacturers of BST that its cost of less than $1 a day per cow would make it *scale-neutral* (Schneider, 1989). Moreover, others argue that the trend towards fewer and larger dairy farms (e.g. 30 per cent decline in the number of dairy farms in the US over a short period of time) was already in existence irrespective of the introduction of BST (Buttel and Geisler, 1989).

However, to be able to offer BST at such low prices to farmers requires significant economies of scale to reduce unit costs of producing BST. It is clearly demonstrated that increasing the scale of production of BST from 0.5 million to 7 million doses per day reduces average costs from $4.23 per gram to $1.97 per gram (Kalter et al. 1984). If the size of the market was limited domestically to the US, a plant with the capacity to produce 7 million doses would cover nearly two-thirds of the US dairy herd (Molnar and Kinnucan, 1989). Even lower unit costs (to the level of $1 per day per cow as claimed above) through larger economies of scale can be achieved as even bigger global (approaching $1 billion annually) and international ($100–$500 million annually) markets are emerging (Schneider, 1989, and UNDP, 1989).

Therefore, there is an economic incentive to manufacture BST under monopoly conditions. It is little surprising that four giant multinationals, Monsanto, Eli Lilly, Upjohn and American Cyanamid, are currently engaged in the development of BST (Schneider, 1989). While this offers the prospects of supplying BST cheaply to dairy farmers worldwide including the Third World (making it more scale-neutral at the user level), it creates a *monopsonistic* market structure in the US with concentrations in *both* the dairy and BST-manufacturing industries.

SCP technology in developing countries such as Nigeria, Venezuela and Cuba which possess the substrate like natural gas can create jobs in the industry and reduce malnutrition in developing countries. For instance, in Nigeria the economic climate is favourable for its acceptance as can be seen from the following: (a) the income elasticity of demand for poultry products is higher than that for beef; (b) the supply–demand projections reveal an excess demand for poultry products; (c) relative prices of other sources of poultry feed (soybean and fish meal) compared with SCP are higher and on the increase; and (d) the ban on the import of poultry products and poultry feed provides the protection and opportunity for import substitution.

In contrast, to protect employment in the soya protein industry in the US, which accounted for two-thirds of world exports, the development of the SCP technology was shunned (Junne, 1990).

Induced Innovation

The theory of induced innovation (Ahmed and Ruttan, 1988) suggests that technological change enables the farmers to use less of relatively higher-priced agricultural inputs. As the agricultural wage rate rose compared with agrichemicals in the US, technological change was clearly labour-saving in its character. With the advant of biotechnology, this trend will continue, as herbicide-tolerant *transgenic* crops lead to the application of higher doses of chemical herbicides with corresponding reduction in labour use.

In contrast, large subsidies on agrichemical use in labour-abundant developing countries have led to high applications of chemicals and lower use of labour. The factor price distortions certainly are labour-saving. For instance, in Indonesia, the farmers, who pay 10 per cent of the cost of fertilisers and pesticides (Barbier, 1989) naturally are induced to overuse this input and save on labour. Structural adjustment measures which remove subsidies and use the private sector for marketing and distribution of inputs and products in Third World countries would induce agrichemical savings and could contribute to the adoption of more labour-intensive biotechnology applications. This is evident from the fact that currently, seeds account for some 20 per cent of the cost of wheat production in Europe, while fertiliser (45 per cent) and pesticides (35 per cent) absorb the remaining 80 per cent (*The Economist*, 1987). With the advent of biotechnology it is believed that seeds could account for nearly 50 per cent of the total cost. This implies that the induced technological innovations process could lead to further substitution of labour by capital.

Global Impact on Employment

While much of the above analysis of the employment impact of biotechnology makes an assessment of what could happen at the sectoral level, a wider global assessment is necessary to identify the impact of biotechnology beyond the sectoral and national boundaries.

Inter-industry repercussions

It is clear from the foregoing discussion that APB is able to develop new crop varieties that are more resistant to disease, drought or pests. New biotechnologies applied to animal husbandry (e.g. BST) enable animals to metabolise feed more efficiently, and their application to the feed industry (e.g. SCP) boosts protein content. This reduces input requirements per unit of output for pesticides, feed (vegetable sources of animal feed), energy for irrigation, etc.

For instance, in Malawi intermediate material inputs used per unit of farmland have declined (by 70–180 kwachas) as a result of the application of APB to the tea crop.[2] There is less need for chemical means of plant protection, as noted earlier in the discussions of Mexico and China. The application of BST to dairy cows reduces feed requirements. Similarly, SCP production in Nigeria reduces the need for soybean meal as poultry feed.

The adoption of biotechnological innovations having significant resource-saving (including labour) effects will therefore generate a series of intersectoral repercussions throughout the economy. The overall cumulative impact of these repercussions, as they spread from one sector to another, may be very great indeed.

An attempt was made using input–output techniques to trace and quantify these direct and indirect impacts on output and on employment through simulation exercise (Lee and Tank, 1991). The results of the simulation reveal that under resource-saving biotechnology the gross domestic product and aggregate employment will decline unless exogenous demand for output from one or more industries can be stimulated or new industries are created. Indeed, as has been observed in China, labour released from farming has been deployed in numerous new and biotechnology-related specialised services (Ahmed, 1991).

The new international division of labour

Biotechnology developments in the industrialised countries, often stimulated by protectionist measures, have disrupted the existing global trading patterns. The new international division of labour is essentially characterised by the unemployment of plantation workers and the loss of the livelihood of small producers engaged in the traditional export sector of developing countries.

   The most quoted example is that of the High Fructose Corn Syrup (HFCS) which threatens the livelihood of 50 million workers engaged in the sugar industry of the Third World (Panchamukhi and Kumar, 1988). Biotechnology developments for numerous other export crops will affect the earnings of the most vulnerable groups of the Third World population. For instance, the substitution of vanilla flavour by biotechnology substitutes threatens the livelihood of 70,000 small farmers in Madagascar (Mushita, 1989). Such adverse effects on the employment of the Third World poor are spreading rapidly to many more export commodities which are substituted by biotechnology-derived products. These include the small cocoa producers in Cameroon, Ghana and the Ivory Coast; coffee producers in Colombia, Burundi, Uganda and Ethiopia. Jobs of 500,000 small coffee producers in Rwanda and another 650,000 in Indonesia are directly threatened by such substitutions (Ahmed, 1991).

   Apart from the North–South trading patterns, biotechnology developments have jolted South–South trade. Biotechnology applications to convert vegetative oils to produce lipids or tailored fats will affect the market shares of 11 vegetative oil crops traded by the countries of the South. For instance, decline in Philippine coconut exports will directly affect the employment of the 15 million Filipinos directly dependent on the coconut industry.

Concluding Remarks

The limited empirical evidence suggests that overall labour intensity in agriculture will not be affected by labour displacement in some agricultural operations (e.g. in plant protection) as it will be made up by increased labour absorption in newly created crop operations. Some attention will have to be given to agrarian reform measures since, in addition to the biotechnology developments, institutional factors like farm size distribution could also determine labour use. Overall rural employment will expand, bringing in newer jobs through a greater blending of skills, primarily that of women scientists, and through the linkage effects stimulated by biotechnologies.

The chapter also reveals that biotechnologies can be specifically designed and better targeted to create more jobs. Induced innovations process will lead to biotechnology developments and applications which will tend to substitute capital for labour in the industrialised countries and tend to have the converse effect for Third World countries, particularly with greater price liberalisations and withdrawal of subsidies on inputs as a result of structured adjustment measures.

The cumulative effect of inter-industry repercussions of the resource-saving biotechnologies is to reduce GDP and depress aggregate employment. A new international division of labour will emerge due to the substitution of traditional Third World exports by biotechnology products in the industrialised countries. Plantation workers and small producers will bear the brunt of job losses that will result from this process of substitution.

NOTES

[1] Genetic manipulation of many species of potatoes is easier because they carry genes on four sets of chromosomes in each cell, as compared with two sets carried by animals and most other plants (*The Economist*, Oct. 1990).
[2] Intermediate material inputs include chemicals (pesticides, fungicides and herbicides) used in land protection and chemical fertilisers, and the fuel/power required to deliver them.

REFERENCES

Ahmed, I. (ed.), *Biotechnology: A hope or a threat?* (Macmillan/ILO, 1991).
Ahmed, I., *Technological change and agrarian structure: A study of Bangladesh* (Geneva, ILO, 1981).
Ahmed, I. and Vernon W. Ruttan, *Generation and diffusion of agricultural innovations: the role of institutional factors* (Aldershot: Gower, 1988).
Barbier, E., The contribution of environmental and resource economics to an economics of sustainable development, in *Development and Change*, Vol. 20, No. 3, July 1989.

Bartsch, W. H., *Employment and technology choice in Asian agriculture* (New York: Praeger, 1977).

Biotechnology and Development Monitor (The Hague), No. 4, September 1990.

Buttel, F. H. and C. C. Geisler, The social impacts of bovine somatotropin: Emerging issues, in Joseph J. Molnar and H. Kinnucan (eds), *Biotechnology and the new agricultural revolution* (Boulder, Colorado: Westview Press, 1989).

Doyle, J., *Altered harvest: Agriculture, genetics and the fate of the world's food supply* (New York: Viking, 1985).

Eastmond, A., Amarelia, R. L. Gonzalez, H. L. Saldana and M. L. Robert, *Towards the application and commercialisation of plant biotechnology in Mexico*, Universidad Autónoma de Yucatán, Mexico, Oct. 1989, unpublished draft.

Eastmond, A., Amarelia and Manuel Robert, Advanced plant biotechnology in Mexico: A hope for the neglected?, in Iftikhar Ahmed (ed.), *Biotechnology: A hope or a threat?*

Halos, S. C., *Biotechnology trends: A threat to Philippine agriculture?*, in I. Ahmed (ed.), *Biotechnology: A hope or a threat?*

Hameed, N. D. A., et al., *Rice revolution in Sri Lanka* (Geneva: United Nations Research Institute for Social Development, 1977).

Johnston, B. F. and P. Kilby, *Agriculture and structural transformation: Economic strategies in late-developing countries* (London: Oxford University Press, 1975).

Junne, Gerd, The impact of biotechnology on international commodity trade, paper presented at the *International Seminar on the Economic and Sociocultural Implications of Biotechnologies*, Vézelay, France, 29–31 October 1990 (Paris: UNESCO, 1990).

Kalter, Robert J. et al., *Biotechnology and the dairy industry: Production costs and commercial potential of bovine growth hormone*, Department of Agricultural Economics, AE Research 84–22 (Ithaca, New York: Cornell University, 1984).

Keya, S. O., J. Freire and E. J. DaSilva, MIRCENs: Catalytic tools in agricultural training and development, in *Impact of Science on Society* (Paris: UNESCO), No. 142.

Lee, Harold L. and Frederick E. Tank, *A conceptual framework for biotechnology assessment*, in I. Ahmed (ed.), *Biotechnology: A hope or a threat?*

Manandhar, A., S. Rajbhandari, P. Joshi and S. B. Rajbhandari, Micropropagation of potato cultivars and their field performance, in *Proceedings of national conference on science and technology* (Kathmandu: Royal Nepal Academy of Science and Technology, 1988).

Mani, Sunil, Biotechnology research in India: Implications for Indian public sector enterprises, in *Economic and Political Weekly* (Bombay), 25 August 1990.

Molnar, Joseph J. and Henry Kinnucan, Introduction: The biotechnology revolution, in Joseph J. Molnar and H. Kinnucan (eds), *Biotechnology and the new agricultural revolution* (Boulder, Colorado: Westview Press, 1989).

Mureithi, L. P. and B. F. Makau, Biotechnology and farm size in Kenya, in I. Ahmed (ed.), *Biotechnology: A hope or a threat?*

Mushita, A. T., The impact of biotechnology in developing countries, in *Development* (Society for International Development, Rome, 1989, 2/3).

Okereke, G. U., Biotechnology to combat malnutrition in Nigeria, in I. Ahmed (ed.), *Biotechnology: A hope or a threat?*

Panchamukhi, V. R. and N. Kumar, Impact on commodity exports, in *Biotechnology Revolution and the Third World*.

Rajbhandari, S. B., Plant tissue culture method and its potential, in *Proceedings of national conference on science and technology* (Kathmandu: Royal Nepal Academy of Science and Technology, 1988).

Schneider, Keith, 5 big chains bar milk produced with aid of BST, in the *New York Times*, 24 August 1989.

The Economist, 15–21 August 1987.

The Economist, 13 January 1990.

The Economist, 13 October 1990.

UNDP, *Programme Advisory Note: Plant biotechnology including tissue culture and cell culture* (New York: UNDP, July 1989).

Unnevehr, L. J. and M. L. Stanford, Technology and the demand for women's labour in Asian rice farming, in *Women in rice farming* (Aldershot: Gower for IRRI, 1985).

Uyen, N. V. and P. V. Zaag, Potato production using tissue culture in Vietnam: The status after four years, in *American Potato Journal*, Vol. 62, 1985.

Zamora, Alfinetta B. and R. C. Barba, Status of tissue culture activities and the prospects of their commercialisation in the Philippines, in *Australian Journal of Biotechnology*, Vol. 4, No. 1, Jan. 1990.

30

Some Social Implications of Diagnostic Applications of Biotechnology

H. David Banta

Introduction

The 'new biotechnology' has initiated a revolution in diagnostic and screening tests. The ability to develop more definitive diagnostic technologies and predictive tools, as well as more cost-effective therapies, such as pharmaceutical products, gives biotechnology the potential to reduce the severity and burden of disease in the population. Biotechnology has also been applied directly to human material to map the structure of DNA and to analyse gene mutations. Direct genetic diagnosis is becoming possible for an increasing number of conditions. In addition, since disease depends on the interaction of the genetic inheritance of the person, with the broader influences within the person's body and in the environment, much more will gradually be understood about the nature of multifactorial diseases such as cancer and heart disease. This will also contribute to early diagnosis, genetic counselling, and prevention.

This chapter acknowledges these important positive outcomes, but will focus on possible problems. The availability of kits for diagnosis raises certain social issues of importance. The field of genetic testing brings up social issues of a more profound nature.

Monoclonal Antibodies and Diagnosis

Monoclonal antibodies are highly specific, homogeneous antibodies. Their greatest immediate application is in diagnosis. Monoclonal antibodies will lead to improvements in available tests, and in many cases make them easier, quicker, and cheaper to perform. One implication is that diagnostic kits are possible, making diagnosis in the clinic or even the home more feasible. In many cases, the test is simplified by the use of this technology. One major advantage of a test based on monoclonal antibodies is that the results can be obtained in a short time and should be quite precise, if the test is performed correctly. The impact of this new technology applied to diagnosis can include early rational management of patients with infectious disease.

By 1987, diagnostic kits based on monoclonal antibodies were available for more than 20 conditions, including kits for pregnancy testing, testing for iron deficiency, and testing for acute heart attack (by measuring blood enzymes). By 1989, products were available to measure susceptibility to blood coagulation (thrombosis), cancer (cancer markers), and susceptibility to emphysema. The U.S. market in 1988 for diagnostic tests based on biotechnology was US$250 million, and growing at more than 15 per cent per year. The greatest volumes were made up of cancer tests ($65 million) and pregnancy tests ($60 million). Kits have been developed for diagnosis of bacterial meningitis, streptococcus A, Chlamydia, Candida infection, detection of bacteria in the urine, and others. By 1987, in Europe, the market for kits to diagnose sexually transmitted disease had risen to US$149 million, and was expected to grow rapidly.

In the future, a number of diagnostic kits could be offered to the general public. The first, a test for pregnancy, has already been marketed. Others include tests for sexually transmitted diseases (e.g. gonorrhea), and hepatitis. Certain screening tests for cancer (e.g., of the uterine cervix and prostate) could be offered by 1995. Other possibilities include skin diseases and periodontal disease. The U.S. market for home diagnostic tests had already reached US$350 million in 1988. In the United States, a firm has applied to the Food and Drug Administration (FDA) for permission to market a kit for diagnosis of the HIV virus (AIDS) directly to the public, but has not received permission. Genetic screening kits are also being developed for home use, especially by U.S. companies. Companies plan to develop and market tests for common genetic diseases such as diabetes.

Diagnosis seems certain to move progressively to less centralized sites. Diagnostic services in the home, using kits and dry chemistries, will almost certainly increase. Clinicians can do more and more tests in the home. In addition, the patient him or herself can do certain tests. Whether home diagnosis by lay people using kits is a beneficial practice or not is a major issue for the future. When dealing with chronic diseases, continual monitoring of a person's biological functions through such means as diagnostic tests is often necessary. The modern trend when dealing with chronic diseases is self-care by the patient. It is not feasible to have a trained health-care professional with every person having a chronic disease all of the time. Increasingly, people with certain chronic diseases monitor their own clinical status and adjust their therapy accordingly. Monitoring can also be done by different providers: relatively simple monitoring by family assistants, more complicated monitoring by nurses, and complicated disease monitoring by physicians. Diabetes is the prototype condition for self-monitoring with professional support and supervision. Several manufacturers sell portable devices or chemically-treated test strips that the diabetic patient can use at home. The patient is able to check his or her blood sugar two to four times a day, and can adjust the amount of insulin needed accordingly. If necessary, consultation with the clinician by telephone can be used to check any decision.

The most important immediate implication of this development is for accuracy of the test results (whether the results are correct or incorrect). In the United States, laboratory functioning is regulated, and required proficiency tests ensure a minimum level of accuracy. Physician office testing, however, is usually not regulated. In Europe, regulation of clinical laboratories is less developed. Voluntary and professional efforts generally ensure a minimum level of quality in hospital laboratories. Practitioners usually do not have their own laboratories. Careful thought needs to be given to how to ensure quality if testing moves more to the office and clinic setting.

A great problem comes with direct marketing of kits to the public. In Europe there is no FDA to require evaluation of diagnostic kits before they are marketed. Governments have limited authority to ensure the correctness of advertising and promotional material. In the case of the HIV test kit in the United States, organizations involved with AIDS requested FDA not to allow direct sales to the public, since correct interpretation of test results and counselling could not be ensured.

The move of testing to the home can promote autonomy and self-care, but it also has dangers:

- Lay people are not taught to perform tests nor to interpret their results.
- No test can be completely accurate.
- Interpretation of test results depends on consideration of the clinical picture and the outcome of other diagnostic procedures.

The implications of false positives and false negatives need to be carefully considered in terms of potential anxiety or reassurance, and the subsequent health-care consequences.

Genetic Diagnosis and Screening

New techniques developed in the research laboratory will enlarge the scope of genetic diagnosis to a wide variety of genetic diseases and susceptibilities. There are said to be about 10,000 disorders with a strong genetic component. Until now only about 200–300 have been related to specific gene defects. The new technology will make it possible to diagnose directly all 10,000 genetic disorders and will provide gene probes to diagnose carriers and embryos carrying the disease. The first applications will be for the intrauterine diagnosis of genetic diseases such as cystic fibrosis and Duchenne muscular dystrophy. Carrier screening for such diseases as phenylketonuria is becoming feasible. Carrier screening in Cyprus, Sardinia, and Greece has led to impressive declines in hemoglobin diseases such as Thalassemia B. Tay-Sachs disease can also be prevented in this way. The gene for cystic fibrosis was identified and cloned in 1989, which means that this disease is now preventable. (One in 20 members of the population carry the gene.) Screening will then be extended to genetic disorders that become clinically manifest at adult age, such as Huntington's disease.

The increasing knowledge of each individual's genetic make-up will bring many social issues to the fore. Who owns the information? Can it be kept in data banks without the individual's permission for purposes of medical research? Does the individual have the responsibility to inform members of the family or potential marriage partners? Does the society have the right to require that others concerned be informed?

Many diseases have a genetic component. In the longer term, genes will probably be identified for susceptibility to a wide variety of cancers (e.g., breast and colon cancer), cardiovascular disease, diabetes, and affective disorders (such as depression). A gene is known to influence Alzheimer's disease. Very often this information will indicate a probability of risk of the disease, especially if the person comes into contact with certain substances, either through environmental and workplace exposures or through personal behaviour. Those at high risk can alter their life styles, participate in intensive screening programmes, and so forth, to prevent development of certain diseases, or they can be followed medically so that disease can be identified in its early stages. In the long run, this capability in itself will probably lead to considerable change in the health status of the population.

It will become possible to screen each neonate for its genetic structure, including such factors as likelihood to develop cancer or to be a carrier. All known genetic disorders could be discovered at any moment of an individual's life, beginning in the uterus and continuing until death. Cells could be kept frozen to be examined later, as new capabilities appear.

The question here is what intervention is justified in the interests of the health of the population. The implications, in terms of individual privacy, medical ethics, and the norms and values of society, could be profound and far-reaching.

During the next decade or two, such testing will be too expensive to be routine. However, automation of tests seems likely, and perhaps within 15–20 years it will be feasible to do a complete genetic profile on each individual. In practice, selection of testing to cover only a small proportion of known genetic diseases and susceptibilities will probably be the norm.

Genetic testing raises a number of complex and interrelated issues. For example, one could possibly envision widespread testing for certain susceptibilities, such as to breast cancer, as a first step in the design of a screening programme. Testing could also be the first step in health-information activities. For example, those susceptible to certain diet-related diseases could be targeted by specific health-education programmes. Such activities would result in an increasing medicalization of society. The complexities make predicting future acceptance and implications difficult.

One important implication of the use of genetic tests for susceptibilities to disease is the use of testing for selection purposes by employers and insurers. The chemical industry could, for example, not hire those susceptible to diseases associated with exposure to certain chemicals. Insurance companies could reject applicants carrying the gene for certain diseases or susceptibilities that might be deemed to be associated with future poor health or early death.

Genetic testing raises a number of problems of individual autonomy and freedom of choice, confidentiality of personal data, and broader social implications. Genetic testing has always been a somewhat controversial field, and ethical analyses and discussions are frequent. Within the health care system, such problems may be dealt with without great disruption. A more serious problem is the use of genetic information outside of the health care system. The question is, should genetic testing be used to enhance autonomy and self-determination, or could it also be used to promote economic ends, such as profit maximization of insurance companies. Genetic information has already been used in job selection in the United States. Insurance companies may soon begin to test potential clients, or at least to ask if they have ever been tested for genetic disease. It has been suggested in the Netherlands that such activities should be forbidden by law.

Conclusions

The new biotechnology provides many new tools for the diagnosis of disease and for the identification of carriers of adverse genes. The point is that these technological tools are not neutral, they have social consequences.

If used wisely, these new tools can improve health. In many cases they can also increase individual choice and autonomy. This chapter concerns another possibility: that they may have adverse consequences. In particular, genetic tests could lead to attacks on individual choice. Stigmatization of those with genetic problems and redefinition of handicapping conditions are certainly possible. And the use of genetic tests by non-health organizations raises great concern.

The hope of this chapter is to encourage early and active discussion of these issues so that the value of the new biotechnology for diagnosis may be maximized. The social consequences of advances in genetics are in themselves so profound and their use so dependent on social and cultural factors that widespread discussion in society is warranted.

ACKNOWLEDGEMENT

The author is particularly grateful to Dr S. Gevers, Academic Medical Centre, Amsterdam, who contributed greatly to the material presented in this chapter.

GUIDE TO FURTHER READING

H. D. Banta and A. C. Gelijns (eds), *Anticipating and Assessing Health Care Technology*, Volume 6, *Applications of the New Biotechnology: The Case of Vaccines*. Report of the Scenario Commission on Future Health Care Technology (Kluwer Academic Publishers, Dordrecht, 1988).

H. D. Banta and A. C. Gelijns (eds), *Anticipating and Assessing Future Health Care Technology*, Volume 5, *Developments in Human Genetic Testing*. Report of the Scenario Commission on Future Health Care Technology (Kluwer Academic Publishers, Dordrecht, 1988).

N. A. Holtzman, *Proceed with Caution: Predicting Genetic Risks in the Recombinant DNA Era* (Johns Hopkins University Press, Baltimore, 1988).

Office of Technology Assessment, Commercial Biotechnology: An International Analysis (US Government Printing Office, Washington, DC, 1984).

Office of Technology Assessment, The Role of Genetic Testing in the Prevention of Occupational Disease (US Government Printing Office, Washington, DC, 1983).

H. Rigter, et al. (eds), The Social Consequences of Genetic Testing, Proceedings of a Conference, 16–17 June 1988, Leidscendam, The Netherlands (Scientific Council for Government Policy, The Hague, in press).

Part VI

Policy

Introduction to Part VI

The main concern of this section is policy – what should be done about biotechnology?

The first point to be made is that the 'engine' of the biotechnology revolution (if revolution it indeed is) is located largely in companies, both large and small, since it is in these companies that biotechnological knowledge is commercialised in the attempt to meet market expressable needs. We have seen, however, that unlike in the case of other new technologies where the technologies have been generated primarily within the private corporate sector, in the case of biotechnology the technologies have originated largely in the university and government health-related sector. This raises important policy questions since it means that knowledge emerging in different sectors must be brought together if successful commercialisation is to occur. How is this best achieved?

One policy question relates to the appropriate relationship between universities and government research institutes on the one hand and the private corporate sector on the other. As this section shows, this question is in practice extremely complex. While some have argued that public research institutions should be made more 'relevant' to the commercial needs of companies, others have suggested that this will lead to an undermining of the basic research that has been an important source of change in the capitalist process of development.

Another policy question relates to the role that government should play in helping to reap maximum benefit from biotechnology. Again the issues are complex. There are those who believe that the role of government should be minimal, leaving, as far as possible, decision-making to companies and other institutions. (Just how far the government should go in attempting to regulate biotechnology is one example of a controversial subquestion in this area.) On the other hand, others have argued that the government has a central role to play. In the present section these kinds of debates are examined explicitly and implicitly in the cases of Europe, Japan, and the United States.

Further policy questions arise in connection with Third World countries. Does government need to play a special role in these countries as a result of their lack of development? And relatedly, what role should be played by international organisations. These issues are also discussed in this section.

31

The Role of Private Companies and the State in the Promotion of Biotechnology: Options for Government

Margaret Sharp

Introduction

The purpose of this chapter is to consider how best governments may promote the development of biotechnology. It is wrong to talk about biotechnology as an industry. It is not an industry but a technology, or rather a set of technologies. It is above all a set of new approaches to old problems, and new ways of doing things – a process technology not a product technology. To promote the development of biotechnology, therefore means promoting the rapid diffusion of these new process techniques. This chapter argues that the key to this diffusion process does not always lie, as some would suggest, with the small-firm sector, which is seen essentially as an American phenomenon. For Europe and Japan, the burden of diffusion rests primarily with the major pharmaceutical and chemical companies. It further argues that the best way to promote the diffusion process lies in strong support for the science base, the building of bridging mechanisms between industry and academia, and a tough but sympathetic regulatory environment.

Three Important Characteristics of the New Biotechnology

The new biotechnology is a prime example of what is termed a new technological paradigm. (For a discussion of this concept, see Dosi, 1988). The standard definition, say that used by the OECD – 'the application of scientific and engineering principles to the processing of materials by biological agents' (Bull, Holt and Lilly,

1982) – emphasises the broader processing base of biotechnology, but fails to capture the extent to which developments in genetic engineering in the early 1970s constituted a radical change of technological trends. Two seminal break-throughs – recombinant DNA and the cell fusion methodology of monoclonal antibodies – have opened up hitherto undreamed of developments in science, particularly in the fields of medicine and agriculture. In doing so this has changed the whole set of basic ground rules, opened up new approaches to old problems, and new fields of development. (See chapter 11 – Applications of Biotechnology: An Overview.)

Three important characteristics of the new biotechnology associated with this shift in ground rules can be identified:

(i) *Pervasive uncertainty.* The shift in ground rules means that techniques are, at least initially, untried and experimental, and the public reaction to the new, genetically engineered products, unknown. Those moving in early to develop these techniques were faced by genuine uncertainty – the risks were not quantifiable. By early 1982 it was clear that the first phase – the cloning and expression of simple proteins – had been mastered, and some of the patentability issues were resolved. But the fast movement of the science base, the problems of scale up, and the remaining question of public acceptability left many issues subject still to considerable uncertainty.

(ii) *The importance of the academic science base.* More than in any other new technology (except perhaps nuclear power in the 1950s), the locus of knowledge lay in the academic sector, and required mechanisms of technology transfer which both accessed that knowledge base and helped translate science into technology. Moreover, the continued fast movement of the science base – for example, current developments in protein engineering – have meant continued dependence on the academic sector, whereas in other new technologies the locus of knowledge has rapidly shifted into the research laboratories of the company sector.

(iii) *The need to mix established disciplines.* In both academic and industrial settings biotechnology demands project teams which span traditional disciplines – protein chemists and biochemists, molecular geneticists and molecular modellers, microbiologists, protein crystallographers, fermentation scientists, and separations process experts. For many firms it has meant accessing new skills and building up new cross-disciplinary in-house capabilities. Putting such teams together takes time and care. The short-term solution has been to buy-in such skills on a temporary basis from the academic sector (from where they originate) or via the new small biotechnology firms.

These characteristics have had important effects on the way in which biotechnology has developed.

The New Biotechnology Firm – an American Phenomenon

The most publicised phenomenon of the development of biotechnology has been the emergence of the new specialist biotechnology firm – the NBF. It has above all been an American phenomenon and the story is well known. Starting with the launch of Genentech in 1976, numbers grew rapidly from 10 in 1978 to 40 in 1980, to 85 by 1981, 150 by 1983, to 250 by 1985 and 300 by 1987 (OTA, 1988). In spite of the many firms which fell by the wayside both in the stock market crash of 1987 and in the normal rough and tumble of small-time corporate life, company start-up has continued and there are today some 400 small biotechnology firms in the US. As *The Economist* commented recently:

Call it a miracle. Or may be madness. Over the past decade hundreds of tiny biotechnology firms have sprung up and new ones are still being created almost every week. . . . Only one firm – Genentech – makes a sustained profit, and even that is disappointingly small. Yet investors continue to pour money into these companies on an heroic scale – $10 billion so far. How much longer can this industry defy gravity? (*Economist*, 1989)

It is easy to see in retrospect why the NBF grew and flourished in the early years of biotechnology. Essentially they fulfilled two functions. First, they were a hedge against uncertainty – a clever way by which the larger firms were able to 'keep a window' on the technology while keeping their own commitment (of money and people) to the minimum. It was only after the uncertainties of the early years were eliminated that the big firms began making substantial investments in in-house capabilities. Secondly, the NBF provided a bridge between academia and industry. They were essentially a means of accessing the scarce skills of the academic sector and putting them at the disposal of industry. Many NBFs were effectively private research laboratories for bright professors, separating academia from commercial activities (which often suited both university and client), and enabling the academic to realise the value of scarce skills. They also enabled cross-disciplinary teams to be rapidly assembled across the various disciplines without upsetting the established institutional framework either within the academic sector or within industry.

Funding for the NBFs came partly from the large companies, but most funding came from the venture capital market – from individuals and institutions who were prepared, in effect, to gamble on their development.

Table 31.1 presents estimates, derived from a UK Government source, of the number of NBFs in the UK, France and West Germany compared with those in the US. Although it would be foolish to pretend that all of these firms were operating at the leading edge of technology – the number of firms of the calibre of Genentech, Cetus, Amgen or Calgene in the US; of Celltech or British Biotechnology in the UK; or Transgene in France, is small compared with the total – nevertheless the message is clear. The small-firm sector has been, and remains, very active in the US; is of growing importance in the UK, but remains negligible in other European countries. And, although we have no figures for it, it remains negligible in Japan.

Table 31.1 Biotechnology companies and sector of interest, 1986

	US	UK	France	FRG
Agriculture	73	15	5	2
Chemical	37	4	1	4
Diagnostics	141	10	3	6
Food	18	12	2	1
Pharmaceuticals	65	9	2	4
Veterinary	54	6	3	0
Total	388	56	16	17

Source: Coleman (1987).

Why should this be? Why has the new biotechnology firm flourished in the US, but not in Europe or Japan? The answer lies partly in institutions, partly in culture, and partly in the quality of the academic base. But it has, as we shall see, considerable implications for policy.

(i) Institutions

In contrast to the US, the venture capital markets in Europe and Japan are underdeveloped. The most active venture capital market is in the UK, where some half dozen specialist funds are active and an estimated total of over $1 billion has been invested since 1980 in biotechnology (*Financial Times*, 1988). The doyen of this market is the Rothschild Fund – Biotechnology Investments Ltd (BIL) – now capitalised at $200 million and the largest specialist fund in Europe. By contrast, the largest German venture capital fund, Techno Venture Management, established in 1984, had an initial capitalisation of only

$10 million (Yuan, 1987). The availability of venture capital, however, is only one part of the equation. BIL, for example, whose investments span biotechnology and medical technology, have not found in Europe the quality of investment they are looking for; 75 per cent of their investments are in the US, only 25 per cent in Europe, and these concentrated almost entirely in the UK. This pattern of investment is mirrored by nearly all the investment funds in Europe, all of which invest a large proportion of their funds for biotechnology in the small-firm sector in the US, and only a very small proportion in the small firms in Europe.

(ii) Cultural

We need therefore to look beyond the availability of venture capital to find an explanation for the absence of the small-firm sector outside the US. An explanation frequently mentioned is the different academic ethos in Europe and Japan from that in the US. In fact there are two different factors at work here. One is cultural. There remains in many countries, particularly in West Germany and Japan, an elitism which consciously or unconsciously puts curiosity-led academic science on a pinnacle and shuns commercial involvement or commercial interest. American science is not without elitism, but the dominant culture, encouraged by a university system in which many pay their way through college by taking on a variety of part-time jobs, is to cash in on opportunities which present themselves.

There are also major institutional differences in the organisation of academic science between countries. The UK comes closest to the US model, with medical research organised (via the Medical Research Council – MRC) by a mixture of institute (e.g., LMB – Laboratory of Molecular Biology at Cambridge) and university-based programmes, but with the universities playing a substantial role in research in both the medical and biological sciences. In West Germany, although the universities maintain a dual role in teaching and research, much premier research is undertaken in specialist research institutes, particularly the Max Planck Institutes. Japan, whose universities were modelled on the German system, with each professor 'king' of his own laboratory, has a surprisingly underdeveloped academic research base and a system which still shuns direct links between industry and university researchers. In France, universities are primarily teaching institutions, and most research is undertaken in the publicly funded research laboratories of CNRS (Centre National de Récherche Scientifique), INSERM (Institut National de Santé et Récherche Médicale) and INRA (Institut National de Récherche Agronomique), or in the privately financed institutes, chief of which, for biotechnology, is the Institut Pasteur, which has provided the powerhouse for much of French work in the biosciences (Sharp, 1989).

These differences in the organisation of scientific research may well be a major factor in explaining the paucity of new biotechnology firms in Europe. A researcher at a CNRS laboratory in France, or at a Max Planck Institute laboratory in Germany, is the full-time employee of that institution. It is not easy also to be the part-time employee of a private firm, and the opportunity costs of leaving to set up an independent small firm, in terms of loss of seniority, pension rights, etc., may be considerable.

Table 31.2 Breakdown of national expenditures on academic and related research by main field, 1987[a]

| | Expenditure (1987 M$) | | | | | | |
	UK	FRG	France	Neths	US	Japan	Average[b]
Engineering	436	505	359	112	1,966	809	
	15.6%	12.5%	11.2%	11.7%	13.2%	21.6%	14.3%
Physical sciences	565	1,015	955	208	2,325	543	
	20.2%	25.1%	29.7%	21.7%	15.6%	14.5%	21.2%
Environmental	188	183	172	27	859	136	
sciences	6.7%	4.5%	5.3%	2.8%	5.8%	3.7%	4.8%
Maths	209	156	175	34	596	88	
and computing	7.5%	3.9%	5.4%	3.5%	4.0%	2.3%	4.4%
Life sciences	864	1,483	1,116	313	7,285	1,261	
	30.9%	36.7%	34.7%	32.7%	48.9%	33.7%	36.3%
Social sciences	187	210	146	99	754	145	
(and psychology)	6.7%	5.2%	4.6%	10.4%	5.1%	3.9%	6.0%
Professional	161	203	67	82	490	369	
and vocational	5.7%	5.0%	2.1%	8.5%	3.3%	9.9%	5.8%
Arts	184	251	218	83	411	358	
and humanities	6.6%	6.2%	6.8%	8.6%	2.8%	9.6%	6.8%
Multi-	6	32	3	1	217	28	
disciplinary	0.2%	0.8%	0.1%	0.1%	1.5%	0.8%	0.6%
TOTAL	2,798	4,037	3,212	958	14,904	3,736	
	100%	100%	100%	100%	100%	100%	100%

Note: a Expenditure data are based on OECD 'purchasing power parities' for 1987 calculated in early 1989

b This represents an unweighted average for the six countries (i.e. national figures have not been weighted to take into account the differing sizes of countries).

Source: J. Irvine and B. Martin (1990): *Investing in the future: An International Comparison of Government Support for Academic and Related Research* (Aldershot: Edward Elgar).

(iii) The Science Base

Tables 31.2 and 31.3 tell their own story – the sheer weight of spending on the life sciences in the US compared with spending in Europe or Japan. This is reflected in research productivity as measured by citations in learned journals (table 31.3), based on the assumption that a much cited paper will have been an influential one. The US contributes over 50 per cent of the highly cited papers in most academic fields close to biotechnology; the UK, in spite of its low spending, contributed around 10 per cent in most areas; and France, West Germany and Japan contributed at around the 5 per cent level. If the small biotechnology firm is an important means of accessing academic science, it is apparent why the US small biotechnology firm has so much more to offer than its European counterparts.

Table 31.3 Citations of scientific publications in fields relevant to biotechnology (country shares)

		USA	Japan	UK	Germany	France
Biomedical	1973	55.3	3.0	11.2	4.8	4.0
	1984	55.1	6.6	9.4	5.7	4.2
Biology	1973	50.4	4.4	14.1	3.2	2.5
	1984	43.5	7.4	12.8	4.8	2.2
Genetics and Heredity	1973	42.5	4.2	14.5	7.0	3.8
	1984	39.6	7.0	10.0	6.4	4.1
Biochemistry and Molecular Biology	1973	54.7	3.9	10.2	5.1	4.2
	1984	49.1	10.1	9.7	5.3	4.9
Biophysics	1973	50.8	0.2	8.6	6.3	0.9
	1984	72.1	1.8	7.4	4.1	0.2
Cell Biology	1973	54.4	2.8	8.5	6.8	3.3
	1984	55.7	5.2	7.2	7.3	2.5
Microbiology	1973	52.7	4.1	15.1	4.0	3.1
	1984	39.8	4.7	17.4	8.2	4.7
Virology	1973	63.5	3.2	9.2	2.4	1.5
	1984	62.3	7.7	8.5	6.1	3.4

Source: CHI/NSF Data Base (held at SPRU, University of Sussex); adapted from Tanaka (1988).

The Importance of Europe's Large Chemical and Pharmaceutical Companies

Table 31.4 Sales and R&D expenditure of the major chemical firms in Europe USA and Japan

	R&D Expenditure 1987/88	Annual Sales 1987/88	% R&D/Sales $ Million
Bayer	1276.70	20635	6.20
BASF	896.10	22354	4
Hoechst	1231.67	20531	6
ICI	641.24	18242	4.10
Rhone–Poulenc	585.36	9344	6.30
Ciba–Geigy	1122.82	10580	10.60
Sandoz	540.27	6026	9
Akzo	451	8631	4.70
Montedison	291	10641	2.73
Enichem	158	–	3.60
Du Pont	1223	30468	4
Dow Chemical	670	13377	5
Union Carbide	159	6914	2.30
Monsanto	557	7639	7.30
Sumitomo Chem	88	5034	3.19
Ashai Chem	234	6621	3.50
Takeda Chem	216	4378	4.90
Mitsubishi Ch	221	7586	5.10

Source: Annual reports, *European Chemical News*.

Table 31.4 lists some of the world's largest pharmaceutical and chemical companies, together with sales and R&D data. These companies already have substantial involvement in biotechnology, although the depth of that involvement varies from company to company. Companies such as Hoechst and Bayer, Dow, Du Pont, Monsanto, Ciba–Geigy, are all now spending upwards of $100m a year of their research expenditures on biotechnology-related research.[1] As multinationals they have the ability to access knowledge on a worldwide basis. Ciba–Geigy, for example, in addition to its Swiss research facilities, which are extensive and include a major research institute for the biological sciences in Basle, has research centres in the US and in the UK; has two joint ventures in biotechnology, one with the NBF Chiron and one with Corning Glass; and has research contracts/linkages with six other new biotechnology firms in the US and three in Europe. Bayer, which has developed close links with the Max Planck Institute for Plant Breeding in Cologne, established its own (internal) Institute for Biotechnology at Mannheim in 1983, but has three separate research centres in the US – the Miles Cutter Laboratories at Elkhart, the Mobay (agricultural/veterinary) Research Centre in Kansas, and the recently (1985) established West Haven Research Centre, which concentrates on molecular biology and immunology. Bayer also has close links with the Universities of Rochester and Yale in the US, and has had a research link-up with Genentech to develop Factor VIII. In this way, international companies are themselves acting as a technology transfer mechanism, transferring skills from one country to another.

How may Governments best Promote Biotechnology?

What then may governments do to promote biotechnology? To date, the pace of diffusion has been largely dictated by developments in the US, and these have been spearheaded by the activities of the NBFs. Although the US government has, through its Small Business Administration Act and similar legislation, facilitated the existence of such small firms, its main promotional policy for biotechnology has been the new substantial support it has given to academic science, and particularly to the life sciences. Just as spending on military and space programmes in the 1950s and 1960s fuelled developments in microelectronics, so spending on medical research and particularly research into cancer (and now AIDS) has fuelled the development of biotechnology. It is notable that, although regularly threatened with cuts by the Reagan administration, the NIH budget escaped unscathed through the decade of the 1980s, if for no other reason than that Congress did not wish to be seen to be cutting 'life saving' medical research (OTA, 1988, pp. 37–38).

In other respects, the US government's policy towards biotechnology might be described as 'laissez faire' – it has taken the view that it is up to business to exploit the science as best it wishes and it has not sought, as it has in the electronics area, to mount programmes of support in important areas of application. The new biotechnology firm has, as we have seen, been the business response, with large firms only recently stepping up their investment programmes to build up substantial research programmes in-house.[2]

The only other government to adopt a similar laissez faire approach has been the Thatcher government in the UK, which as early as 1981 made it clear that, in its view, if biotechnology promised such riches it was up to the business community to exploit it. But it singularly failed to establish the preconditions for such a policy to work. Funding for the academic science base in the life sciences (table 31.2), already lower than competitors, actually fell in proportionate terms between 1983 and 1987 (from 34.4 to 30.9 per cent) (ABRC/SPRU, 1986) and was also lower in absolute terms, because the UK devotes a smaller proportion of a smaller GDP to academic research than any comparable country. Moreover, although the 1980s has seen the development of a substantial venture capital market in the UK, many would claim that, compared with its US counterpart, it remained unduly risk averse, as the story of Celltech, the most prominent of the UK's NBFs, illustrates only too well.[3] Likewise, the large-firm sector in the UK, dominated by the chemical company ICI, has tended to be more conservative in its approach to biotechnology than its counterparts in other countries.

In contrast to the UK, France and Japan have both established national programmes to promote biotechnology. In Japan, biotechnology was designated a 'next generation basic industry' by the influential Ministry of International Trade and Industry (MITI) in 1980 and it established a ¥26 billion ($80m) 10-year programme with 14 large companies, mainly from the chemicals sector, forming the Biotechnology Industry Association, and subsequently a looser federation of 400 companies in BIDEC – the Bio-Industry Development Centre. But MITI does not have responsibility for either agriculture, education or health services, and inter-departmental rivalry has plagued MITI's efforts to promote a national programme. In particular, the relatively weak base of academic research in the life sciences and poor linkage between universities and industry have limited indigenous capabilities and this explains the active role of many of Japan's major companies in collaborative ventures with US NBFs. These companies are now investing heavily in building up basic research capabilities in-house and using these links to gain knowledge and competence and to train researchers. Company spending, not government spending, is now the driving force behind biotechnology development in Japan.[4]

The French government was more successful than MITI in mobilising efforts behind a co-ordinated programme and in 1982 launched its Mobilisation Programme for biotechnology, pulling together the efforts of ministries, universities and research institutes. Total government spending of approximately $150m p.a. was to be matched by equivalent spending from industry, the whole aimed at increasing the French contribution to the global 'bio-industries' from 7 to 10 per cent. Unfortunately the rhetoric exceeded achievements, and a combination of economic crisis and change of governments meant much reduced commitments with a revamped Mobilisation Plan in 1987 concentrating on technology transfer, particularly between the research institutes and industry. With their major agricultural interests the French have been very keen to develop the agro-food sector in biotechnology, an interest now reflected in some of the European Community programmes (Sharp, 1989).

The governments of West Germany and the Netherlands have both adopted a corporatist approach towards biotechnology. The West German federal government, which in fact introduced its first programme to promote biotechnology as long ago as 1972, is advised in its priorities in relation to both academic and industrial research support by representatives from the major chemical and pharmaceutical industries, and, not surprisingly, these same firms are the recipients of the bulk of the industrial funding. Even so, its funding, amounting to approximately $100m p.a., is marginal compared with the sums which these firms are themselves now putting into biotechnology – Hoechst and Bayer are each reckoned to be spending at least $150m p.a. in this area of research.[5] As indicated earlier, one of the major issues in West Germany remains technology transfer from research institute to industry. This has not proved a problem in the Netherlands, where their Innovation Action Programme in Biotechnology (IOP-B) brings together academics, businessmen and government officials to identify priorities and target funding. Being a small country, the aim has been to develop a niche strategy based on two areas, agriculture and pharmaceuticals, and to concentrate resources accordingly. The degree to which the programme has succeeded in integrating industry's needs with the public research infrastructure is impressive. Generally, however, the last decade has seen a squeeze on resources going into academic research, and there are worries, as in Britain, that the cuts have adversely affected research capabilities. In both West Germany and the Netherlands, there are also fears among industrialists that the rising influence of the Green Party, which wishes to see the prohibition of all experiments involving genetic engineering, will result in over-stringent regulations. Already, for example, BASF and Bayer are planning to move their biotechnological research to the US.

What then can governments do? This quick review of policies put forward by different governments, together with the earlier analysis of the respective role of large and small firms, suggest that today the role of government is limited. In essence the message is that the development of biotechnology has now reached the stage when it is out of the hands of small firms or governments and into the hands of the large corporation. The annual R&D spending on biotechnology of one of these companies far exceeds that of any NBF; it often also exceeds the total of government spending. As multinational companies they can shift resources around the globe to maximise the advantages gained from different locations.

In such circumstances what can any one government do to ensure that its own country gets as much benefit as possible from this new technology? Here's a check list of measures which are worth thinking about.

(i) *More support for basic science.* It is clear that the magnet attracting international firms to invest in US biotechnology research has been the sheer quantity and quality of its research base.

(ii) *Institutional arrangements matter.* The second lesson to be learned is that institutional arrangements matter. Research institutes may have advantages for researchers, but it is clear that they pose problems for technology transfer which have not yet been solved.

(iii) *Bridging mechanisms also count.* The NBF in the US has proved an important bridging mechanism. Outside the US, where the NBF plays only a marginal role, mechanisms have to be established which bring academic research and industry together. There has now been a diversity of experiments, and governments should learn from these.

(iv) *The regulatory environment is important.* The regulatory environment assumes far greater importance for biotechnology than for other new technologies. Fears about the safety, controllability and the long term impact of genetic engineering are understandably strong, and the regulatory authorities need to establish codes of conduct which are adhered to and can be enforced, but which are not so stringent that they throttle research creativity.

(v) *Competition also counts.* Many of the major European-based chemical and pharmaceutical companies have been relatively slow compared with their US and Japanese counterparts to develop capabilities in biotechnology. The reason for this may lie in part with the absence of the small-firm sector and in part with the fragmented regulatory framework, which enables such firms to insulate themselves from the full rigour of the market. 1992 will help, but a tough competition policy is also necessary.

(vi) *Venture Capital helps.* Although it has been argued that promoting the small-firm sector will not, *per se*, help diffuse biotechnology, the presence of the small-firm sector is important both as a vehicle for new ideas and, from time to time, to price the complacency of the large companies.

(vii) *Concentration is useful.* All companies, large and small, find it useful to assess in advance the resources that will be available. Given the vital importance of the public-sector research base to the development of biotechnology, an approach such as that of the Netherlands government, which seeks to co-ordinate public and private-sector activities, finds a ready response from industry.

NOTES

[1] I am indebted to Ilaria Galimberti, a doctoral student at SPRU, for the information contained in this paragraph, which she has collected from annual reports, investment analyses and press reports about the companies mentioned.

[2] This is shown most clearly in the figures quoted in the recent OTA report on investment in biotechnology. Corporate investment in biotechnology shot up from approximately $200 million in 1983 to over $1.2 billion in 1985 (OTA, 1988, pp. 82–83).

[3] Celltech was established in 1980 as a government-initiated small biotechnology firm expressly to promote technology transfer from the MRC research institutes. Its struggles for viable financing make it clear that the City of London never clearly understood the NBF phenomenon. See Dodgson (1990) for a full account of the saga.

[4] The information in this paragraph derives from Lewis (1984), Brock (1987), and personal interviews conducted in 1988.

[5] Same source as note 1.

REFERENCES

ABRC/SPRU (1986). *International Comparisons of Government Funding of Academic and Academically Related Research* (obtainable in mimeo from SPRU, University of Sussex, Brighton, BN1 9RF).

Brock, M. (1987). The Involvement of Government in the Development of Biotechnology in Japan, M. Litt Thesis, University of Oxford.

Bull, A. T., Holt, G. and Lilly, M. (1982). *Biotechnology: International Trends and Perspectives* (Paris: OECD).

Coleman, R. F. (1987). National Policies and Programmes in Biotechnology, paper presented at Canada–OECD Joint workshop on National Policies and Priorities in Biotechnology, Toronto, Canada, 7–10 April. Reproduced in OECD, *Biotechnology: The Changing Role of Government* (OECD, 1988).

Dodgson, M. (1990). *Celltech: the story of a new UK biotechnology firm,* DRC Special Working Paper No. 1, SPRU, University of Sussex, UK.

Dosi, G. (1988). The Nature of the Innovative Process, Chapter 10 in Dosi, G., Freeman, C., Nelson, R., Silverberg, G. and Soete, L. (eds), *Technical Change and Economic Theory* (London: Frances Pinter).

The Economist (1989). The Money Guzzling Genius of Biotechnology, 13 May, pp. 91–92.

Financial Times (1988). *Supplement on Biotechnology*, p. II – Ventures that may appeal to the heart of the City, 28 May.

Irvine, J. and Martin, B. (1990). Investing in the future: An International Comparison of Government Support for Academic and Related Research (Aldershot: Edward Elgar).

Lewis, H. W. (1984). Biotechnology in Japan. Report for the National Science Foundation (mimeo from NSF).

OTA (1984). *Commercial Biotechnology: An International Analysis*, US Congress Office of Technology Assessment, January.

OTA (1988). *New Developments in Biotechnology (4): US Investment in Biotechnology*, US Congress Office.

Sharp, M. (1985). *The New Biotechnology: European Governments in Search of a Strategy*, Sussex European Paper No. 15, Science Policy Research Unit, University of Sussex.

Sharp, M. (1989). Biotechnology in Britain and France: The Evolution of Policy. Chapter 5 in Sharp and Holmes (eds), *Strategies for New Technologies*, Case Studies from Britain and France (London: Philip Allan, 1989).

Tanaka, M. (1988). Industry–university relations in the case of new biotechnology in Japan. Saitama University. Mimeo.

32

Safety Issues in Genetic Engineering: Regulation in the United States and the European Communities

Rogier Holla[1]

In several countries, the new developments in biotechnology have given rise to intense debate about the risks involved for public health and the environment. Many countries have introduced regulation to guarantee appropriate risk assessment prior to the use of genetic engineering techniques in contained systems and to the release into the environment of genetically modified organisms.

This chapter seeks to give a brief overview of the development of the regulation of genetic engineering research and biotechnology products in the United States and the European Communities.

History of the Safety Concerns: The Berg Letter

Reservations about the potential hazards of recombinant DNA (rDNA) research were first voiced by scientists in 1974. In the US, a group of prominent scientists involved in rDNA research, led by Paul Berg, formed a Committee of the US National Academy of Sciences (NAS). Berg was a professor at Stanford University and a pioneer of rDNA technology. In 1971, he planned to conduct an experiment involving the transfer of DNA from an animal tumour virus which can infect the common human intestinal bacteria Escheria Coli. When cancer specialists heard about it, they warned Berg that if such a virus were to escape from his laboratory, it might survive in the human intestinal bacteria and result in a cancer epidemic. Berg decided to abandon the experiment and to establish a NAS committee to discuss the safety issues.[2]

The committee addressed an open letter to two leading scientific journals, *Science* and *Nature*.[3] The letter carried four recommendations: (1) scientists should accept a voluntary moratorium on certain types of rDNA research; and (2) 'weigh carefully' plans for certain other types of experiments; (3) the National Institutes of Health (NIH) were requested to establish a committee of experts to evaluate risks and develop guidelines; (4) an international meeting of scientists should be organized to review scientific progress and evaluate risks.

The moratorium was accepted. It was enforced through peer pressure, strengthened by the fact that the elite of the molecular biotechnology world were its initiators. Such a moratorium is probably unique in the annals of science. The third recommendation resulted in the NIH Guidelines, which laid the foundation for research guidelines all over the world. The fourth recommendation led to the Asilomar Conference.

The Asilomar Conference

In February 1975, a conference for scientists involved in rDNA research took place at the Asilomar Center in Pacific Grove, California.

One of the major objectives of the Asilomar Conference was 'to discuss appropriate ways of dealing with the potential biohazards of rDNA molecules'. The most vocal group of scientists opposed to rDNA research, however, was not represented at the conference. Rather than attend, they sent an open letter which criticized the 'molecular biology community' for usurping decision-making authority and called for direct involvement of the general public. One of the few lawyers present described the major part of the conference as 'virtually indistinguishable from an ordinary scientific meeting held for the purpose of exchanging information'.[4] The latest information on rDNA techniques, experiments, and results was exchanged. Many of the conferees hoped that a consensus could be reached about the safety procedures for allowable rDNA experiments, and about the experiments that would then be initiated.[5]

On the last morning a conference statement was accepted without much discussion. It lifted the moratorium and described safeguards for rDNA experiments. This statement served as the basis for the now universally used NIH Guidelines.

The National Institutes of Health Guidelines

The NIH, which funded the major part of academic rDNA research in the US, decided to develop guidelines for federally supported research. Although the NIH had no enforcement powers other than the threat to withdraw funds, the Guidelines were widely accepted. They conciliated promoters of genetic engineering by permitting the research to proceed, and the opposition by guaranteeing a review procedure that seemed satisfactory.

A Recombinant DNA Advisory Committee (RAC) was tasked to draft the guidelines. The first version of the 'Guidelines for Research Involving Recombinant DNA Molecules' was issued in July 1976. It focused on the prevention of accidental escape of rDNA organisms. One of the principal substantive mechanisms was to link the degree of containment with the amount of (hypothetical) risk. Four different containment levels were established.

The Guidelines also established review procedures. The RAC became the most important review institution. It was established as an advisory committee, with the authority to approve experiments being given to the director of the NIH. Originally, the RAC consisted only of scientists, but later lay-people were added; it must be said, though, that the guidelines and proposals reviewed were extremely technical. During the Carter Administration (1977–1981), the RAC could realistically be called a forum for active debate about risks and benefits of biotechnology. Membership included representatives who had warned of the potential dangers of genetic engineering, as well as people favouring the developments. During the Reagan Administration (1981–1989), the number of critical RAC members declined substantially.

The Guidelines were only applicable to institutions receiving NIH support. It was, however, not necessary that the experiments themselves were funded: if an institution received NIH support for any project, its genetic engineering research fell within the scope of the Guidelines. This made the Guidelines applicable to practically all genetic engineering research undertaken at universities, including those projects funded by private firms.

The NIH Guidelines had a substantial impact on biotechnology regulation throughout the world. Denmark, Ireland, Norway, Sweden and Switzerland copied the Guidelines. West Germany, the United Kingdom and the Netherlands developed guidelines based on those of the NIH.

Since the first version was issued in 1976, the NIH Guidelines have been revised and relaxed several times.[6] A revision in April 1982 lifted the prohibition of environmental release. Field tests could be initiated after review by the RAC and joint approval by the director of the NIH and a local review institution.

Foundation on Economic Trends v. Heckler

The first approvals by the NIH of deliberate release experiments, in 1983, refuelled the public debate over the risks of genetic engineering. One of the approved field tests involved genetically engineered bacteria, designed to play a role in bringing down the temperature at which ice crystals would be formed in crops, and thus reduce frost damage. This 'ice-minus experiment' was the first field test of a genetically engineered micro-organism which was ready to proceed, and it received the full attention of the opposition to genetic engineering. The Foundation on Economic Trends, led by activist Jeremy Rifkin, challenged the NIH approvals in court and sought a preliminary injunction.

The District Court of the District of Columbia ruled that the NIH had violated the National Environmental Policy Act, which requires federal decision makers to compile an Environmental Impact Statement (EIS) prior to approval of all 'major federal actions significantly affecting the quality of the environment'.[7] The NIH should have prepared an EIS before introducing the possibility of field tests in the Guidelines or before approval of the experiment. A preliminary injunction was granted.

However, the court explicitly exempted field tests undertaken by private firms from the injunction. These experiments were not governed by the NIH Guidelines because they did not involve any NIH funding. The judgment thus drew attention to the outright absence of a review mechanism for commercial research.

Jurisdictional Confusion

The Environmental Protection Agency (EPA) attempted to plug the regulatory gap. In 1983, officials of the EPA stated their belief that they had authority to regulate biotechnology under the Toxic Substances Control Act (TSCA) and the Federal Insecticide, Fungicide, and Rodenticide Act (FIFRA). To their opinion 'genetically engineered organisms' fell under TSCA's broad definition of 'chemical substance'. With this statement, however, the EPA entered spheres of competence that traditionally belonged to other agencies. The Department of Agriculture (USDA) and the Food and Drug Administration (FDA) protested and claimed regulatory authority over biotechnology research and products in their areas of interest.

The need for greater coordination of the regulation of rDNA research became increasingly necessary. The jurisdictional dispute received considerable criticism from industry and commerce. Spokesmen from industry maintained that the regulatory uncertainty hampered investment in biotechnology R & D.

The US-Coordinated Framework for Regulation of Biotechnology

As the opportunities for economic growth in the biotechnology-related sectors became clearer, concern over the regulatory situation also increased in the White House. The Executive Office of the President formed a working group composed of representatives of fifteen agencies. The working group reviewed federal biotechnology regulation and published a proposal for a 'Coordinated Framework for Regulation of Biotechnology' in December 1984. The final regulatory framework was proclaimed in June 1986.[8] It divided regulatory responsibility among the different agencies on the basis of a number of criteria.

The most important agencies in this complex network of review procedures are the Environmental Protection Agency, the Food and Drug Administration, the National Institutes of Health, the Animal and Plant Health Inspection Service (APHIS, part of the USDA), the Science and Education Division of the USDA, and the National Science Foundation (NSF).

The most significant criteria for jurisdiction are related to the intended use of the organisms, the source of funding for research, whether the organisms contain genetic material from other species, and the categories: micro-organisms – plants – animals. In many cases more than one agency has regulatory competence. When this occurs, a 'lead-agency' becomes responsible for coordination of the review processes. Which agency assumes this task is again determined on the basis of certain criteria. Another mechanism to enhance cooperation is the Biotechnology Science Coordinating Committee, consisting of representatives of all agencies involved. It has advisory capacity with respect to coordination of agency activities and development of definitions.

US Congress and the Regulation of Biotechnology

Congress has not been involved in the development of the Coordinated Framework, which thus has no legislative status. It can be seen as a policy declaration of the agencies that clarifies regulatory competence, enacted with the mediation of the White House Working Group. The US Congress has extensively deliberated about possible new legislation for biotechnology. After the moratorium and the

Asilomar Conference, scientists showed aversion to the imposition of any non-voluntary regulations on genetic engineering, and organized an intensive lobby in Washington against legislation. A lobby of environmental organizations, however, pushed for strict regulation and new legislation. Some Senators and Representatives were strongly in favour of a new Biotechnology Regulation Act. Congressional Committees undertook several hearings. Numerous bills have been proposed to amend existing statutes or enact new ones, but none of these proposals has been accepted.

State and local-level regulation

None of the federal laws on which the Coordinated Framework is based 'preempts' regulation at other levels of government. The willingness of non-federal governments to regulate biotechnology has been striking. Some states and several counties and municipalities have passed laws supplementing federal regulation or imposing more stringent specifications.

Regulation in European Countries

At present there is a drastic divergence of biotechnology regulation in European countries, ranging from the absence of specific regulations, through voluntary or mandatory guidelines, to comprehensive legislation. In practically all countries a variety of health and environmental protection laws exists, in part enacted to regulate the 'old' biotechnology. Legislation usually covers risks relating to foodstuffs and food additives, pharmaceuticals, chemicals, pesticides, etc. But these laws usually focus on the commercialization of products. They were not enacted to deal with monitoring and risk management of genetic engineering.

Greece, Italy, Luxembourg, Portugal and Spain have not developed specific regulations with respect to environmental applications of genetic engineering. In view of the forthcoming EC harmonization, these countries will probably not develop licensing systems, but will implement the proposed EC Directives once they have been adopted.

Other European countries have adopted guidelines and regulations in a variety of forms. Most countries have issued guidelines for genetic engineering research, based upon the NIH Guidelines, for voluntary application. The United Kingdom has adopted mandatory guidelines; in Germany the guidelines are compulsory or voluntary depending on how the research is funded.

Many European countries have established national advisory committees. The United Kingdom has its Advisory Committee on Genetic Manipulation (ACGM), Germany has its 'Zentrale Kommission für die biologische Sicherheit', France its 'Commission du Génie Biomoléculaire'. Also Ireland, the Netherlands and Belgium have advisory bodies. These committees are responsible for drafting guidelines and reviewing research proposals on a case-by-case basis in their respective countries.

Denmark adopted a comprehensive statute regulating environmental applications of biotechnology in 1986. This Environment and Gene Technology Act[9] establishes a licensing system for the development of biotechnology-derived products, which includes a procedure for risk assessment and inspection. Deliberate releases are in principle prohibited. However, the Minister of the Environment may approve releases in 'special cases'. Before such an exemption is given, an assessment of the possible harmful effects on the environment must be conducted (on a case-by-case basis). Detailed conditions may be prescribed for each release.

In Germany, a Parliamentary Commission issued a report in 1987 examining the adequacy of existing laws that apply to biotechnology.[10] The report recommended that existing voluntary guidelines be made compulsory for all research, and that a five-year moratorium be established on field tests with genetically modified microorganisms. The German Parliament ('Bundestag') rejected the moratorium in October 1989. In the mean time, a Genetic Engineering Act ('Gentechnik Gesetz') has been proposed. This proposed Act, in line with proposals for European Community-level regulation, would establish mandatory review procedures for all deliberate releases.

The European Communities and the Regulation of Biotechnology

Against the backdrop of these differing national positions in Europe, the Commission of the European Communities[11] (EC) became increasingly interested in the harmonization of biotechnology regulation. Its concerns are: (1) since health and environmental impacts of biotechnology might easily transcend national frontiers there is a need for coordination of risk management; (2) harmonized regulation would prevent market fragmentation and thus provide a more attractive environment for innovation; (3) risk-assessment coordination and data exchange would mean a more efficient use of resources in the advancement of regulations and their adjustment to increased scientific knowledge and technological change.

In 1982, following a proposal from the Commission, the Council of the EC issued a Recommendation concerning 'the registration of work involving recombinant deoxyribosenucleic acid'.[12] A notification procedure was recommended for registering laboratories wishing to undertake work involving recombinant DNA techniques, as well as research projects envisaged, safety evaluations, and control measures.

A Communication from the Commission to the Council in 1983 explicitly recognized the necessity for a coherent European regulatory response to biotechnology. In 1984 the Commission established the Biotechnology Steering Committee (BSC) to coordinate the implementation of the Commission's initiatives in biotechnology. It consists of the Directors-General of the Commission services principally concerned with biotechnology: Internal Market and Industrial Affairs; Competition; Agriculture; Environment, Consumer Protection and Nuclear Safety; Science, Research and Development; and Information, Market and Innovation.

The BSC is supported by the Concertation Unit for Biotechnology in Europe (CUBE), which acts as its secretariat, monitors developments in biotechnology and provides information and assessments. Some interservice working groups cover specialized areas for the BSC. The Biotechnology Regulations Interservice Committee (BRIC) deals with regulatory aspects. The BRIC drafted a proposal for an EC regulatory framework for biotechnology.

A Regulatory Framework for the European Communities

In March 1988, the Commission proposed two Directives[13] to the Council: (1) on the contained use of genetically modified microorganisms, and (2) on the deliberate release into the environment of genetically modified organisms.[14]

The proposed Directive on the contained use of genetically modified micro-organisms (GMMOs) addresses issues such as physical and biological containment levels, accident control, and waste management in industrial applications. It seeks to adopt working practices, safety precautions for routine releases (e.g. as waste or in airborne emissions), and the prevention of accidental releases of GMMOs and the limitation of consequences of any such accidents.

The proposed Council Directive on the deliberate release to the environment of genetically modified organisms[15] seeks to establish a case-by-case notification and endorsement procedure. Every EC Member State is required to designate a competent authority (CA), or authorities. Its composition and its position in the national administration are left to the discretion of the Member States. Prior to a release, a notification has to be submitted to the CA of the respective Member State. This notification must include a technical dossier supplying information necessary for risk evaluation and a statement evaluating the impacts and risks for people and the environment. The CA will review and evaluate the notification and decide upon endorsement. No release may be carried out before it has been endorsed.

For experimental releases each Competent Authority is fully responsible for the releases carried out in its Member State. Placing on the market genetically modified organisms for a given use requires agreement with the other Member States before the product may be endorsed, because the product enters the common market.

Coordination of Regulation in OECD Member States

In the framework of the Organisation for Economic Cooperation and Development (OECD), useful work has been done towards international development of risk-assessment methods and safety recommendations. In 1986 the report *Recombinant DNA Safety Considerations* was issued,[16] which made safety recommendations for large-scale industrial practice, and environmental and agricultural applications. The report continues to have a major impact on regulatory approaches worldwide.

The Focus of Regulatory Consideration for the 1990s

In the 1970s safety precautions focused on accidental escape of pathogenic rDNA micro-organisms, and on infection of humans in contained systems. In the 1980s the scope of concern shifted towards industrial use and the effects of deliberate release of genetically engineered organisms. Regulatory activity has resulted in registration of genetic engineering activities, guidelines for research and industrial practice, and review procedures for certain categories of research and marketing of products.

We should beware the thought, however, that genetic engineering is sufficiently regulated with adequate case-by-case review procedures, and not forget the forest when looking at the trees. With respect to the environment, biotechnology might lead to favourable as well as undesirable developments, and we should not leave the choice between them to mere market mechanisms. To give an example: genetic engineering techniques might be used to increase resistance against certain diseases. This would lead to reduced use of pesticides. However, the techniques could also be used to increase resistance against pesticides or herbicides. This, in contrast, would stimulate the use of pesticides and herbicides.

With the increase of scale of biotechnology applications, the scope of regulation should be extended to oversee the effects of biotechnology as a whole, and ensure a responsible, integrated approach and adequate risk management.

NOTES

[1] European University Institute, Florence – Department of Law. The author would like to thank Carla Bergman and Ken White for their editorial suggestions.

[2] Judith Swazey, James Sorenson and Cynthia Wong, Risk and Benefits, Rights and Responsibilities: A History of the Recombinant DNA Research Controversy, *Southern California Law Review*, Vol. 51 (1978), pp. 1019 ff., esp. pp. 1021–1022.

[3] *Science*, Vol. 185, 26 July 1974, p. 303; *Nature*, Vol. 250, 19 July 1974, p. 175.

[4] Roger B. Dworkin, Science, Society, and the Expert Town Meeting: Some Comments on Asilomar, *Southern California Law Review*, Vol. 51 (1978), pp. 1471 ff., esp. p. 1473.

[5] Swazey et al., *Risk and Benefits*, p. 1033.

[6] The latest revision has been published in the *Federal Register*, Vol. 51, 7 May 1986, pp. 16958 ff.

[7] Foundation on Economic Trends v. Heckler, *Federal Supplement*, Vol. 587 (1984), pp. 573 ff.; the judgment was vacated in part and affirmed in part by the D.C. Circuit Court of Appeals, *Federal Reporter, 2nd Series*, Vol. 756 (1985), pp. 143 ff.

[8] Executive Office of the President, Office of Science and Technology Policy, Coordinated Framework for Regulation of Biotechnology', *Federal Register*, Vol. 51, 26 June 1986, pp. 23302 ff.

[9] Act no. 288 of 4 June 1986.

[10] Enquete Kommission des 10. Deutschen Bundestages, *Chancen und Risiken der Gentechnologie* (Bonn: Deutscher Bundestag, 1987).

[11] The 'European Communities' (EC) embrace the European Economic Community (EEC), the European Coal and Steel Community (ECSC), and the European Atomic Energy Community (Euratom). Present members of the EC are: Belgium, Denmark, the Federal Republic of Germany, France, Greece, Ireland, Italy, Luxembourg, the Netherlands, Portugal, Spain, and the United Kingdom. Most of the organs of the European Communities are integrated. These joint organs include the Commission, the Council, the European Court of Justice, and the European Parliament.

[12] Recommendation 82/472/EEC, *Official Journal of the EC*, No. L 213 of 21 July 1982, pp. 15 ff.

[13] A Directive is binding as to the result to be achieved, upon each Member State, but leaves to the national authorities the choice of form and methods (Art. 189 of the EEC Treaty). Member States are obliged to implement Directives in their own legal system. Directives are adopted by the Council of the EC upon proposal of the Commission.

[14] Document COM (88) 160 (Brussels: Commission of the EC, 1988).
[15] For a more comprehensive evaluation of this proposed Directive, see: Rogier Holla, Ecological Risk Assessment and European Community Biotechnology Regulation, in Lev R. Ginzburg (ed.), *Assessing Ecological Risks of Biotechnology* (Stoneham: Butterworths, 1989).
[16] OECD, Paris, 1986.

GUIDE TO FURTHER READING

David Bennet, Peter Glasner and David Travis, *The Politics of Uncertainty; Regulating Recombinant DNA Research in Britain* (London: Routledge and Kegan Paul, 1986).
Jeffrey N. Gibbs, Iver P. Cooper and Bruce F. Mackler, *Biotechnology and the Environment: International Regulation* (New York: Stockton Press; Basingstoke: Macmillan, 1987) (mainly US regulation, furthermore short overviews of regulation in the UK, Australia, Austria, Canada, Denmark, West Germany, France, Ireland, the Netherlands, New Zealand and Switzerland).
Review and Analysis of International Biotechnology Regulations, prepared by Arthur D. Little, Inc. for a consortium of US government agencies (US Department of Commerce, 1986) (regulation in Japan, France, Germany and the European Communities).
M. Chiara Mantegazzini, *The Environmental Risks from Biotechnology* (London: Frances Pinter, 1986) (regulation in EC Member States).

33

Activities of the Commission of the European Communities in Biotechnology R&D

D. de Nettancourt

Introduction

Molecular biology, one of the foundations of modern biotechnology, was placed on European rails at the time R. K. Appleyard, from Euratom, launched the first transnational programme in Radiation Protection and became, in July 1964, the first executive secretary of EMBO, the European Molecular Biology Organisation. The Radiation Protection Programme was a starting point of pioneering efforts at Community level on the analysis of the replication, translation, transmission, mutation and repair of the hereditary material in living organisms; EMBO, the origin of the European Molecular Biology Conference (EMBC) and, later on, of EMBL (European Molecular Biology Laboratory), provided the training basis to molecular biology in Europe through courses, workshops and exchange fellowships.

Some 15 years later, the 'New Biotechnology' expressed its European dimensions when D. Behrens, Managing Director of DECHEMA in the Federal Republic, founded the European Federation of Biotechnology (EFB) and when the Commission of the European Communities implemented its first R&D programme (BEP) entirely devoted to the applications of biomolecular engineering to agriculture and to industry.[1]

BEP, which covered the period 1982–1986 with a modest budget of 15 million ECU, was followed by several other programmes (BAP, BRIDGE) and biotechnology-related programmes (ECLAIR, FLAIR) aiming at the concertation of national R&D strategies in the

Community, the stimulation of basic biotechnology and the development of a strong basis for agro-industrial technology.

It is the purpose of the present chapter to define each of these programmes, to provide a brief inventory of scientific and technical achievements and to outline recent plans for the expansion of Community R&D in biotechnology such as they have been expressed by the Commission in a recent proposal to the Council of the European Communities.

The First Community Programmes for Biotechnology R&D

While several research activities (such as Radiation Protection, Environment, Energy, Agricultural Research) implemented by the CEC displayed a strong biological component, it was with BEP (Biomolecular Engineering Programme) that a first systematic effort was made to exploit the fundamentals of enzyme and genetic engineering. The objective of this programme was to pool, through cost-shared research contracts, the main competences and infrastructures of the Community, dispersed throughout the Member States, into transnational ventures for the development of the new methods and molecular tools required for the control, transformation and exploitation of biological systems. With 53 specific cooperation agreements for exchange of equipment, sharing of staff, and joint research on specific issues such as the development of cytoplasmic male sterility, the characterisation of important genes in cultivated plants, the upgrading of plant products or genetic engineering in lactic acid bacteria, BEP, in spite of its meagre resources, acted as a strong catalyst of transnational collaboration in Europe.[2]

BAP (Biotechnology Action Programme, 1985–1989) took over from BEP, with a budget of 75 million ECU and efforts focused upon a larger number of objectives, ranging from contextual measures for the development of Community infrastructures in bio-informatics and culture collections, basic biotechnology, to the assessment of risks and the concertation of policies in the Member States and the Community as a whole.[3] Among the very diversified achievements of BAP which were reported in 1988, one may list, as examples:

- the characterisation and isolation of numerous microbial and plant genes with strong importance for agriculture or industrial applications;
- the discovery, as a step towards the rational control of plant growth, of a plant hormone receptor;
- the sequencing, now near completion and for the first time in the world, of an entire chromosome in a eukaryotic cell (yeast);

- the transfer and expression of soybean leghemoglobin genes in Lotus corniculatus, a forage legume, under control of nitrogen-fixing rhizobia which normally do not recognise soybean as host;
- cloning for the very first time in the world of a regulatory gene from cultivated higher plants: the gene is 'opaque-2' which controls zein synthesis in maize;
- preliminary encouraging results on the ability of engineered red blood cells to produce and release antineoplastic drugs from encapsulated pro-drugs;
- the isolation of thermotolerant methylotrophic bacillus species with promising biotechnological potential;
- the development of a kit to distinguish between vaccinated and infected pigs (pseudo-rabies);
- a successful case of co-factor regeneration – a process which self-perpetuates a transformation reaction – at pilot plant phase;
- the launching of MINE, the Community project for harmonising and computerising data about 150,000 strains of microorganisms in European culture collections for on-line availability.

European Laboratories Without Walls: the Tool for Transnational Cooperation in Basic Biotechnology

However, the most successful achievement of BAP certainly was the development of the concept of 'European Laboratories Without Walls' (ELWWs) and the creation during the late 1980s of 34 such ELWWs throughout the Community. An ELWW is an open-ended, transnational, usually multidisciplinary, association of cooperating European groups (laboratories from universities, institutes or industries) with a common commitment to a specific target-oriented research. Within each ELWW, the information, data and materials arising from cost-shared contractual Community research circulate freely, creating a rapid mobility among research staff. Particular importance is attached to exchanges between university and industrial partners within the ELWW. Regular meetings are held in order to evaluate and discuss the results obtained and to plan future experiments.[4]

The structure of each ELWW depends upon the nature of the problem which it attempts to tackle and upon the area of biotechnology covered by its activities. In protein engineering, for instance, there are six ELWWs involving 17 universities or institutes (an average of 2.8 per ELWW) and ten industrial participants (average 1.7 per ELWW). The largest ELWW constructed in the framework of BAP concerns the sequencing of chromosome III in yeast (figure 33.1); it involves 35 laboratories and 16 associated industries. The most recent creation is

Fig. 33.1 The yeast chromosome III of sequencing network

the ELWW which concentrates on new in vitro approaches to pharma-co-toxicology and metabolism in cancer chemotherapy: five labora-tories participate with the close collaboration of one industry and a specific expression of interest from several others.

European Economic Interest Groupings (EEIG)

ELWWs are flexible, rapidly evolving entities which completely lack a legal basis. It is often desirable, once the major research objectives of an ELWW have been reached and commercial applications are in sight, to officialise relationships between research partners or to es-tablish specific agreements with industrial laboratories. The 'Euro-pean Economic Interest Grouping' (EEIG) is a new instrument, directly incorporated into Community law, which facilitates cross-frontier cooperation between firms in the Community wishing to undertake joint activities such as, for example, research and develop-ment, operation of specialised services, quality control of substances.[5]

Post-graduate Training in Basic Biotechnology

In parallel to research work implemented by means of cost-shared contracts, BEP and BAP have also supported training activities which give junior and senior scientists from the Member States the possi-bility of acquiring specific knowledge and know-how in one or sev-eral of the complex disciplines composing modern biotechnology. The programme provides these scientists with access to the very best centres of excellence throughout the Community and covers a wide range of research sectors (enzymology, plant cell research, bio-infor-matics, genetic engineering of industrial microorganisms, animal cell technology and risk assessment). In the framework of BAP, more than 400 applicants have been selected and have received grants for train-ing through research for periods ranging from several weeks to a maximum of two years. In addition, 13 summer schools have been organised in 1988 and 1989 in Greece, Portugal and Spain for provid-ing specific training (practical and applied) in molecular biology, genetic engineering and process engineering.[6]

The Problem of Risks Possibly Associated
with modern Biotechnology

In BAP, a specific sector of activities has been created which concen-trates exclusively on biosafety research. The transnational projects presently implemented under this heading were initiated in 1989 and will end in December 1990; they deal with the assessment of risks possibly originating from microorganisms under physical containment, depollution bacteria, plant interacting bacteria, transgenic plants and genetically engineered viruses. Current activities involve the development of specific monitoring techniques, the standardisation of model ecosystems and limited field-trial experiments, and the study of the stability and possible transfer of genes from released organisms. Fifty-five laboratories distributed in 16 transnational projects participate in what represents the largest integrated effort ever implemented in the world on the subject.[7]

BRIDGE: Biotechnology Research for Innovation, Development and Growth in Europe

BRIDGE will succeed BAP in 1990. This more ambitious programme (100 million ECU for the period 1990–1993) is subdivided, as was BAP, into two actions. Action I for research and training,[8] and Action II for accompanying activities grouped under the heading 'Concertation'.[9] Ninety per cent of the total budget is to be devoted to Action I, which includes four sectors:

- information infrastructures for the processing and analysis of biotechnological data;
- enabling technologies (protein design/molecular modelling, biotransformation, genome sequencing);
- cellular biology (physiology and molecular genetics of industrial microorganisms, basic biotechnology of plants and associated organisms, biotechnology of animal cells);
- pre-normative research (in vitro evaluation of the toxicity and pharmacological activity of molecules, safety assessment associated with the release of genetically engineered organisms).

Action I in BRIDGE will be executed by means of training contracts and cost-shared research contracts. Two types of projects are foreseen for the implementation of the research activities: N-projects for the integration within ELWWs of research efforts in all areas of the programme where bottlenecks result from gaps in basic knowledge; T-Projects for the removal, through a significant investment of skills and resources, of important bottlenecks resulting from scale and structural constraints. To date, four areas have been identified for T-projects:

- sequencing of the yeast genome;
- molecular identification of new plant genes;

- characterisation of lipases for industrial applications;
- biotechnology of lactic acid bacteria.

Other T-projects are now being discussed which could also be adopted for implementation in BRIDGE or in a revised form of the programme.

Agro-industrial Technology: ECLAIR and FLAIR

ECLAIR (European Cooperative Linkage of Agriculture and Industry through Research; 80 million ECU for the period 1989–1993) is the first multi-annual programme of the Community for biotechnology-based agro-industrial research and technological development.[10] It contributes, through such research development work, to the enhancement of Europe's competitiveness in several economic activities based on the life sciences and biotechnology. In 1988, contractual arrangements involving a wide range of different projects (such as production of vaccines, control of insect pests through the development of resistant transgenic plants or of insect parasitic nematodes, development of new antifungal proteins, improvement of shelf-life for fresh fruits and vegetables) have been negotiated with some 150 public and private organisations in the Community. A second selection round is foreseen in 1990.

FLAIR (Food-Linked Agro-industrial Research; 25 million ECU for the period 1989 to mid-1993) is a programme concentrating on research and development on food quality, hygiene, safety, toxicology, nutritional value and wholesomeness.[11] A first choice of projects to be supported in concerted and cost-shared actions took place in 1989.

A Complement to Current Activities

The Commission of the European Communities proposed on 2 August 1989 a new framework programme for research and technological development which foresees an additional and significant effort in the life sciences and technologies (1000 million ECU) for the period 1990–1994. The Commission suggests, for this action line, that it should complement some of the Community R&D activities in biotechnology, agriculture, agro-industry, health and medicine, tropical agriculture, tropical health, implemented under the current framework programme (1987–1991). With regard to basic biotechnology, the research actions are to concentrate on the understanding of biological information and the control of transformation systems and should include, among other priorities, the analysis of genomes from representative species, macromolecular modelling, essential aspects of immunology and the development of testing methods as a prenormative scientific basis for Community regulations.

Other Strategic Objectives of the Community in the Field of Biotechnology

Five action priorities, in addition to research and training, have been defined by the Commission, which concern respectively the pricing of new materials, the preparation of regulations adapted to the requirements and possible consequences of modern biotechnology, the protection of intellectual property, demonstration projects at the interface between agriculture and industry and the concertation of the national and Community policies which seek the implementation of these priorities.[12]

NOTES

[1] For references to early efforts, including the assessment of the importance of biotechnology for Europe, the preparation of BEP, the activities of the EFB and studies in the framework of FAST (Forecasting and Assessment in Science and Technology, DG XII, CEC) see D. de Nettancourt, Community research in basic biotechnology: an example of transnational scientific cooperation, Proceedings of the IDF/UKDA seminar on dairy research, Bulletin of the International Dairy Federation, in press.

[2] Full details on BEP and the results of Community research are to be found in E. Magnien (ed.), *Biomolecular Engineering in the European Community* (Martinus Nijhoff Publishers, 1172 pages, 1986).

[3] The final activity report of the research and training action was published in the spring of 1990 and assessed by nine invited experts at the 5th European Congress on Biotechnology organised by the EFB in July 1990 in Copenhagen.

[4] A detailed description of the features, functions, performances and typical evolution of ELWWs has been provided by E. Magnien, A. Aguilar, P. Wragg and D. de Nettancourt, A new tool for biotechnology R&D in the Community, *Biofutur*, November 1989, 84, pp. 17–34.

[5] For a detailed description, see The European Economic Interest Grouping, CEC, April 1989, 6/89, pp. 1–21 (Office for Official Publications of the E.C., L.2985 Luxembourg).

[6] For further information on the subject, see P. Fasella, Proceedings of the 8th International Biotechnology Symposium, Paris, 1988, and D. de Nettancourt, Training in biotechnology R&D at Community level, *Chimicaoggi*, March 1989, pp. 9–10.

[7] I. Economidis, Research and training activities of the European Communities in Biotechnology: biosafety. European forum on risk

management in biotechnology, ADEBIO, Grenoble, 1989. Abstracts of contributions, J. Defaye, H. de Roissart, P. M. Vignais (eds).

[8] Implemented by the Biotechnology Division (responsible: D. de Nettancourt) within the Biology Directorate (responsible: F. van Hoeck) of the Directorate-General XII for Science, Research and Development, Commission of the European Communities.

[9] Implemented by the Concertation Unit for Biotechnology in Europe (responsible: M. F. Cantley), Biology Directorate, DG XII, CEC.

[10] Implemented by the Division for Agro-industrial research (responsible: F. Rexen), Biology Directorate, DG XII, CEC.

[11] Ibid.

[12] For a review, see M. F. Cantley, Managing an invisible elephant, *Biofutur*, 1989, 84, pp. 8–16.

GUIDE TO FURTHER READING

D. de Nettancourt. Research and training activities of the European Communities in Biotechnology, *Biotech-Forum*, 5, 1988, 334–341.

Evaluation of the Biomolecular Engineering Programme, BEP (1982–86) and the Biotechnology Action Programme, BAP (1985–89). Report of a panel of 7 independent experts (Luxembourg: Office for Official Publications of the European Communities, 1988, ISBN 92-825-8901.3).

D. de Nettancourt. Training in biotechnology R&D at Community level, *Chimicaoggi*, March 1989, 9–10.

E. Magnien, A. Aguilar, P. Wragg and D. de Nettancourt. European Laboratories Without Walls. A new tool for biotechnology R&D in the Community, *Biofutur*, Nov. 1989, 17–29.

A. Vassarotti and E. Magnien (eds), *Biotechnology R&D in the European Communities: I. A catalogue of BAP achievements. II. Detailed final reports of BAP contractors*. (Paris: Elsevier Editions Scientifiques, 1990, in preparation).

34

The Strengths and Weaknesses of the Japanese Innovation System in Biotechnology

Martin Fransman and Shoko Tanaka

Introduction

In recent literature on technological innovation, use of the concept of a 'national innovation system' has become widespread (see, for example, Nelson, 1988; Freeman, 1987, 1988; Fransman, 1988, 1990, 1991; and Nelson and Rosenberg, forthcoming). The concept is particularly appropriate in the case of industries for which the knowledge-base extends beyond the private corporate sector itself to include functionally differentiated institutions such as universities and government laboratories. The concept is also useful in the case of biotechnology since 'new biotechnology', including recombinant DNA, cell fusion, and protein engineering, originated primarily in the university and government health-related sectors and was subsequently diffused from here to private companies. The concept of a 'national system', which takes account of the interactions of institutions such as these and the way in which they shape the process of technical change, is therefore potentially helpful.

The aim of this chapter is to elaborate on the concept of a national system of innovation in the case of Japanese biotechnology – the Japanese Innovation System in Biotechnology (JISB) – and to identify the additional insights that are derived from the use of this concept. These insights are summarised in the conclusion to this chapter.

The National Innovation System

In most accounts of national innovation systems the concept of a 'national system' has not been rigorously specified. This has less to do with the justifiability, or indeed desirability, of proceeding without a rigorous specification than with the difficulties presented by the sheer complexity of a national innovation system. Although, in the Hegelian sense, the truth is in the whole, the whole is a complex whole which is extremely difficult to specify analytically.

In the present chapter the way that has been selected around the problem of complexity involves abstraction. This abstraction involves the identification of a number of *primary institutions* which together play a central role in the generation and use of knowledge in the technology area concerned. It is worth noting at once that national innovation systems are technology-specific since primary institutions influence different technologies in different ways.

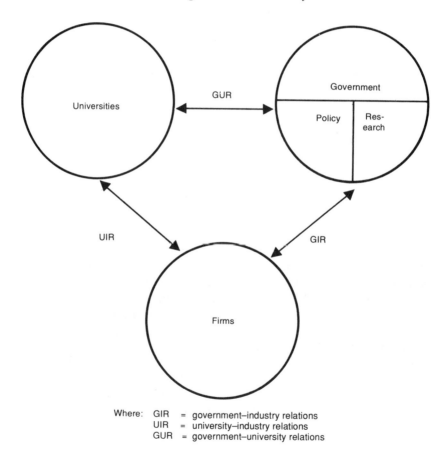

Figure 34.1 An overall 'model' of JISB

The primary institutions distinguished in this chapter are: firms run for profit, universities, government research and policy-making bodies. In addition there are a number of *supporting institutions* which also influence the generation and use of technological knowledge, although less directly so. The latter institutions are abstracted from, in the present chapter, in order to render tractable the problem of complexity already referred to. They would, however, need to be analysed in a more detailed account of the national innovation system. Supporting institutions include: institutions of education and training, legal and regulatory structures, industrial relations systems, language and culture, etc.

A simplified 'model' of the primary institutions comprising the national system of innovation is presented in figure 34.1. As already mentioned, the main focus of the national innovation system is on the generation and use of *knowledge*. Unlike information which is the basis for probabilistic decision-making, knowledge varies over time and space and therefore is held with a degree of uncertainty. While information is 'closed', knowledge is always 'open'. Furthermore, knowledge is selectively structured (implying that it could be structured in different ways), is unevenly distributed across decision-makers, and is always costly to transfer. One way of seeing 'the economic problem' – significantly different from the traditional focus on the allocation and distribution of given scarce resources – is in terms of the problems involved in the coordination, generation, and use of knowledge (see Hayek, 1945). The relevance of the national innovation system as defined is obvious.

Of the three primary institutions in the national innovation system, for-profit firms constitute the site where knowledge relevant (or potentially relevant) for commercial purposes, that is technological knowledge, is integrated with needs expressed through the market; or, to put it more simply, where 'seeds' are integrated with 'needs'. Unlike universities or government research institutions, which may also be involved in the generation of 'seeds', for-profit firms *specialise* in the *integration* of 'seeds' and 'needs'. That is, 'seeds' are not created or used for their own sake, but rather for the purpose of meeting needs, now or in the future. To put this slightly differently, knowledge-creation and use is subordinated to value-creation in for-profit firms. It is this which distinguishes for-profit firms from universities and government research institutions in terms of their respective role in the generation and use of knowledge. While commercial needs may also be expressed in the research of universities and government research establishments, the matching of 'seeds' with market-expressed 'needs' is not the central raison d'être of these institutions.

As figure 34.1 implies, the concept of a national innovation system raises a number of difficult and controversial issues. Two of these are of particular importance. The first is the nature of the relationships between 'science', 'technology', and economic processes. The second, even more controversial, issue is the nature of the relationship between the economy and the polity. More specifically, since government plays an important role in the generation and use of knowledge it is necessary to understand why the government behaves in the way that it does. We shall return to both these issues in our analysis of the Japanese Innovation System in Biotechnology.

The Japanese Innovation System in Biotechnology (JISB)

A. Firms

In terms of the corporate private sector in Japan, new biotechnology has been the preserve of large, established firms. Start-up firms based on new biotechnology, so-called new biotechnology firms (NBFs), have been insignificant, unlike in the United States and to a lesser extent Britain where they have played an extremely important role in the diffusion of knowledge from the university and government health-related sectors to the private sector. Later we shall return to examine whether the absence of Japanese NBFs has been a source of weakness in JISB.

The large Japanese firms which have entered the field of new biotechnology may be divided into two groups. The first are firms with a base in 'old' fermentation-related biotechnology, such as food, beverages and pharmaceuticals (significantly antibiotics), which later moved into new biotechnology. The second group includes chemicals firms, which by and large had not been involved in fermentation technologies, and others in sectors such as manufacturing, engineering/construction, energy/mining, and instrumentation. Some idea of the relative importance of these firms is given in table 34.1, taken from Dibner and White (1989), p. 295. As can be seen from table 34.1, of the 244 firms included in Dibner and White's sample, the largest group of firms involved in biotechnology was from the chemicals sector (56), followed by food and beverages (41), pharmaceuticals (38), and manufacturing (26).

In the mid-1970s, when the new biotechnologies such as recombinant DNA and cell fusion were invented, the large Japanese companies referred to here had very little ongoing research in fields closely related to these new technologies. They were therefore confronted with the problem of developing capabilities in these areas.

This problem became all the more pressing from the early 1980s when efforts made, first by the Japanese Ministry of International Trade and Industry (MITI) and later by other ministries and government agencies, highlighted the strategic significance of new biotechnology. (We shall return later to a more detailed analysis of the role played by the Japanese government.) On the other hand, although they lacked capabilities in the *core knowledge and techniques* necessary for new biotechnology, they were strongly positioned in terms of *complementary assets* such as bioprocessing (in the case of the fermentation-related firms), marketing and distribution capabilities. Furthermore, in most cases they possessed the necessary *financial capabilities* as a result of the close relationship that they had with the main banks which supported them.

Table 34.1 Japan's companies, averages by industry group

Category	No.	Employees	R&D Budget ($ million)	Sales ($ million)	Year
Agriculture	5	7,125	26.27	5,406.31	1954
Biotechnology	9	69	NA	42.72	1976
Chemicals	56	3,024	77.00	1,405.08	1940
Energy/Mining	12	6,554	91.22	6,480.83	1929
Food	41	2,270	20.95	903.01	1924
Engineering/ Construction	15	5,116	35.91	3,975.12	1923
Instruments	11	2,649	82.49	944.62	1941
Manufacturing	26	22,247	583.30	7,613.92	1919
Other Categories	17	4,099	49.93	1,619.73	1924
Pharmaceutical	38	2,360	73.09	776.69	1928
Steel	5	27,638	170.58	7,062.65	1943
Trading	9	5,788	NA	82,692.31	1927
All Companies	244	5,845	126.43	2,654.00	1930

Source: Dibner and White (1989).
Japanese biotechnology-related companies were sorted by their industry group, as shown. The number of companies in each classification, average employee number per company, average 1988–1989 R&D expenditures and average 1988–1989 sales are shown for each group and for all companies. The general trading companies are not included in the analysis for all companies (bottom line). Dollar values were calculated at a conversion rate of 130 yen per dollar.

In general, large Japanese firms pursued three closely related strategies in order to develop new capabilities in the core knowledge and techniques in new biotechnology. The first strategy involved taking steps to build these capabilities in-house by recruitment (usually of newly qualified graduates in the relevant disciplines) and training (often by sending staff to do post-graduate research in Western universities). Since these firms were on the whole reluctant to acquire the capabilities either by expanding their boundaries through merger or acquisition or through the recruitment of experienced personnel from other firms, the first strategy necessarily implied a relatively long lead time. In turn this necessitated two further related strategies: cooperative links with Western and with other Japanese firms. Through these links Japanese biotechnology-related companies were often able to gain access to core knowledge and techniques or other complementary assets in advance of having developed them internally.

Data provided by Dibner and White (1989) on 244 Japanese biotechnology-related firms provide an overview of the kinds of links that were established. Of the approximately 200 links between 1981 and 1988 on which they had data, 48% were marketing agreements, 28% were licensing agreements, 13% research contract agreements, 9% involved equity participation, while 2% related to acquisition. Of the licensing and marketing agreements 74% were with US firms while 8% were with Japanese firms (p. 284). Through these links Japanese companies were able to begin selling, and in some cases producing, biotechnology-related products before having developed internally the underlying core knowledge and techniques. Through research contracts they were able to access the core knowledge of other companies while they were in the process of developing this knowledge in-house. The relatively small number of research contracts (13% of the total) was an indication both of the longer-term goal in Japanese companies to rely increasingly on an internal knowledge-base, and in some instances the reluctance of other companies to sell their knowledge in this form.

Was the relative absence of new biotechnology firms a source of weakness in JISB? The seminal 1984 report on biotechnology of the US Office of Technology Assessment (OTA, 1984) is unequivocal in the view that NBFs give the US at least a temporary competitive advantage, thus implying that their absence in Japan is a source of initial disadvantage. However, the report is careful to stress that this US advantage may only hold in the 'current research-intensive phase of biotechnology's development':

Because NBFs were founded specifically to exploit perceived research advantages, they are providing the United States with a commercial edge in the current research-intensive phase of biotechnology's development. Through their R&D efforts, NBFs are contributing to innovation, expansion of the US research base, technology diffusion, and encouragement of technical advances through the increased domestic competition they create. All of these contributions provide the United States with a competitive advantage. (p. 11)

Despite this perceived competitive advantage, the OTA report concluded that 'Japan is likely to be the leading competitor of the United States.' (p. 7). One of the reasons for this conclusion was that 'Japanese companies in a broad range of industrial sectors have extensive experience in bioprocess technology. Japan does not have superior bioprocess technology, but it does have relatively more industrial experience using old biotechnology, more established bioprocessing plants, and more bioprocess engineers than the United States.' (p. 7).

Three points may be made in commenting on this assessment, all of which require modification of the OTA's conclusions. The first, as illustrated in the data quoted from Dibner and White, is that national innovation systems are not as self-contained as implied in the OTA report. More specifically, it is clear that large Japanese firms have been able to reap at least important short-term gains from the relationships they have established with US NBFs. There is no clear evidence suggesting that Japanese firms have had greater difficulty than their American counterparts in forging links with US-based NBFs. This is illustrated by the licensing and research agreements established between many of the top ten NBFs (not all of which are American) and large Japanese companies. Examples include Genentech (taken over in 1990 by Hoffman La Roche) and its agreements with Mitsubishi Chemical, Daiichi Pharmaceutical, Fujisawa Pharmaceutical, Kyowa Hakko, and Takara Shuzo; Biogen's agreements with Green Cross, Suntory, Yamanouchi Pharmaceutical, CST Research, Fujisawa Pharmaceutical, Meiji Seika, Shionogi, and Sumitomo Pharmaceutical; and Genetics Institute's agreements with Chugai Pharmaceutical, and Yamanouchi Pharmaceutical. (For recent listings of such agreements see Dibner and White, 1989, and Roberts and Mizouchi, 1989.) To the extent that these kinds of links have been established, Japanese firms have also been able to benefit from the knowledge-base and technology transfer provided by these NBFs. This strikingly underlines the point that although national innovation systems have their own internal coherence (which provides the justification for viewing them as systems), they are nevertheless open systems, connected to other national systems through the global flow of knowledge and information.

Secondly, the OTA's conclusions regarding the advantages bestowed on Japanese biotechnology-related firms as a result of their experience and capabilities in old biotechnology needs to be modified to the extent that Japanese chemical companies on the whole have not had particularly significant experience with fermentation technologies. Indeed, for firms such as Mitsui Petrochemical it was only in the 1960s that such experience was gained, primarily as a result of an abortive attempt to develop single cell proteins. In the case of Mitsubishi Kasei, the largest Japanese biotechnology-related company, the company is involved in six basic technology areas: carbon chemistry, inorganic chemistry and metallurgy, organic chemistry, polymer chemistry, electronics, and biotechnology. Although there is some connection between organic chemistry and biotechnology in the production of pharmaceuticals, in general biotechnology represents something of a departure from the company's main areas of technological specialisation. This is true of most of the other chemicals and petrochemicals companies. As noted above, chemicals companies made up the largest single group of biotechnology-related companies in Japan, constituting 23% of Dibner and White's sample of 244 such firms.

Thirdly, it must be kept in mind that even the largest Japanese biotechnology companies are much smaller than their American (and European) counterparts and in areas such as chemicals and petrochemicals have not in the past had the degree of success enjoyed by the latter. This is seen, for example, by comparing the largest Japanese chemicals company, Mitsubishi Kasei, with the third largest American chemicals company, Monsanto, both of which have made significant commitments to biotechnology. In 1985, when both companies were beginning to accumulate capabilities in new biotechnology and a year after the OTA report was published, Mitsubishi Kasei's total sales were $4,445 million, or 66% of those of Monsanto; its total R&D expenditure in the same year was $155 million, 33% of Monsanto's; and its R&D in biotechnology (for 1986) was $72.2 million, or 27% of Monsanto's research budget in the life sciences for the same year.

The first point noted here implies that although American NBF's have undoubtedly speeded the commercialisation of university-based biotechnology knowledge, by providing through their business linkages important advantages to non-American firms they have ensured that the American Innovation System has not been able to capture all of the resulting benefits. Taken together, the second and third points imply that the competitive advantages which Japanese firms enjoy in terms of their fermentation-related capabilities are seen in a slightly different light when account is taken of the chemicals companies and the relative size of Japanese biotechnology-related companies. This does not imply, however, that Japanese companies in food, beverages, and pharmaceuticals are not benefiting from their expertise in fermentation and bioprocessing nor that in the future Japanese biotechnology-related companies will not become a significant threat to American and European companies. Indeed, our own analysis of patents, below, suggests that Japanese firms are rapidly increasing their relative strength in biotechnology. (For an extremely optimistic forecast of the future competitive strength of Japanese chemicals companies, based partly on their ability to take advantage of complementary information technologies, see Freeman, 1990.)

B. Government–Industry Relations

The next feature of JISB to be addressed is the relationship between industry and government (see figure 34.1 above). In discussing this relationship it must be kept in mind that the aim of the analysis is not the government–industry relationship per se, but rather, the implications of this relationship for the generation and use of knowledge. These implications will be brought out in this section and summarised at the end of the section.

The Distinctiveness of Japanese Government–Industry Relations

The OTA (1984) report referred to earlier was clear in its judgement that the competitive strength of Japanese biotechnology owed a good deal to the role played by the Japanese government. While, as already noted, the report states that Japanese strength in bioprocessing is one reason why Japan is likely to be the leading competitor to the US, the second reason is that 'the Japanese Government has targeted biotechnology as a key technology of the future, is funding its commercial development, and is coordinating interactions among representatives from industry, universities, and government' (pp. 7–8). Is there compelling evidence to support this judgement?

In order to examine this question it is necessary to delve more deeply into the relationship between industry and government in Japanese biotechnology. It is clear from even a cursory comparison of the role of government in Japanese industry with that in countries like the United States and United Kingdom (though less so in countries like France) that in Japan there is a far closer link between industry and government. This is evident, for example, in the debates that have taken place in the US and UK on the question of industrial policy. In

Japan the role for such policy has not needed to be debated. This does not mean that the Japanese government is a direct representative of the interests of Japanese industry. As in other capitalist countries, the Japanese state is 'relatively autonomous' from industry, and although it does represent many of the interests of industry, it is also in charge of regulation and the representation of other sometimes conflicting interests. Nevertheless, it is true that in Japan there is a far closer *direct* link between industry and government, with both the bureaucracy and the ruling Liberal Democratic Party in power continuously since 1955. (This link has recently been examined in an illuminating way by Aoki, 1988, chapter 7.) The precise nature of this link will not be discussed further here, comment being confined to the form the link has taken in the case of Japanese biotechnology.

Government–Industry Relations in Japanese Biotechnology

As in other industrial sectors, Japanese biotechnology-related firms have over the years forged a *direct* link with the Ministries and Government Agencies most directly concerned with their activities. This is seen most clearly in the case of the Japanese chemicals companies which by the late 1970s were closely involved with the Basic Industries Bureau of the Ministry of International Trade and Industry (MITI), which had the task of working closely with the chemicals industry in order to assess its needs and define areas where government assistance, within the tight overall budgetary constraints, was desirable. In order to carry out this task effectively, the Basic Industries Bureau needed accurate information on the state of the industry in Japan and also worldwide. This information gathering and analysing role was well established within MITI and was recognised to be important by Japanese companies which realised that their own information was constrained by their resources, their current activities, which bounded their 'vision', and the competitive relationship that limited inter-company flows of information. It was therefore widely accepted by Japanese companies that government had an important role to play in collecting and analysing information and using this information in a joint effort with industry to define overall priorities and strategies, particularly in the medium to long term. These joint efforts were cemented by a number of institutional practices which included the formation, often with government assistance, of industry associations, the establishment of government–industry consultative committees, and the practice of *amakudari*, by which retired government bureaucrats 'descended from heaven' to lucrative posts in large private firms. From the point of view of information flows, all of these practices served to improve the information basis of policy-making in both the private and government sectors. Furthermore, direct links were not confined to connections between government and the bureaucracy. Private companies, both directly and through their representative organisations such as the Keidanren, also forged direct linkages with the ruling Liberal Democratic Party that itself worked closely with the bureaucracy and played an important role in resolving inter-Ministerial conflicts of the kind that will shortly be discussed.

The life science in JISB in the 1970s and new international trends

The government machinery in JISB first began to highlight the industrial strategic significance of the life sciences in the early 1970s. In April 1971 the Prime Minister's Council for Science and Technology recommended the promotion of the life sciences. In 1973 the Science and Technology Agency established the Office for Life Science Promotion.

At the same time, global events began to highlight the potential significance of the life sciences in general and the emerging new biotechnology in particular. Drawing on Watson and Crick's model of the double helix in 1953, the first gene was cloned in 1973 by Boyer and Cohen, giving birth to the new genetic engineering technique of recombinant DNA, while in 1975 the first hybridoma (fused cell) was created by Milstein and Kohler. In 1976 the first so-called new biotechnology firm was set up to exploit recombinant DNA technology, Genentech, a spin-off from university-based research. When Genentech shares were first sold on Wall Street in 1980 they set the record for the fastest price increase, rising from $35 to $89 in 20 minutes. In 1981 the initial public sale of shares by Cetus, another NBF, established a new Wall Street record for the highest amount raised in an initial offering, amounting to $115 million. By the end of 1981, over 80 NBFs had been established in the US. In the same year Du Pont, the largest American chemicals company, allocated $120 million to R&D in the life sciences, followed by Monsanto, the third largest, which committed a similar amount. While events such as these emphasised the opportunities that were perceived to exist in the field of new biotechnology, the changes that were beginning to occur in the property rights regime signalled the potential threats that faced companies which delayed entry into this new technology area. In 1980, in Diamond V. Chakrabarty, the US Supreme Court ruled that microorganisms could be patented under existing law, and in the same year the Cohen/Boyer patent was issued for the technique relating to the construction of recombinant DNA.

The response of MITI and large Japanese companies

Like their American and European counterparts, Japanese chemicals companies were 'pushed' into biotechnology by falling rates of profit in their traditional areas of business as much as they were 'pulled' into this new technology area by expected rates of return. For many of these companies the concentration on oil-based products such as bulk chemicals and plastics led to difficulties when these products began maturing in the early 1970s and when the level of international competition in petrochemicals began increasing. Profitability was further hit by the oil price rise in 1979 and 1980.

In 1980 a Biotechnology Forum, with the objective of consolidating interest in biotechnology, was established by five Japanese chemicals firms: Asahi Chemical, Kyowa Hakko, Mitsubishi Chemical, Mitsui Toatsu, and Sumitomo Chemical. At around this time these and other Japanese companies began to do their first research in new biotechnology-related areas. Reflecting both the increasing sense in Japan of the importance of biotechnology and the initiating role played by MITI, with its direct links with Japanese companies with a potential interest in biotechnology, particularly the chemicals companies, MITI entered the field of biotechnology in 1981. Background motivating factors included the recession in the international chemicals industry, MITI's long concern at the relative weakness of Japanese chemicals companies, and the Ministry's preoccupation, in the wake of the 'second oil shock', with energy policy. After initial discussions with around 50 companies, MITI in 1981 eventually invited 14 companies to join in the biotechnology part of the Next Generation Base Technologies Development Programme (the other parts consisting of future electronic devices and new materials – see Fransman, 1990, for a detailed analysis of the former). The companies chosen were: Ajinomoto, Asahi Chemical, Daicel Chemical, Denki Kagaku Kogyo, Kao Soap, Kyowa Hakko, Mitsubishi Chemical, Mitsubishi Chemical Institute of Life Sciences, Mitsubishi Gas Chemical, Mitsui Petrochemical, Mitsui Toatsu, Takeda, Toyo Jozo, and Sumitomo Chemical. Significantly, all of the participants in this research programme had important interests in the chemicals area, while 11 of them (that is excepting Ajinomoto, Kao Soap, and Toyo Jozo) were primarily chemicals companies. This underlined the fact that it was the chemicals companies that were the main 'constituents' in this part of MITI's domain. (This programme is analysed in more detail later.)

In order to further consolidate the link in this new technology area with its 'constituents', MITI established in 1982 its Bioindustry Office, significantly located within the Basic Industries Bureau which had long held responsibility for the chemicals, steel, and non-ferrous metals sectors. In June 1982 the Bioindustry Office established the Committee for the Development of Bioindustry, a government–industry consultative committee charged with surveying the situation in the industry and producing a 'vision' for the future. In the same year the Bioindustry Office established a new industry association, the Bioindustry Development Center (BIDEC), initially under the auspices of the Japanese Association of Industrial Fermentation (which was itself established in 1943). BIDEC would gather information on the biotechnology industry, facilitate the flow of non-proprietary information between its member companies, coordinate the presentation of the industry's views to government, and provide feedback on government programmes. Of BIDEC's approximately 150 members in the mid-1980s (including several foreign-owned companies), 28% were from the chemicals sector, 23% from the electrical, machinery, and construction sectors, and 20% from the food sector.

The response of other ministries and agencies

Three other government ministries and one agency also had an interest in the emerging biotechnology because of its application to areas under their jurisdiction: the Ministry of Health and Welfare (MHW); the Ministry of Agriculture, Forestry, and Fisheries (MAFF); the Ministry of Education, Science, and Culture (MESC); and the Science and Technology Agency (STA).

Until the early 1980s, MHW and MAFF tended to relate to the firms in their 'constituency' primarily through the regulatory processes that were under their control, such as safety requirements and procedures for the approval of new drugs and chemicals. In some instances the relationship between Ministry and firm was extremely close. For example, Yoshimasa Umemoto, the President of Takeda, the largest pharmaceutical company in Japan, 'came to Takeda in 1977 after a brilliant career that included service as Vice-Minister at the Ministry of Health and Welfare and at senior posts in the Environmental Agency and the Cabinet. . . . Mr Umemoto, having spent his career in public service, much of it at administrative groups that define some of the essential parameters of our market, has already been with us for nearly a decade' (Takeda, Annual Report, 1986, p. 4).

From the early 1980s, however, the nature of the relationship between MHW and MAFF and their 'constituency companies' changed in a number of important respects. Several factors were responsible for these changes, as Masami Tanaka, a senior official in MITI who played a leading role in the initial activities of MITI's Bioindustry Office, amongst others, has shown (Tanaka (undated)). To begin with, by the late 1970s, having completed the process of 'catch up', Japanese firms in many areas had reached the international technology frontier. Rather than learning-by-following, they now had no option but to seek the way forward in the same way that their Western competitors were doing. This meant that *oriented-basic research* was given a new priority in Japan. Furthermore, having become increasingly competitive, Japanese firms found greater difficulty in gaining access to technology from their Western competitors. In addition, increasing international pressure was being put on Japanese government and companies to make a greater contribution to the international stock of knowledge from which they had so greatly benefited. For these reasons research was accorded a greater significance in Japan from the late 1970s. At the same time it was widely acknowledged that government had a particularly important role to play in facilitating longer-term research in uncertain areas where for-profit firms, in the absence of government support, might not do as much research as was desired.

With its biotechnology activities launched from 1981 MITI had demonstrated that cooperative research programmes, bringing together private companies, government researchers and policy makers, and university academics, provided an important way for government bureaucrats to expand both their political influence (through their enhanced role in national strategic policy-making) and the resources under their control (through the extra funds made available by the Ministry of Finance for research). This lesson was not lost on bureaucrats in MHW and MAFF, who saw the opportunity to do likewise in an emerging technology area that had potentially great significance in their own fields of application. As Tanaka notes, their interest, like that of MITI, was enhanced by the severe budget constraints that were imposed by the Ministry of Finance in a period of significant government budget deficit.

Following in MITI's wake, the Ministry of Health and Welfare, MHW, established a Lifescience Office in 1983, and a government–industry consultative committee, the Cooperation Group for Industrial Policy Related to the Pharmaceutical Industry. In April 1985, on the recommendation of the Pharmaceutical Industry Policy Forum advising the Director of the Pharmaceutical Affairs Bureau of MHW, the Japan Association for Advanced Research of Pharmaceuticals was formed. In April 1986 this Association became the Japan Health Services Foundation (HSF), which, according to its official literature, had two main aims: 'The first is to function as the central organisation, in cooperation with the Ministry of Health and Welfare, to develop the *Japanese pharmaceutical and other health science-related industries* by applying advanced technologies such as genetic engineering.' (Emphasis added.) The second aim was 'to act as a catalyst for triangular cooperation among national research institutes, universities and private laboratories'. In the late-1980s HSF had a membership of about 140 companies, including the major Japanese firms involved in the pharmaceuticals area and a number of foreign firms such as Ciba–Geigy, Du Pont, Hoechst, Pharmacia, and Upjohn. The Foundation funded joint research projects as a result of a grant of about Yen 1,050 million ($5.8 million) from the 'Funds for Research into Basic Health Science for Better Aging' from the 1986 budget of MHW. These funds were allocated to private laboratories, universities, and government research laboratories on the basis of competitive grant applications. Three fields of research were selected: biotechnology, new materials for artificial organs and related medical equipment, and research on the immune system.

In 1984 the Ministry of Agriculture, Forestry and Fisheries, MAFF, set up its Biotechnology Division and its government–industry consultative committee, the Cooperation Group for Promoting R&D in Biotechnology. Like MITI, but unlike MHW, it established a number of research associations involving private firms, including the Bioreactor Research Association for Food Industry Biotechnology (1984), the Research Association for Pesticides (1984), and the Research Association for Analysing Structures Relating to Agricultural Genes (1986).

The Science and Technology Agency (STA) was in a somewhat different position from MITI, MHW, and MAFF. Although the STA had played a leading role in biotechnology in the 1970s, the Agency's official responsibility, based as it was in the Prime Minister's Office, was to coordinate policy. This meant that the Agency as such did establish research associations like MITI and MAFF, even though it had had its Life Science Office since 1973 and was closely linked to the Lifescience Committee of the Council for Science and Technology. Nevertheless, from an early period the STA had established a stake in technology-related research. In the 1950s, at a time when there was concern in Japan at the dependence on imported technology, proposals were made by Keidanren and the Japanese Academy of Sciences, amongst others, to address the problem. This eventually resulted in the establishment of the Research Development Corporation of Japan (JRDC) in May 1961 under the supervision of the STA. The birth of JRDC was accompanied by a tough battle for control between the STA and MITI, which argued that it had responsibility for projects aimed at increasing technological capabilities in the country. Keidanren, the Ministry of Finance, and politicians from the Liberal Democratic Party were also involved in the battle, which was ultimately resolved in STA's favour. According to one participant, 'MITI was extremely nasty to JRDC for several years after the decision.'[1] The evolution of JRDC's activities mirrors the growth of science and technology in Japan. Until the late 1970s, the JRDC's main activity involved commercialising technologies that were 'incubated' in universities and government research institutions. From the late 1970s, however, under the circumstances referred to earlier, the role of JRDC began to change significantly. With basic research a much higher priority in Japan, JRDC launched its Exploratory Research for Advanced Technology (ERATO) programme. In 1989, 48% of JRDC's expenditure went to the development of new technology, while 36% went to the ERATO programme. (The ERATO programme is analysed in more detail below.)

The Ministry of Education (MESC) differed from the other four government organs insofar as its links with private industry were of far more of an arms-length nature, partly as a result of a traditional belief amongst many in the Ministry that in matters of education a close link with industry was undesirable. As a consequence, the Ministry and institutions under its control, including universities, tend to have a looser relationship with private firms than MITI, MHW, MAFF and the STA, although initiatives have been taken by MESC to link university-related research more closely with industry. The implications for university–industry relations are considered in greater detail below.

Inter-ministerial conflicts

The expansion of several organs of government into the same new technology area inevitably produced conflicts. For example, in the mid-1980s MAFF introduced a new five-year programme to develop cell fusion techniques for modifying microbial and plant cells. The firms that agreed to cooperate in this research programme included Asahi Chemical, Hitachi, Kagome, Kikkoman, Kirin Brewery, Kubota, Kyowa Hakko, Meiji Seika, Mitsui Toatsu, the Nihon Kinoku Centre, the Plantech Research Institute, Sapporo Brewery, and Suntory. These 'constituents' reflected MAFF's jurisdiction over food and other agriculture-related areas. Thus six of these thirteen firms had food as their primary business, while another five were closely related to food or agriculture. Their participation in this MAFF programme was a reflection of the close direct links that they had formed with this Ministry.

However, five of these companies – Asahi Chemical, Kirin Brewery, Kyowa Hakko, Mitsui Toatsu, and Suntory – also participated in one of the two major MITI programmes launched at this time (the Next Generation Base Technologies Programme referred to earlier and the Protein Engineering Research Institute discussed later). Similarly, companies such as Takeda and Ajinomoto were involved in both the MITI and the MHW programmes.

Although the areas of prime application differed in the research programmes of the different Ministries, there was a significant degree of overlap in the generic technologies that were being developed. As Masami Tanaka shows, this led to opposition from many of the firms involved, which felt that they were being pressurised to contribute resources (albeit subsidised resources) in an inefficient way (as a result of the multiplicity of programmes and the degree of overlap involved) in order to further the competing ambitions of rivalrous bureaucrats. Furthermore, as Tanaka notes, 'a Ministry which may have a close relationship to a specific firm, will often be reluctant to have that firm participate in another agency's programme' (p. 25). The opposition of the companies was even expressed through the business federation, the Keidanren.[2] Politicians from the Liberal Democratic Party also became involved in the conflict. By no means above the fray, some of them had also developed a close interest in high-technology policy questions and became involved in the allocation of funds to research, partly because of the possibilities raised for related resources to be allocated to their constituencies. Nevertheless, these politicians ultimately decided to avoid the minefield that is interministerial politics. As Tanaka delicately puts it:

The Diet members [of the LDP] realise that, in theory, politicians are the master of policy ideas and that government agencies [i.e. bureaucrats] are the master of routines and technique. They have also appreciated, however, the fact that bureaucrats are frequently capable of moulding not only techniques but also policies. Although Diet members obviously became aware of interministerial rivalries, they accepted fragmented policies initiated by [government] agencies [in the area of biotechnology]. They preferred government rivalry rather than risking Diet-induced confusion through attempted co-ordination. It may indeed appear to them that competition for new policy initiatives means an expansion of public expenditure. This could favour them as it may provide them with more benefits to distribute among their supporters and thus enhance their chances for re-election. (pp. 27–8)

On occasion, however, inter-ministerial conflict produced favourable consequences from the point of view of research in Japan. One notable example is the so-called 'telecom wars' that raged between MITI and the Ministry of Posts and Telecommunications (MPT) from 1981 to 1985 (discussed in illuminating detail by Johnson, 1989). In this case the convergence of computing and communications technologies in the form of expanding computer networks and the ambitions of MPT bureaucrats to upgrade their Ministry from a 'business' to a 'policy' agency provided the basis for the conflict. While, in the process of this conflict, MPT officials came to the conclusion that their interests would best be served by the part-privatisation of NTT, the national telecommunications operator which was under their direct control (a conclusion shared by other powerful business and political interests in Japan), they wished at the same time to expand their influence and policy-making role by establishing new research facilities with funds received from dividends from government-owned NTT shares. This, however, was opposed by MITI, which was in the process of attempting to increase even further its profile in the area of advanced research and which succeeded in gaining the support of the powerful Ministry of Finance in its struggle with MPT. A major role in resolving this three-ministry dispute was played by the Liberal Democratic Party and its ZOKU (caucus) politicians who were most closely involved in the policy areas concerned. The outcome was the establishment of the Japan Key Technology Center (*Kiban Gijutsu Kenkyu Sokushin Senta*) in October 1985, which was funded largely from dividends from government-owned shares in NTT and Japan Tobacco Inc. From the research point of view, the main benefit that flowed from the establishment of the Center was the securing of research funding from sources independent of the Ministry of Finance's annual budget and thus freed from the tight constraints on this budget at a time of significant budget deficit. Although formally under the joint control of MITI and MPT, there is a de facto split down the middle of the Center, with the two ministries taking sole charge of the areas under their own jurisdiction. Biotechnology falls under the control of MITI and the first major project which was established in this area was the Protein Engineering Research Institute (PERI), which is analysed in more detail below.

Inter-ministerial conflict and the creation and use of knowledge in JISB

The significance of inter-ministerial conflict is seen in a different light when viewed from the point of view of the creation and use of knowledge in JISB. From this point of view such conflict has had beneficial consequences to the extent that it has increased *variety* in the national system. It is being increasingly understood that the generation of variety is an important determinant of longer-term performance within national systems, variety and selection being the two crucial processes of evolutionary change within such systems.[3] Inter-ministerial conflict has contributed to variety by increasing the alternative sources of research funding and therefore the range of possible priorities for research, by increasing the number and sectoral composition of firms involved in government-initiated research, by increasing the range of differentiated institutions with their distinctive competences involved in research (such as government research institutes dedicated to different areas of research), by increasing the areas of application of biotechnology research, and possibly by increasing the overall quantity of national resources allocated to research. In these ways inter-ministerial conflict has had the unintended consequence of increasing the vitality of JISB.

Government-initiated Research Programmes

The Japanese government, for the reasons mentioned, has played an important role in initiating research programmes in biotechnology involving cooperation between industry, universities, and government research institutes. In this section three of the more high-profile of these programmes will be briefly analysed: MITI's Next Generation Base Technologies Development Programme and Protein Engineering Research Institute, and the STA's ERATO programme.

The Next Generation base technologies development programme

From the point of view of JISB the main question regarding all of these programmes relates to *their effect on the creation and use of knowledge*. In the case of the Next Generation (NG) programme it must be recalled that research began at a time when the companies were just beginning to make a firm commitment to new biotechnology. Furthermore, significant uncertainty still surrounded the commercial applicability of the main technologies. In this context MITI officials, in consultation with the fourteen participating companies, established a research programme involving 'coordinated in-house research', rather than 'joint research in joint research facilities', which was the modus operandi for the Protein Engineering Research Institute (PERI). Accordingly, research was done under the NG programme by the participating companies working in-house and sharing some, but not necessarily all, of the knowledge that resulted.[4]

The fourteen companies were divided, on the basis of consultation, into three separate research groups: the Bioreactor Research Group (consisting of Daicel Chemical, Denki Kagaku Kogyo, Kao Soap, Mitsubishi Chemical, Mitsubishi Gas Chemical, and Mitsui Petrochemical); the Large-Scale Cell Cultivation Research Group (containing Ajinomoto, Asahi Chemical, Kyowa Hakko, Takeda, and Toyo Jozo); and the Recombinant DNA Research Group (consisting of Mitsubishi Chemical Institute of Life Sciences, Mitsui Toatsu, and Sumitomo Chemical).

As noted earlier, the choice of companies was heavily biased in favour of the chemicals sector, the main companies from which had a long-established close relationship with MITI. MITI's twin concern with the longer-term competitiveness of these companies and the effects of the latest oil shock, which were still reverberating, influenced the choice of applications areas. In particular, the research of the Bioreactor Research Group was heavily oriented towards the development of energy-saving biological processes which it was hoped would substitute for more energy-intensive conventional processes. Thus Mitsubishi Chemical was involved in the screening of microorganisms for the production of muconic acid, used in the production of high-performance resin polyesters such as nylon, from benzoic acid, an inexpensive raw material. Kao Soap was also involved in the selection of microorganisms, in this case for the microbial oxidation of higher alkyl compounds. Similarly, Daicel selected microorganisms for the production of acetic acid. The Bioreactor Research Group was the only one of the three involved in a degree of joint research in joint research facilities emerging from their joint work with, and use of expensive bioreactor facilities in, MITI's Fermentation Research Institute. Members of the Bioreactor Research Group met each two months in order to report on their results, in addition to the annual conferences held by each of the three groups to more formally announce research findings.

An example of the research done in the Large-Scale Cell Cultivation Research Group is the research done by Takeda with applications in the field of pharmaceuticals. In this research Takeda researchers developed hybridomas (fused cells) which could produce human monoclonal antibodies against tetanus toxin or the surface antigen of the hepatitis B virus. A major aim of this project was to produce these antibodies on an industrial scale in biofermentors. In the Recombinant DNA Research Group all three of the members were involved in the cloning of various types of DNA in host systems such as *E. coli*, and *B. subtilis*. For example, Sumitomo Chemical was involved in the genetic engineering of *E. coli* with enhanced monooxygenase activity, with large-scale industrial applications in mind in areas such as the oxidation process of various industrial chemicals and the removal of hydrocarbons from industrial waste.

From the point of view of the creation and use of knowledge, however, a number of questions are raised. For example, how much additional knowledge was created in these projects (that is, knowledge which would not have resulted had the project not been undertaken)? To what extent was knowledge, created in one firm, shared with other participating firms? (By avoiding the need to duplicate efforts, the sharing of knowledge provides one of the major benefits of cooperative research programmes.) How useful was the knowledge in terms of its ultimate commercialisation?

Important though they are, not enough information yet exists in order to answer the first and third questions. Some light can, however, be thrown on the second question.

In a widely-quoted study, Herman Lewis (1984) of the US National Science Foundation passed judgement on the extent of research cooperation in the three research groups:

bringing about cooperative research or even getting the companies within each group to agree upon a common project has been the biggest problem of the [research] association. During the first two years of the association's history this has not been accomplished. Within each of the three research groups there has been what is generously labeled 'cooperative competition' which is a euphemism to describe consensus on the broadest general terms but disagreement on all specifics. Within this framework, the three research groups can be characterised with respect to the degree of communication within the group. The recombinant DNA Research Group displayed the least desire to exchange information or ideas between the companies within the group, while the Large Scale Cell Cultivation Research Group seems to be willing to exchange certain kinds of information. The Bioreactors Research Group gets an intermediate score. (p. 50/1) [Accordingly, Lewis concluded that the Research Association for Biotechnology which supervised the three projects] has not yet succeeded in pooling the resources of the participating companies to carry out long term cooperative research. (p. 53)

One way of examining the extent to which knowledge, created in one participating company, was shared with other participating companies is to analyse the extent of joint patents, defined as patents where the inventors come from more than one of the participating companies. In order to examine such patents the present authors obtained a list of all the patents, granted and applied for, emerging from the NG programme up to January 1987. The data are presented for the three research groups in tables 34.2, 3, and 4.

Table 34.2 Patents granted and pending for the Bioreactors Research Group of MITI's Next Generation biotechnology programme

Firm	No. of Single Patents	No. of Joint Patents[1]	Joint Patent Holders[1]
Daicel Chemical	23	0	–
Denki Kagaku Kogyo (DKK)	0	12	MPI, FRI
Denki Kagaku Kogyo (DKK)	–	1	FRI
Fermentation Research Institute (FRI)	0	12	MPI, DKK
Fermentation Research Institute (FRI)	–	1	DKK
Kao Soap (KS)	13	0	–
Kao Soap (KS)	–	1	MGC
Mitsubishi Chemical Industries (MCI)	8	–	–
Mitsubishi Chemical Industries (MCI)	–	11	MGC
Mitsubishi Gas Chemical Co. (MGC)	4	–	–
Mitsubishi Gas Chemical Co. (MGC)	–	1	KS
Mitsubishi Gas Chemical Co. (MGC)	–	11	MCI
Mitsui Petochemical Industries (MPI)	0	12	DKK, FRI
Total	48	25[2]	

Source: Information obtained by the authors.

NOTES:

[1] All patents in this programme are owned by MITI. 'Joint Patents' here refer to patents where the inventors come from more than one company.

[2] There is 'double counting' in this column. This results from showing in the table the joint patent holders for each of the firms.

Table 34.3 Patents granted and pending form the Recombinant DNA Research Group of MITI's Next Generation biotechnology programme

Firm	No. of Single Patents	No. of Joint Patents
Mitsui Toatsu Chemicals	9	0
Mitsubishi Chemical Institute of Life Sciences	5	0
Sumitomo Chemical	8	0
Total:	22	0

Source and notes: see table 34.2.

Table 34.4 Patents granted and pending for the Large-Scale Cell Culture Research Group of MITI's Next Generation biotechnology programme

Firm	No. of Single Patents	No. of Joint Patents
Ajinomoto Co.	8	0
Asahi Chemical	8	0
Kyowa Hakko Kogyo	6	0
Takeda Chemical	3	0
Toyo Jozo Co.	5	0
Total	30	0

Source and notes: see table 34.2

A number of conclusions emerge from these tables. To begin with, it is only in the Bioreactors Research Group that joint patents (formally owned by MITI) have resulted. Neither in the Recombinant DNA Research Group nor in the Large-Scale Cell Cultivation Research Group had there been any joint patents. Secondly, in the Bioreactors Research Group the joint patents have been granted to two groups of companies. The first group consists of Mitsui Petrochemical, Denki Kagaku, and the Fermentation Research Institute, which received 12 out of a total of 25 joint patents held by the Bioreactors Group. The second group consists of Mitsubishi Chemical and Mitsubishi Gas Chemical, which received 11 out of the 25 joint patents. Of the two remaining joint patents, one was held by Denki Kagaku and the Fermentation Research Institute, and the other was held by Kao Soap and Mitsubishi Gas Chemical. Thirdly, the ratio of single to joint patents is almost two-to-one. Fourthly, the firm holding the highest number of patents, Daicel Chemical, was engaged in only a limited amount of joint research (to the extent that patents serve as an accurate indication of joint research). Fifthly, most of the patents were from the Bioreactors Research Group, which accounted for 73 or 58% of the total of 125 patents.

This analysis of patent data suggests a number of conclusions. The most important of these is that joint research, to the extent that this is indicated by joint patents, existed only in the Bioreactors Research Group. It is reasonable to assume that such joint research has been facilitated by the *complementary*, rather than competitive, relationship that exists between the cooperating firms. On the basis of interviews we established that neither Mitsui Petrochemical nor Denki Kagaku were involved in direct competition with one another, and furthermore, they were both closely related to the Mitsui group of companies. (Mitsui Petrochemical has an ongoing research programme in the area of biotechnology with other Mitsui group companies, including Mitsui Toatsu, Oji Seishi, and Daiichi Engei.) Similarly, Mitsubishi Chemical and Mitsubishi Gas Chemical are complementary firms which are part of the Mitsubishi group

In contrast, competitive relationships are more prevalent in the other two research groups. To take one example, *all* the eight firms in these two research groups were involved in the development of pharmaceuticals. It is reasonable to suppose that these competitive relationships provide at least part of the explanation for the absence of joint patenting, and by implication the absence of joint research, in these two groups..

However, joint patents provide only one indicator of the sharing of knowledge between firms. In particular, the patent data might well obscure more informal flows of knowledge between the participating firms. We therefore supplemented our patent analysis with interviews with several of the participant firms which revealed that there were indeed instances of more informal knowledge-sharing which were regarded as important by the recipient firms. For example, as an unintended spin-off from its research under the NG programme, Takeda developed an innovative medium (brandnamed GIT medium) for culturing animal cells. This medium, containing partially purified growth factors from the serum of adult cattle, was a highly economical substitute for the conventional medium, which contained expensive fetal-calf serum. Unlike the several serum-free media which were already being marketed and which were effective only in cultivating specific cell lines, the GIT medium is almost equivalent to that supplemented with fetal calf or bovine serum with respect to its applicability to a wide range of cells. Ajinomoto also developed an innovative medium under the NG research programme, namely a non-serum medium for animal cell culture. Although there were similar mediums developed in the US and Europe, the one developed by Ajinomoto was reported to be the first which could be heated and sterilised.

For present purposes it is significant that the data on the Takeda and Ajinomoto mediums were shared with *all* the other firms in the Large-Scale Cell Cultivation Research Group *before* the technological knowledge was commercialised. Similarly, Toyo Jozo, Asahi, and Kyowa Hakko shared their findings on growth factors of cells under various conditions. Kyowa Hakko developed a new apparatus for cell culture under the programme and this was made available to, and was used by, some of the other firms.

It may therefore be concluded that although most of the research under the NG programme was undertaken in-house in the participating companies, and although there was limited joint creating and sharing of knowledge of the kind that leads to patentable output, there was a fair amount of inter-firm flow of knowledge which benefited the participants. These flow as facilitated by the regular meetings of a formal and informal kind that took place under the auspices of the project. On the whole it seems likely that a significantly greater inter-firm flow of knowledge occurred as a result of the NG programme than would have taken place in its absence. Nevertheless, the underlying competitive relationship between some of the companies grouped particularly in the Large-Scale Cell Culture and the Recombinant DNA research groups, and the commercial implications of some of the research being done in these groups, served to constrain the extent of knowledge flow between firms.

Although in this way the NG programme helped to increase inter-firm flows of knowledge in JISB, more information is required on the quality and relevance of the research output from this programme in order to evaluate its contribution. For example, after a time a good deal of dissatisfaction was expressed by some of the participants within the Bioreactors Research Group regarding the relevance of much of its research. These reservations emerged in a changed climate where the importance of energy-saving biological-based processes was not as great, given the fall in relative oil prices, as was originally envisaged when the programme was initiated in the aftermath of the second oil shock. According to some of the critics of this project,[5] MITI erred by refusing to terminate the project earlier in the light of the changed circumstances. In the event, however, the Bioreactors Research Group was terminated before the completion of the NG programme as a whole, an extremely rare event in the history of MITI research programmes.

We shall return later to this programme in an overall assessment of the role of government research programmes.

The protein engineering research institute (PERI)

PERI was established under the auspices of the Japan Key Technology Center (JKTC), the political origins of which were examined earlier in this chapter. The main activity of JKTC involves the financing of joint R&D companies, where 70% of the funds are provided by the Center while 30% are contributed by the mainly corporate members of the company.

The original idea for a national research programme in the area of protein engineering came from a number of individuals in MITI who, in conjunction with various formal and informal advisory bodies, had responsibility for making policy in the area of biotechnology. They included Mr Hosokawa and Mr Masami Tanaka.

By 1985 it was acknowledged (a) that protein engineering was a potentially important new generic technology, and (b) that Japanese companies and universities lagged seriously behind their Western counterparts in this field. This was underlined in a report published in 1985 under the US-sponsored Japanese Technology Evaluation Program. The report identified three technologies as being essential for protein engineering: recombinant DNA technology, protein structural analysis, and computer graphics. It was concluded that the first technology was 'the easiest to acquire and [that] Japan has as good a capability in these areas as the United States' (p. 6–5/6). However, in the area of protein structural analysis, including crystallography, it was noted that 'Japan has a relatively small pool of trained experts . . . and it is not easy to rapidly expand this base.' (p. 6–6). There were only two major centres for protein structural analysis in Japan, the Institute for Protein Research in Osaka, and the Faculty of Pharmaceutical Sciences at the University of Tokyo. 'Neither Director [of these two institutions] knew of any significant effort in protein engineering anywhere in Japan as of late 1984' (ibid.). In the case of both hardware and software it was concluded that the 'computer graphics equipment and programs are being imported into Japan and are usually a generation behind that now being used in the United States' (ibid.). The overall conclusion of the report was that 'Japan currently ranks fourth to the United States, United Kingdom, and Western Europe in protein engineering. *There is not much activity at the present time at either the universities or industries*' (p. 6–7, emphasis added).

In 1985 MITI officials began discussions with a number of companies and eventually, in April 1986, PERI was established by five companies: Kyowa Hakko, Mitsubishi Chemical, Takeda, Toa Nenryo, and Toray. Of these companies only the latter two had not taken part in the earlier Next Generation programme. Later a number of other firms joined, including Ajinomoto, Fujitsu, Kanegafuchi, Kirin Brewery, Showa Denko, Suntory, Toyobo and two foreign firms, Nihon Digital Equipment (DEC), and Nihon Roche, a subsidiary of the Swiss pharmaceutical company; 17 billion Yen was budgeted for the project over a period of ten years (ending in March 1995) with 30% provided by the mainly corporate membership. The original five company members own a slightly higher proportion of the equity and send a greater number of researchers to the Institute. The President of PERI is Dr Masakazu Ito, Chairman of Toray Industries, while the General Manager is Dr Morio Ikehara, Professor Emeritus of Osaka University. PERI is divided into five departments which work jointly on topics of common interest. These are: 1. Structural Analysis (13 researchers); 2. Structural-Function Correlation And Design Of New Proteins (13 researchers); 3. Synthesis (12 researchers); 4. Purification And Characterization (9 researchers); and 5. Database Analysis and Computer System (11 researchers). Of the 58 researchers, 5 are postdoctorates, 43 researchers, and 10 assistants. Of these 43 researchers, one third have Ph.Ds, 30 are from the participating companies, and 13 from universities. Of the postdoctorates, one is from Birkbeck College, London University, associated with Professor Blundell, whose work in protein engineering has an international reputation.

Unlike the Next Generation programme, PERI is an example of 'joint research in joint research facilities'. Although it was not consciously modelled on the MITI Institute For New Generation Computer Technology (ICOT), PERI bears striking organisational similarities to ICOT. (See Fransman, 1990, for a detailed analysis of ICOT.)

The following are the main features of PERI, analysed in terms of the creation, diffusion, and use of knowledge:

(i) *Structural separation of generic research and applications research.* The role of PERI itself is to undertake generic research into protein engineering. Although the boundary between generic research and applications research is sometimes difficult to draw, it is intended that the latter research is done inside the laboratories of the individual member companies, rather than in PERI. Typically, the researchers from the member companies spend about three years in PERI after which they rotate back to their companies and are replaced by other company researchers. At times, parallel research groups are established inside the companies, whose research closely follows that being done in PERI, but with concentration on applications. In this way there is a relatively smooth transfer of generic knowledge from PERI, where joint cooperative research is undertaken, to the member companies, where *private* applications-oriented research is done.

(ii) *PERI's generic research and the world stock of knowledge.* Like similar Japanese organisations such as ICOT, PERI is able to tap effectively the world stock of knowledge. As the major centre for protein engineering research in Japan, and with a concentration of resources in a single location exceeding that of comparable research programmes in other Western countries, PERI has an extremely high international profile. This has greatly facilitated knowledge exchanges with the other major centres, and with individual researchers, elsewhere in the world. These exchanges are assisted by the open nature of PERI,

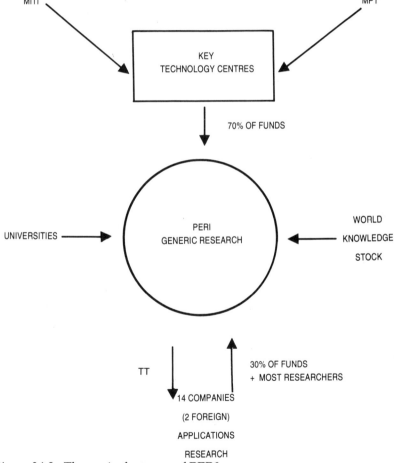

Figure 34.2 The main features of PERI

where research publications and results are made easily available and where visitors are readily received and PERI researchers sent to other international centres. However, the applications research which is done inside the member companies and which draws on the generic research done in PERI, as in the case of any research done in any company in any country, normally remains private.

(iii) *Interfacing with Japanese universities.* PERI also provides a suitable vehicle for interactions with Japanese universities. As will be discussed in more detail later, the Japanese Innovation System has faced a significant problem as a result of the factors which restrict (but do not entirely prevent) the participation of Japanese university researchers in industry-related research. PERI, however, provides an institutional mechanism for overcoming some of these restrictions. Although full-time employees of Japanese national universities cannot take up full-time jobs in PERI, this restriction does not apply to retired academics and to academics from private universities. Furthermore, university researchers can be involved in PERI-related research and receive resources from PERI in undertaking this research in their universities. In addition, the activities of PERI are regularly reviewed by a committee of academics who also provide important feedback on the Institute's research. In these ways Japanese universities make an important contribution to the research of PERI.

(iv) *Subsidising the development and diffusion of new generic technologies.* With 70 per cent of the research funds coming from external sources, the companies participating in the PERI programme receive a significant subsidy. Although in view of the centrality of protein engineering it is likely that the companies would in any event, even without this or similar projects, have acquired protein engineering capabilities, it is probable that they have done so faster and more cost-effectively as a result of this project. The lag in Japanese protein engineering in 1985 relative to other Western countries described in the Japanese Technology Evaluation Program's report discussed above bears testimony to the degree of uncertainty that individual companies would have faced in making the decision to invest in developing capabilities in this area. The subsidy that in the event they have received is likely to have significantly speeded their entry into the field of protein engineering.

The following are the other benefits that PERI has provided: (a) *Avoidance of overlap.* In the absence of a cooperative project like PERI it is likely that the individual companies would have engaged to a considerable extent in overlapping research. One of the main benefits of PERI, however, has been the avoidance of overlap in developing capabilities in the companies in the generic protein engineering technologies.

(b) *Combining distinctive competences.* A further benefit has resulted from the combining in PERI of the distinctive competences of researchers from various companies, from universities, and from government research laboratories, with their idiosyncratic strengths and skills. It is unlikely that isolated research, done inside these different kinds of organisations, would have achieved the same synergistic result.

(c) *Sharing expensive skill-intensive equipment.* One of the major benefits provided by PERI has followed from the sharing of expensive equipment, such as a supercomputer and nuclear magnetic resonance equipment, and the shared development of generic software tools. This sharing has reduced significantly the cost per company or research laboratory of access to equipment and software, thus providing a significant benefit.

(d) *Increased diffusion of generic protein engineering technology.* In the absence of cooperative research there would have been a strong tendency for the companies to privatise their knowledge and limit its spread to other companies. In PERI, however, with the joint participation of companies (which in some cases are competing) in joint research laboratories the diffusion of generic protein engineering technologies has been increased.

(e) *Focusing national attention on protein engineering as a strategic technology.* One of the important spin-off benefits from the PERI project has been the raising of awareness in Japan of the significance of protein engineering as a strategic technology. It is likely that a number of other companies, together with the members of PERI, have as a result of this project devoted additional attention and resources to protein engineering.

(f) *Overcoming some of the problems of technology transfer.* With the current absence of a rigorous theoretical underpinning for protein engineering, there is a significant degree of *tacitness* in this area of research. In general, the greater the degree of tacitness, the greater the difficulty in transferring knowledge from one location to another. The interdisciplinary nature of protein engineering, which necessitates cooperation between researchers from diverse areas such as molecular genetics, protein structural analysis, and computer science, further complicates the problem of knowledge transfer. The organisation of PERI, however, with company researchers who have been involved in the generic research returning to their companies after their secondment to the Institute, provides an effective solution to this difficult problem.

It is still too early for an assessment of PERI's output. There are, however, three indirect indications which are of some use. The first is the assessment made by the Japan Key Technology Center's academic committee, which is charged both with the selection and with the assessment of Center projects. This committee uses an 'internal' method of assessment whereby the performance of a project is judged according to the extent to which the objectives of the project, stated in the project proposal, have been achieved. According to information informally supplied to one of the authors by a senior member of this committee, PERI has been very positively assessed and is regarded as one of the most successful of all the Center's projects. The second indication comes from two senior British academics, closely involved in British national research programmes in protein engineering, who have on two occasions been exposed to the research results of PERI. In their opinion, the research currently being done in PERI is of the same standard in many fields as the frontier research being done in Britain. Thirdly, at least one of PERI's researchers has been suggested as a possible Nobel Prizewinner. He is Dr Kosuke Morikawa, a department director at PERI, who has devised techniques using X-ray crystallography which enable the precise drawing of DNA and RNA molecules (*Financial Times*, Survey of Japanese Industry, 3 December 1990, p. III).

Together these indirect sources of assessment suggest that substantial progress has been made in PERI since the project started. To the extent that this is indeed correct, a substantial revision is needed to the state-of-the-art in Japanese protein engineering summarised in the 1985 Japanese Technology Evaluation Program's report.

The exploratory research for advanced technology (ERATO) programme

The circumstances surrounding the establishment of the ERATO programme were discussed earlier. From the point of view of the present concern with the creation and use of knowledge in JISB the ERATO programme derives its significance from two factors. First, it is a programme designed to bring university science and industrial technology closer together, thus increasing the connectivity between the primary institutions of JISB. Secondly, the programme aims at founding some projects which are highly uncertain, and which may therefore otherwise be underfunded, and furthermore it aims to provide researchers with a considerable degree of freedom in carrying out their research. It is these two factors which give the programme its significance, even though in total financial terms the programme is not particularly large.

Each ERATO project lasts five years and receives approximately £2–3 million per year. About twenty projects have so far been funded.

The plan is to have four new projects beginning each year so that there are approximately twenty running at any time. The ERATO programme is also of interest in terms of the methods used for project selection and evaluation of results. A major criterion in the selection of projects is the creativity and scientific standing of the project leader/s. While the overall aim of the ERATO programme is to support uncertain research which may yield returns only in the longer run (although the programme does have a portfolio of projects which vary in terms of the degree of uncertainty involved), at the project selection stage more attention is given to the creativity and standing of the project leader/s than to the details of the proposal which they put forward. In making a judgement on questions of creativity and standing, particular attention is paid to the attitudes and assessments of younger scientists. The ERATO staff of six have their own networks in order to collect information and make judgements on these matters, they regularly attend scientific conferences in order to assemble further information, and have conducted surveys for the same purpose. At the time the ERATO programme was established, in 1981, its founder, Mr Genya Chiba, waged some tough battles, particularly with the Ministry of Finance, in order to establish the principle that there should be no evaluation of the project after its completion at the end of the five-year period. In Chiba's view, which ultimately prevailed, evaluation at project completion stage did not make sense in the case of oriented basic research projects of the kind the ERATO programme was designed to support. By definition, the outcome of these kinds of projects only becomes apparent in the longer term (perhaps from about five years after the project has ended). Furthermore, it was feared that the pressures resulting from early evaluation might deflect the researchers' attention from the longer-term goals of their research. The refusal to evaluate on project termination, however, placed increased importance on the correct initial choice of project leader/s.[6]

In 1988 the ERATO programme was evaluated by the US National Science Foundation-sponsored Japanese Technology Evaluation Program (JTECH (1988)). The evaluating panel noted that the ERATO programme 'has a tendency to select high-risk, adventuresome areas that are not in vogue but that would become highly visible if successful' (p. 10). The general conclusion of the panel was that the programme was 'a success' and that the 'overall scientific quality of the . . . programme was high, but varied considerably from project to project' (p. 10).

In the area of biotechnology the panel evaluated three projects. The Bioholonics Project (1982–87) directed by Professor D. Mizuno involved parallel processing with many units (holons) working synergistically. The aim of this project was 'to simulate and utilize the organization of information and processes found in biological systems to develop novel approaches to the treatment of diseases, such as cancer and atherosclerosis' (p. 7). The general conclusion of the panel was that the 'quality of this research is comparable to similar efforts in the United States'. Furthermore, the development of improved tumour necrosis factor (TNF-S) 'is something of a breakthrough' (p. 47).

The second project was the Superbugs Project (1984–89) directed by Professor K. Horikoshi. The purpose of this project was 'to search the world for unusual microorganisms such as those which thrive under extreme conditions of pH, temperature, salinity, or in the presence of biotoxic materials. The aim of the project is to establish a "new biotechnology" using knowledge acquired in studies of the unique properties of these superbugs' (p. 54). A notable feature of this project was the substantial participation by 16 Japanese companies, 5 Japanese universities, and foreign researchers from 6 countries. In evaluating this project the panel concluded that it was 'a guaranteed success project (p. 61) . . . The area of research was well chosen in the sense of gaining success . . . but we don't think that it is truly basic knowledge with a basic technology or technical impact, and it is certainly not a high-risk project' (p. 62). The panel was of the opinion that the 'quality of the research is about comparable to that of the United States or elsewhere in Japan' (p. 63).

The third project was the Bioinformation Transfer Project (1983–88) directed by Professor Osamu Hayaishi. The general aim of this project was 'understanding how the brain is wired and how information is processed, transferred, and stored' (p. 49). It was hoped that this would lead to new 'information technology with important applications in the field of medicine, as well as computers' (ibid.). More specifically the project was concerned with a class of biological compounds called prostaglandins (which are produced by nearly every cell in the body) and their role in the central nervous system. Seven pharmaceuticals companies sent researchers to work on this project. The panel concluded that this project 'has been extremely successful. It is being directed by one of the best known biochemists in the world, who has attracted an impressive group of productive scientists' (p. 52). The panel noted that the ERATO directors had 'picked a strong leader with a good track record who had the respect of younger scientists, and then gave him the freedom and flexibility to pursue all aspect of a given research area. Serious evaluation of the project is to be delayed until five years after the termination of the five-year project' (p. 53). It was concluded that this 'ERATO project has clearly established Prof. Hayaishi as the world's expert on prostaglandins, in general, and, more specifically, for their role in the central nervous system' (p. 52).

More general comment on the ERATO programme and its significance in JISB is undertaken in the following section of this chapter.

The Role of Government in JISB

In assessing the role of government in JISB it must be reiterated that our concern is primarily with the creation and use of knowledge. From this point of view it must be kept in mind that for-profit firms, the main site in national innovation systems where 'seeds' are integrated with 'needs', face a number of constraints in their ability to create knowledge. These constraints follow largely from the fact that, as pointed out in the introductory sections of this chapter, knowledge-creation in for-profit firms is subordinated to value-creation. Accordingly, knowledge which is supposed by for-profit firms to have little chance of being transformed into value is unlikely to be a target for investment. Similarly, areas where a significant degree of uncertainty attaches to the chances of creating commercially relevant knowledge may also receive less investment than is desirable. It is here that government, relatively immune from the need to itself generate value (while still being bound by the national value-creating process), has an important role to play. It is clear that in all three of the government-initiated research programmes considered here government support is likely to have hastened the entry of Japanese firms into new areas of biotechnology. As a result, knowledge in these areas was created sooner, and possibly a greater amount of knowledge was created, than would have happened otherwise.

Furthermore, the 'vision' of for-profit firms is bounded by their knowledge, which in turn is a function of their activities. That is, firms 'see' what they know; and they know what they do. This 'bounded vision' may at times mean that firms fail to perceive technological trends, such as the emergence of new technologies or the fusion of existing technologies one or more of which they initially know little about, that may have longer-term significance for their operations. It is likely that this was the case for at least some Japanese firms in 1981 when MITI established its Bioindustry Office. Although there was a good deal of publicity at the time given to the breakthroughs in the new biotechnologies and to the establishing of the new biotechnology firms, many Japanese firms would have found difficulty in developing a view (vision) of the significance and implications of new biotechnology for their operations. From this point of view the task of the Bioindustry Office in developing, with the aid of the government–industry consultative committee and other formal and informal links with industry, a 'vision' for the area of new biotechnology, taking practical factors such as potential market implications into account, was a useful activity supplementing in a productive way the normal planning activities of firms. MITI later played a very similar role in the area of new materials.

As a result of these activities it is likely that the importance of new biotechnology very quickly permeated the 'visions' of private companies. It is likely that in Japan a greater range of firms outside the 'obvious' areas of fermentation-related industry and chemicals entered the field of new biotechnology than in other Western countries, and that this was largely due to the demonstration effect following the highlighting of the strategic significance of this technology area by MITI and the other ministries. However, while this may have led to entry by a broader range of firms as well as more rapid entry, and while on the whole this may have been beneficial for the national system, it is also possible that government activities gave rise to 'excessive entry'. This is a point that has been made by Mr Mitsuru Miyata, editor of *Nikkei Biotech*, who noted that by 1988, of the approximately 30 Japanese firms that did research on alpha interferon, only two succeeded in marketing their research: Takeda to Hoffman La Roche, and Yamanouchi to Biogen and Schering Plough.[7] While the marketing of research may be too stringent a criterion for company benefit, and while companies may have benefited in other ways from this addition to their stock of knowledge, it remains possible that too many firms entered biotechnology, as a result of the government-induced 'biobuzz', in the mistaken view that they would be able to benefit.

Government also played an important role in facilitating gains from the 'economies of research cooperation', such as avoiding the duplication of research, blending distinctive competences, and sharing expensive non-divisible equipment. While economies such as these provide firms with benefits, the competitive relationship between them might limit their ability to cooperate in research. Here government may be able to play an important facilitating role. The case of PERI provides a good example. It was seen that in 1985 Japanese firms and universities were relatively weak in this area. In the absence of government support this almost certainly would have slowed the entry of Japanese firms into the area of protein engineering by increasing the uncertainties involved. In the event, the PERI project was able to encourage firms to rapidly build their knowledge bases in this area.

The provision by the Japan Key Technology Center of a 70% subsidy and the reaping by PERI of important economies of research cooperation which were earlier analysed in detail, together with the effective tapping of the international stock of knowledge in this field and the transfer of knowledge to applications areas in the participating firms, resulted in significant benefits.

In order to get more perspective on the role of government it is worth examining the relevant figures for 1986, a year when the projects examined in this chapter were in full swing. Government-funded expenditure on biotechnology in 1986 is shown in table 34.5. The Ministry of Education accounted for 46% of the total biotechnology budget, mainly in the form of university-related expenditure. Of the other ministries and agency, which made far more targeted expenditures in biotechnology, the STA accounted for 24%, MITI for 15%, MHW for 8%, and MAFF for 7%.

Table 34.5 Government-funded research on biotechnology 1986

Government Ministry/Agency	Expenditure on Biotechnology (Million Yen)	% of Total Biotechnology Expenditure
Ministry of Education, Culture and Science (of which:	20,000	45.7
Cancer Research)	(8,944)	(20.4)
Science and Technology Agency	10,408	23.8
Ministry of International Trade and Industry	6,753	15.4
Ministry of Health and Welfare	3,400	7.8
Ministry of Agriculture, Forestry and Fisheries	3,110	7.1
Total:	43,661	99.8*

Source: *Nikkei Biotech* 13/1/86.

*Difference due to rounding.

In order to get an idea of the importance of the figure for total government expenditure on biotechnology R&D it is worth comparing this total figure with the amount spent on R&D by six of the largest Japanese biotechnology-related firms in the same year. The firms chosen for this purpose are Ajinomoto, Kyowa Hakko, Mitsubishi Chemical, Mitsui Petrochemical, Sumitomo Chemical, and Takeda, all of which participated in at least one of the projects examined in this chapter. Net sales, total R&D, and R&D in biotechnology for these firms are shown in table 34.6.

Table 34.6 Net sales, total R&D, and R&D in biotechnology, 1985 and 1986

Firm	Net sales $m		R&D $m		R&D as % of net sales		R&D in biotech	R&D in biotech as % of Tot. R&D
	1985	1986	1985	1986	1985	1986		
Ajinomoto	2,838	2,862	–	88.7	–	3.1	NA	NA
Kyowa Hakko	–	1,286	–	51.4	–	4.0	NA	NA
Mitsubishi Chemical	4,445	4,504	155	194.0	3.5	4.3	72.2	37
Mitsui Petrochemical	1,825	1,840	73.5	82.3	4.0	4.5	NA	NA
Sumitomo Chemical	3,288	–	105.2	–	3.2	–	5.6	5
Takeda	3,038	3,082	176.0	184.0	5.8	6.0	NA	NA

Source: Company Annual Reports, Nikkei Sangyo Shimbun.

In 1986 total government expenditure on R&D in biotechnology came to Yen 43.8 billion, which at the exchange rate of $1.80 then prevailing was equivalent to $243 million. This amount is equal to 35% of the total R&D expenditure of the six companies referred to here. MITI's expenditure on biotechnology was 5% of the total R&D expenditure of these six firms. Excluding the expenditure by the Ministry of Education, total government expenditure on biotechnology was about 19% of the R&D spent by the six firms.

Figures on expenditure on the major government-funded cooperative research programmes in existence in 1986, including the three examined in this chapter, are presented in table 34.7. PERI received 27% of the total amount spent on major government-funded cooperative research programmes, ranking first; the Next Generation programme received 9%, ranking fourth; while ERATO received 8% and ranked sixth. Interestingly, *around one third of total government R&D expenditure in biotechnology went to cooperative research programmes.*

Table 34.7 Expenditure on major government-funded cooperative research programmes in biotechnology, 1986

Government Agency/Project Title	Expenditure (billion Yen)	%	Rank
Ministry of International Trade and Industry			
Protein Engineering Research Institute (with Japan Key Technology Center)	3.8	27.4	1
Research Association for Alternative Petroleum Development	1.6	11.6	3
Next Generation Biotechnology Programme	1.28	9.2	4
Aqua-Renaissance National R&D Programme	1.07	7.7	7
Science and Technology Agency			
Special Coordination Fund for Promotion of Science and Technology	2.2	15.9	2
Research Development Corporation of Japan – of which:			
Exploratory Research for Advanced Technology (ERATO)	1.15	8.3	6
Development of New Technology	1.25	9.0	5
Ministry of Health and Welfare			
Foundation for Promotion of Human Sciences	0.95	6.9	8
Ministry of Agriculture, Forestry and Fisheries			
Cooperative Projects with Private Companies	0.625	4.5	9
Total:	13.92	100.0	

Based on these figures a further indication of the significance of government support for research for individual companies can be given. The MITI Next Generation programme has a total budget of Yen 20 billion over a ten-year period starting in 1981. This is the equivalent of $111.1 million over the ten-year period (at the exchange rate of $1.80 then prevailing); $11.1 million per year, or an average of $792,857 per firm per year for the fourteen firms in this project. For Mitsubishi Chemical, the largest biotechnology-related firm, the amount of $792,857 represents 0.0051 per cent of the company's total R&D expenditure for 1985. In 1986 Mitsubishi's short-term borrowings amounted to $1.57 billion. For Kyowa Hakko, the smallest firm in the present sample of six, the amount of $792,857 represented 0.0154 of the company's total 1985 R&D expenditure.

Universities in JISB

Universities play a particularly important role in national innovation systems in the case of biotechnology. The reason is that biotechnology is still very much a science-based activity. This means that universities make a significant contribution to biotechnology and its development in industry. This is not to conclude that this feature of biotechnology will remain unchanged into the future. It is possible that with time the core knowledge and techniques of biotechnology will stabilise and that competitiveness will come to depend less on the science base and more on complementary industry-based technologies such as bioprocessing. Be this as it may, it remains the case that at present the science base is very important and that this implies a particularly significant role for universities.

How important has been the contribution of Japanese universities in JISB? Once again, the seminal OTA report of 1984 passed judgement on this question. According to this report, 'Neither Japan nor the European competitor countries identified in this assessment have as many or as well-funded university/industry relationships as the United States does' (p. 17). Nevertheless, the report goes on to qualify the competitive advantage that this bestows on the US: 'In Japan, the ties between university applied research departments and industry have always been close. Additionally (p. 17), the Japanese Government is implementing new policies to encourage closer ties between basic research scientists and industry' (p. 18).

This implicitly somewhat negative view of the role of Japanese universities compared with their American counterparts is shared by many Japanese analysts. For example, one well-known analyst, Masahiko Aoki, has made the following, rather defensive, comment about Japanese universities:

Japanese universities have been downgraded by foreign observers as less dynamic and less innovative in comparison with the advanced industrial laboratories. I recognize that Japanese national universities lag behind first-rate American research universities both in research and graduate-level education except in a few fields. . . . Having admitted this, however, I would like to note the following two points: First the Japanese have made substantial progress in some scientific fields in recent times, and the role of the university in it [sic] is not entirely negligible. . . . Second, under the condition of low mobility of researchers and engineers between firms, laboratories of university professors have been playing nonnegligible roles as information clearinghouses. Professors of major faculties of engineering have decisive roles in allocating new graduates among leading firms. (Aoki, 1988, p. 252)

As noted earlier, the Ministry of Education has far more of an arms-length relationship with industry compared with the other three biotechnology-related ministries, MITI, MAFF, and MHW, which have close and direct links with industry. This is reflected in the greater separation between universities and industry. This separation is evident, for example, in the restrictions that are placed on university teachers working in national universities, limiting their ability to be employed by industry (even though there are many ways in which they are still able to do industry-related research and contribute to industry in other ways).

Further criticisms have been made in various Japanese quarters of the university system in addition to criticisms of the restrictions on close university–industry links. These include criticisms of the internal organisation of universities, in particular the KOZA system in national universities whereby a full professor is placed in charge of the basic teaching/research group consisting, in addition, of one or more associate professors, assistant professors, and post-graduate students. It has been argued that at least in some cases the power and authority of the full professor stifles the creativity of younger academics. Further criticisms have been made of limited government spending on university research and of the absence of a significant, genuinely competitive system for university-based research grants. Furthermore, it may be argued that other features of the Japanese Innovation System, beyond the immediate control of either the Ministry of Education or the universities, have served unintentionally to weaken university research. One example is the tendency for Japanese companies to have a marked preference for hiring company researchers with a Master's (or even Bachelor's) degree, rather than a Ph.D. Many companies argue that Masters graduates have the necessary general education that is required and that they can be more readily moulded to the requirements of the company than Ph.D students, who tend to be older and who may have acquired narrower 'professional' interests as a result of their specialised Ph.D research which do not conform with company needs. Furthermore, with the life-time employment system, companies have an incentive to provide in-house training, in the knowledge that they are unlikely to lose researchers to other employers. While, under these conditions, Japanese companies may have been able to generate an internal pool of researchers whose experience and interests coincide closely with company priorities, one unintended cost has been the weakening of university research by reducing the demand for university-based Ph.D places by Masters graduates, who face strong incentives to join companies on completing their Masters' degrees. Furthermore, the rapid growth of Japanese companies in the postwar period, the rapid increase in their R&D expenditures, the higher salaries that they pay relative to the university sector, and the stagnating government expenditure on university-based research (in real terms), have turned companies into relatively attractive locations for research. This has meant that a greater proportion of Japanese researchers have, under the incentive conditions mentioned, been allocated to the corporate sector than in many Western countries. On the one hand, this has served to weaken university-based research; on the other, it has meant that the research done in companies has closely matched market-expressed needs (companies being the prime site where 'seeds' are matched with 'needs'), and, furthermore, has been done under the normal conditions of corporate confidentiality. A further unintended consequence, of course, has been the international political tensions that have been generated as a result of Western complaints that the Japanese research system, relatively greatly represented by the corporate sector, is not as 'open' as in Western countries, where university research is correspondingly more important.

While criticisms such as these have become part of the conventional wisdom on Japanese universities, they do not provide enough information to answer the key question: to what extent have Japanese companies been disadvantaged as a result of the nature of the Japanese university system? In order to answer this question more directly the authors undertook a study of the relative importance of various sources of knowledge external to the firm for six of the largest Japanese biotechnology companies. The external sources included were: *joint research* with competing and non-competing companies, and with Japanese and non-Japanese universities; and *licensing* from Japanese and non-Japanese companies. The results of this study are summarised in table 34.8.

Table 34.8 Transfer of technology in six major Japanese biotechnology companies

External Channels of Technology Transfer	Firms:	A	B	C	D	E	F
1. Any *joint research* with competing companies (in biotechnology)?		No	No	No	Yes	No	Yes
2. Any *joint research* with non-competing companies?		Yes	Yes	Yes	Yes	Yes	Yes
3. Any *joint research* with Japanese universities (or technology transfer to the company)?		Yes	Yes/ No	Yes	Yes	Yes	Yes
4. Any *joint research* with non-Japanese universities?		No	No	Yes	Yes	Yes	Yes
5. *Licensing* and other technology purchase *from* Japanese companies?		Yes	No	No	No	Yes	No
6. *Licensing* and other technology purchase *from* non-Japanese companies?		Yes	Yes	Yes	No	Yes	Yes
7. Rank the importance of these 6 channels for your company.		3 6 5	6 2 3	3 4 6	3 2 4	3 4 6	4 3 1

Source: Author's interviews.

A number of important conclusions emerge from table 34.8. The first is that for four out of the six companies *Japanese universities* emerged as the most important source of biotechnology external to the firm. For the sixth company, Japanese universities were the second most important source, while for the second company they were third. Secondly, the second most important source of biotechnology knowledge was foreign universities. For one firm they were the most important source; for two firms, the second most important source; while for the fourth firm they were the third most important source. Thirdly, the third most important source was licensing agreements with non-Japanese companies. Four out of the six firms placed this source in the top three most important sources of external biotechnology knowledge. Fourthly, it is relevant to note that for the six companies taken as a whole, joint research with either competing or non-competing companies was not as important as the other three sources referred to in this paragraph. Nevertheless, for the second and fourth firms, joint research with non-competing firms was the second most important source. In only one case, the sixth firm, was joint research with a competing company included amongst the top three sources of knowledge external to the firm, in this case the third most important source. This reinforces the *a priori* hypothesis that competing companies are less likely to cooperate in sharing knowledge. Only two out of the six companies reported any joint research with competing companies, while all of them undertook some joint research with non-competing companies. Fifthly, in only one case was licensing from a Japanese company included amongst the top three sources of external knowledge.

Is there a contradiction between the contention that Japanese universities have been relatively weak in the area of new biotechnology, and the finding in the present survey that Japanese universities have been the major source of external biotechnology knowledge for a sample of the largest Japanese biotechnology-related companies? For a number of reasons it is argued that there is no contradiction. To begin with, it is usually suggested that Japanese universities tend to be relatively weak in *frontier basic research*. In areas of applied research, for example in fields such as fermentation and applied microbiology, it is widely acknowledged that Japanese universities have made some important contributions. Furthermore, from a company perspective it is likely that *intra-frontier research* in universities *in areas where the company has a commercial interest* is of greater importance than frontier basic research. The reason is simply that the products and processes being commercialised by biotechnology-related companies usually depend more on intra-frontier research than on the latest frontier basic research, which only bears fruit in the longer term. In any event, the large Japanese companies can 'plug into' frontier basic research wherever in the world it takes place, through their own research activities (now increasingly globalised) which serve as an 'entry ticket' into global basic research which is by and large open and easy to access. For these reasons it is suggested that the present finding of the importance of Japanese university research for Japanese companies is consistent with the assessment that in general Japanese universities are not as advanced as their Western counterparts in frontier basic research.

Table 34.9 Patents by priority country for Japan and the US in 13 pharmaceutical areas

Product Country:	81	82	83	84	85	Year 86	87	88	89	90	Tot.
1 – J	0	1	16	12	18	15	19	18	8	4	111
– US	0	4	11	25	12	26	25	38	18	18	177
J/US	–	0.25	1.46	0.48	1.50	0.58	0.76	0.47	0.44	0.22	0.63
2 – J	0	7	9	15	9	15	15	17	10	8	105
– US	0	8	3	15	8	25	24	29	21	18	151
J/US	–	0.88	3.0	1.0	1.13	0.6	0.63	0.59	0.48	0.44	0.7
3 – J	0	1	14	12	21	21	19	19	7	6	120
– US	2	3	7	11	20	21	16	24	22	16	142
J/US	0	0.33	2.0	1.09	1.05	1.0	1.19	0.79	0.32	0.38	0.85
4 – J	–	0	18	16	20	22	49	27	14	27	193
– US	–	3	14	7	31	22	53	52	37	65	284
J/US	–	0	1.29	2.29	0.65	1.0	0.93	0.52	0.38	0.42	0.68
5 – J	6	5	6	11	13	2	0	0	1	–	44
– US	13	20	27	23	27	0	0	0	2	–	112
J/US	0.46	0.25	0.22	0.48	48	–	–	–	0.5	–	0.39
6 – J	4	11	0	6	4	9	14	5	8	13	74
– US	2	0	1	1	5	9	6	9	4	10	47
J/US	2.0	–	0	6.0	0.8	1.0	2.33	0.56	2.0	1.3	1.58
7 – J	15	10	26	22	30	33	24	24	25	17	226
– US	9	12	6	5	10	7	25	16	15	26	131
J/US	1.67	0.83	4.33	4.4	3.0	4.71	0.96	1.5	1.67	0.65	1.73
8 – J	–	1	2	6	1	11	25	14	24	29	113
– US	–	0	0	9	10	8	7	13	31	49	127
J/US	–	–	–	0.67	0.1	1.38	3.57	1.08	0.77	0.59	0.89
9 – J	3	10	10	6	2	14	29	28	40	32	174
– US	0	0	1	3	2	18	9	15	6	10	64
J/US	–	–	10.0	2.0	1.0	0.78	3.22	1.87	6.67	3.2	2.72
10 – J	8	1	5	4	9	5	7	13	18	8	78
– US	24	21	34	28	35	25	22	22	21	24	256
J/US	0.33	0.05	0.15	0.14	0.26	0.20	0.32	0.59	0.86	0.33	0.31
11 – J	1	3	3	4	1	11	6	6	9	2	46
– US	11	1	10	14	11	24	17	27	24	36	175
J/US	0.09	3.0	0.3	0.29	0.09	0.46	0.35	0.22	0.38	0.06	0.26
12 – J	–	–	0	0	0	14	17	18	22	30	101
– US	–	–	5	0	0	4	22	20	21	14	86
J/US	–	–	0	–	–	3.5	0.77	0.9	1.05	2.14	1.17
13 – J	0	2	25	38	94	164	183	253	240	270	1269
– US	33	34	67	117	183	210	252	300	397	415	2008
J/US	0	0.06	0.37	0.33	0.51	0.78	0.73	0.84	0.61	0.65	0.63

Source: Derwent Publications.

Products :

1	=	Alpha Interferon	8	=	Epidermal Growth Factor
2	=	Beta Interferon	9	=	Superoxide Dismutase
3	=	Gamma Interferon	10	=	Brain Peptide
4	=	Interleukin-2	11	=	Human Growth Hormone
5	=	Hepatitis B Vaccine	12	=	Granulocyte Colony Stimulation
6	=	Erythropoletin			Factor
7	=	Urokinase	13	=	Monoclonal Antibodies

Table 34.10 Total patents to 1989*

Product	Priority Country							
	Japan	US	UK	FRG	France	Italy	Sweden	Switz.
1	75	93	17	16	8	0	0	2
2	80	72	11	8	2	0	1	0
3	87	75	15	20	8	0	1	2
4	108	120	8	19	5	1	1	1
5	49	84	16	13	23	0	0	4
6	47	38	3	6	1	0	0	0
7	300	74	17	14	9	2	0	6
8	62	35	7	2	1	2	0	1
9	120	33	1	5	4	3	1	0
10	53	152	24	9	19	7	1	6
11	43	118	11	3	10	6	0	0
12	31	37	4	4	0	0	0	0
13	782	981	176	113	88	4	20	14

Source: Derwent Publications.
* data include only the first 43 weeks in 1989.
Products: see notes to table 34.9.

To the extent that the present empirical material and correspond-ing argument are correct, not only for the six sample firms but for Japanese biotechnology-related firms more generally, *substantial revision is necessary to the conventional wisdom that Japanese universities constitute the weak link in the Japanese innovation system in biotechnology.*

The Performance of JISB: a Patent-based Analysis

In the present chapter we have analysed the main features of JISB. A further problem remains in establishing how well, comparatively speaking, JISB has performed relative to other national systems of innovation. Unlike in the case of information technology and

microelectronics, where the major products have long been commercialised and where, therefore, indicators of performance such as market share, profitability, and rates of return on investment are
available, in biotechnology very few products have reached the commercial stage. In order to examine this problem we analysed international patenting in the area where biotechnology applications are
most advanced, namely pharmaceuticals. With the help of researchers
in this field we identified 13 pharmaceutical products where new
biotechnology research has been concentrated. (The products are
listed in table 34.9.)

Table 34.11 Top non-Japanese companies

Product	Top three companies (no. of patents)		
	1	2	3
1	Schering (12)	Boehringer (9)	Genentech (8)
2	Cetus (23)	Searle (6)	Genentech (4)
3	Bioferon (7)	Genentech (7)	Biogen NV (6) & Cetus (6) & Schering (6)
4	Cetus (27)	Hoechst (7) Immunex (7)	
5	Merck (11)	US Dept Health (10)	Inst. Pasteur (7)
6	Genetics Inst. (3) & Kirin-Amgen (3)	Biogen NV (2) & U. Washington (2) & New York U (2) & Zymogenetics (2)	
7	A Med. Cardio. (6) & Beecham (6) & Behringwerke (6)		
8	Chiron (2) & Genentech (2) & Ethicon (2) & US Sec. Commerce (2) Gen. Hospital Corp. (2)		
9	Chiron (5)	Bio-Tech Gen. (3)	Survival Tech. (2) & Yeda R&D (2)
10	Searle (12)	Squibb (10)	Burroughs Well (7)
11	Genentech (13)	Eli Lilly (9)	Monsanto (8)
12	Immunex (6)	Genetics Inst. (4)	Kirin-Amgen (3) & Schering (3)
13	US Dept. Health (41)	Sloan-Kett'g (36)	Tech. Licence Co. (31)

Source: Derwent Publications.
Products: see notes to table 34.9.

Table 34.12 Top Japanese companies*

Product	Top three companies (no. of patents)		
	1	2	3
1	Green Cross (12)	Toray (10)	Takeda (6)
2	Toray (30)	Kyowa Hakko (10)	Green Cross (5)
3	Toray (14)	Green Cross (12)	Suntory (8)
4	Ajinomoto (29)	Takeda (21)	Toray (7)
5	Kagaku Oyobi (7)	Chemo-Sero (6)	Green Cross (5)
6	Snow Brand (9)	Chugai (6)	Green Cross (4) & Toyobo (4)
7	Green Cross (43)	Asahei (7) & Terumo (7) & Wakamoto (7)	
8	Wakunaga (17)	Hitachi Chem (5) & Nihon Chem (5)	
9	Takeda (8) & Toyo Soda (6) Ube (8)		
10	AIST (4) & Otsuka (3) & Daiichi Seiyaku (4)	Takeda (3)	
11	Rikagaku Ken. (5)	AIST (3) & Nakajima (3) & Sumitomo Chem (3)	
12	Chugai (13)	Kyowa Hakko (2)	[seven firms (1)]
13	Green Cross (42)	Teijin (39)	Kyowa Hakko (29)

Source: Derwent Publications.
*data covers 1981–1988 (up to the 45th week in 1988).
Products: see notes to table 34.9.

Because of differences in national patenting regulations, analysts frequently restrict analysis to patents taken out in the US, the largest market in the world. For present purposes, however, this is inadequate since it does not cater for those Japanese firms which do not expect to market their products in the US. Accordingly, we had no option but to examine patent data by priority country (that is, the country which is the first priority for the organisation, usually based in the same country, taking out the patent).[8] Since national patent regulations differ (particularly between Japan and the US over the period we are concerned with), the figures on absolute numbers of patents are non-comparable. However, making the assumption (which may not be correct) that the differences in regulations and processing of patents have remained constant over the period, we have examined the *ratio* of Japanese to US patents in these 13 product areas in order to establish whether there are any clear trends. The data are given in table 34.9. However, as table 34.9 shows, there are no clear signs of Japan over time becoming stronger or weaker relative to the US in these product areas. In table 34.10 the total number of patents by priority country (*note, which are not directly comparable*) for the 13 products are shown for a number of countries.

Table 34.13   Top ten non-Japanese patenters

Company	Patents	(Monoclonal antibodies)
US Dept. Health	63	41
Cetus Corp.	61	0
Genentech	47	0
Sloan-Kettering	44	36
Inst. Pasteur	39	30
Schering Corp.	35	0
Immunex Corp.	33	20
Akad Wissenschaft	33	27
CNRS	31	29
Tech. Licence Co.	31	31

Source: Derwent Publications.

Table 34.14   Top ten Japanese patenters

Company	Patents	(Monoclonal antibodies)
Green Cross (P)	125	42
Toray (C)	78	8
Takeda (P)	61	9
Ajinomoto (F)	57	9
Kyowa Hakko (C)	51	29
Teijin (O)	41	39
Wakunaga (P)	34	17
Otsuka (P)	30	8
Snow Brand Milk (F)	29	17
Asahi Chem. (C)	27	14

Source: Derwent Publications.
Notes: C = Chemicals
    F = Food/Beverage
    O = Other
    P = Therapeutics.

The present patent data, however, is extremely revealing when analysed in terms of the top three patenting organisations. This data is shown in tables 34.11 to 34.14. As these tables make clear, a significant difference between Japan and other Western countries, particularly the US, in patenting by kind of organisation is that *in Japan patents in this area tend to be taken out by large companies, while in Western countries it is government-related organisations or new biotechnology firms that tend to dominate.* This reinforces the point made that the *'centre of gravity' for Japanese innovative activity in biotechnology is currently located more in large companies than is the case in most Western countries.*

Conclusion

Because of the complexity of the empirical material and analysis of this chapter no attempt will be made here to summarise the strengths and weaknesses of the Japanese Innovation System in Biotechnology (JISB). For this the reader is referred to the body of the present chapter. Rather, attention will be focused on the usefulness of the concept of a national innovation system.

While, as we have seen, history and politics shape the actual institutions that make up the national innovation system as well as their interactions, it is knowledge – its creation and use – that provides the conceptual cement, holding the system together. The effectiveness with which these institutions, both directly and through their interactions, influence the creation and use of knowledge will be an important determinant of the performance of the national system. In the case of JISB we have seen that significant complementary roles have been played by the system's primary institutions: for-profit firms, universities, and government research laboratories and policy makers. The 'whole' that is made up by these institutions *and their interactions* is greater than the sum of the institutions 'added up' individually. It is through the analysis of the interactions of the institutions that the concept of the national innovation system produces its additional benefits.[9] In the case of JISB, for example, we have seen how both universities and government research programmes compensate for the shortcomings of for-profit firms, where knowledge-creation is subordinated to value-creation, in the generation and use of knowledge. Furthermore, we also saw how government research programmes have compensated for some of the weaknesses in the interaction between Japanese universities and Japanese industry. Ultimately, differences in national economic performance, and therefore, differences in the economic wealth of nations, are largely attributable to differences in the institutions of their national innovation systems and to the interactions of these institutions. The strength of the concept of a national innovation system is that it enables these institutions and their interactions to be analysed more rigorously.

ACKNOWLEDGEMENTS

The authors would like to express their appreciation to many individuals and organisations, too numerous to mention here, for the help and insights they have given us into Japanese biotechnology over the last five years. Martin Fransman would like to acknowledge financial support from the European Commission's Concertation Unit for Biotechnology in Europe (CUBE), and intellectual stimulation and support from its Director, Mark Cantley. Financial support from the British Department of Trade and Industry is also gratefully acknowledged. We are particularly grateful to the numerous individuals from some of the major Japanese biotechnology companies who gave unstintingly of their time and knowledge. Particular thanks are due in this regard to company officials from Ajinomoto, Kyowa Hakko, Mitsubishi Kasei, Mitsui Petrochemical, Sumitomo Chemical, Sumitomo Pharmaceutical, Suntory, and Takeda. MITI officials, particularly Mr Furudera and Mr Tanaka, were extremely helpful in providing information and the benefits of their deep insights into Japanese biotechnology. Mr Furudera is to be thanked particularly for the information and data that he supplied. Professor Ikehara and his staff at PERI on numerous occasions shared insights with us and generously gave further details. Mr Mori at BIDEC and his colleagues were also particularly helpful in giving information and arranging contact with companies. We are also extremely grateful to Dr Brian Stockdale and Derwent Publications for the access that they have given us to patent data and for other related information. Needless to say, none of these individuals or organisations are to be held responsible for our data, analysis, and conclusions.

NOTES

[1] Interview with one of the authors.

[2] See Federation of Economic Organisations (Keidanren), *Proposal for Promotion of Basic Research in Life Sciences* (Tokyo: Keidanren, 1988), pp. 2–3, quoted in Tanaka (undated).

[3] See Nelson and Winter (1982) and Metcalfe (1988, 1991).

[4] The concepts 'coordinated in-house research' and 'joint research in joint research facilities' are further developed in Fransman (1990) in an analysis of some of the major government-initiated cooperative research projects in Japanese information technology.

[5] This is based on the authors' interviews. MITI officials were also strongly criticised for their lack of openness regarding the difficulties of this part of the NG programme.

[6] Interview with Mr Genya Chiba by one of the authors.

[7] Lecture at Institute For Japanese–European Technology Studies, University of Edinburgh, 1988.

[8] We would like to express our gratitude to Derwent Publications Limited for the patent data that they generously provided and to Dr Brian Stockdale of Derwent for his unstinting help. It should be noted that it is possible that patents taken out by a foreign subsidiary are included in the data by priority country. For example, a US subsidiary's patents taken out in Japan might be reflected in the patents attributed to Japan.

[9] Any analytical concept, to be justifiable, must produce additional conceptual advantages that would not result without the use of that concept.

REFERENCES

Aoki, M. (1988). *Information, Incentives, and Bargaining in the Japanese Economy.* New York: Cambridge University Press.

Dibner, M. D. and White, R. S. (1989). *Biotechnology Japan.* New York: McGraw-Hill.

Fransman, M. (1991). What We Know About the Japanese Innovation System and What We Need To Know, in Inose, H., Kawasaki, M. and Kodama, F. (eds) (1991), *Science and Technology Policy Research. What Should Be Done? What Can Be Done?* Tokyo: Mita Press.

Fransman, M. (1990). *The Market and Beyond: Cooperation and Competition in Information Technology in the Japanese System.* Cambridge: Cambridge University Press.

Fransman, M. (1988). The Japanese System and the Acquisition, Assimilation and Further Development of Technological Knowledge: Organizational Form, Markets, and Government, in Elliot, B. (ed.) (1988), *Technology and Social Process.* Edinburgh: Edinburgh University Press.

Fransman, M. (1988). Corporate Strategy and Technology Transfer in the Japanese Biotechnology-Creating System, in *Proceedings of the Biosymposium Tokyo '88.* Tokyo.

Fransman, M. (1986). *Biotechnology – Generation, Diffusion and Policy: An Interpretive Survey.* Maastricht, the Netherlands: United Nations University.

Freeman, C. (1987). *Technology Policy and Economic Performance, Lessons from Japan.* London: Pinter.

Freeman, C. (1988). Japan: A New National System of Innovation?, in Dosi, G. et al. (eds) (1988), *Technical Change and Economic Theory.* London: Pinter.

Freeman, C. (1990). Technical innovation in the world chemical industry and changes of techno-economic paradigm, in Freeman, C. and Soete, L. (eds) (1990), New Explorations in the Economics of Technical Change. London: Pinter.

Hayek, F. A. (1945). The Use of Knowledge in Society', American Economic Review, Vol. 35, pp. 519–30.

Japanese Technology Evaluation Program (JTECH) (1988). JTECH Panel Report on the Japanese Exploratory Research for Advanced Technology (ERATO) Program. McLean, Virginia: Science Application International Corporation.

JTECH (1985) – see US Department of Commerce.

Johnson, C. (1989). MITI, MPT, and the Telecom Wars: How Japan Makes Policy for High Technology, in Johnson, C., Tyson, L. A. and Zysman, J. (eds) (1989), Politics and Productivity. The Real Story of Why Japan Works. Cambridge, Mass.: Ballinger.

Keidanren (Federation of Economic Organisations) (1988). Proposal for Promotion of Basic Research in Life Sciences. Tokyo: Keidanren.

Lewis, H. W. (1984). Biotechnology in Japan. Washington, DC: National Science Foundation.

Metcalfe, J. S. (1988). The diffusion of innovations: an interpretive survey, in Dosi, G. et al. (eds) (1988), Technical Change and Economic Theory. London: Pinter.

Metcalfe, J. S. (1991). Technology Policy in an Evolutionary World, paper presented to the Second NISTEP Conference, Tokyo, January 1991.

Nelson, R. and Winter, S. (1982). An Evolutionary Theory of Economic Change. Cambridge, Mass.: The Belknap Press.

Nelson, R. (1988). Institutions supporting technical change in the United States, in Dosi, G. et al. (eds) (1988), Technical Change and Economic Theory. London: Pinter.

Nelson, R. and Rosenberg, N. (eds) (forthcoming). National Innovation Systems. Cambridge: Cambridge University Press.

Pisano, G. P., Shan, W. and Teece, D. J. (1989). Joint Ventures and Collaboration in the Biotechnology Industry, in Mowery, D. C. (ed.) (1989). International Collaborative Ventures in US Manufacturing. Cambridge, Mass.: Ballinger.

Roberts, E. B. and Mizouchi, R. (1989). Inter-firm technological collaboration: the case of Japanese biotechnology, International Journal of Technology Management, Vol. 4, No. 1, pp. 43–61.

Tanaka, M. (undated). Government Policy and Biotechnology in Japan: The Pattern and Impact of Rivalry Between Ministries (mimeo), Tokyo.

US Congress, Office of Technology Assessment (1984). Commercial Biotechnology: An International Analysis. Washington D.C.: Office of Technology Assessment.

US Department of Commerce (1985). Biotechnology in Japan (Japanese

Technology Evaluation Programme). Washington, DC: US Department of Commerce.

35

Industrial Biotechnology Policy: Guidelines for Semi-Industrial Countries[1]

Francisco C. Sercovich

I Introduction

Entry into biotechnology (BT) manufacturing does not follow entry into BT research as naturally as is sometimes assumed. The transition is not an easy one even for firms engaged in commercial production of BT-based R&D services. Entry into BT production, marketing and distribution means having to cope with things such as the paucity of off-the-shelf technological and manufacturing solutions, and fierce competition from established firms trying to retain their market shares.

Save a few exceptions, like that of *in vitro* diagnostic kits, the customary reference to low barriers to entry into BT in the literature should be taken with a grain of salt since it applies to pre-competitive entry only. The passage from the lab to the industrial arena is less trivial than many enthusiasts admit.

Furthermore, entry into BT as an industrial activity cannot be dealt with as a purely firm-specific phenomenon. For an emerging, generic technology-based industry, it also refers to a whole set of interacting agents, which calls for the often neglected systemic aspects of entry.

Particularly in developing countries (DCs), the accumulation of basic BT knowledge does not trickle down easily into the economic sphere. This diminishes its potential for wealth creation. The passage from the realm of the scientifically possible, through that of the technically feasible, on to that of the economically profitable is much smoother in the industrial countries (ICs), where, for this reason, bio-policy often entails industrial policy, although it may not be called so.

But in order for a workable transition from scientific effort to the market to occur, a wide variety of capabilities and institutions have to be in place, such as a reasonably well articulated risk-capital market; an enterprise sector permeable to the scientific culture; a scientific sector permeable to the enterprise culture; and corresponding sets of institutions and legal codes.

Although policy interventions are justified on grounds of indivisible investments in R&D, uncertainties and non-appropriabilities, clearly they cannot substitute for an efficient interface between the scientific and the industrial systems, the availability of entrepreneurial and management skills or the necessary interactions among the agents of innovation. In DCs, external dis-economies lead to mis-allocation of resources, e.g. by deterring out-sourcing, thus detracting from the effectiveness of the innovative process.

Actually, not even in the ICs is the trickle down effect taken for granted. Market failures (and national rivalries) lead to active government promotional and stimulatory involvement. Although market failure is nowhere as pervasive as in DCs, in most of them the industrial policy content of BT policies is not very readily identifiable, to say the least.

DCs have a lot at stake on the issue of what standards are set to define entry into BT. After so many short-lived incursions into industrialization, they cannot afford taking false steps into such a critical cluster of generic technologies by adhering to loose guidelines. This refers not just to scientific quality. It concerns especially industrial, engineering, organizational and entrepreneurial standards. These can in no way be satisfied if due attention is not paid to a set of key dimensions such as gaps in technological mastery, polyvalent engineering skills and scale-up-related issues. Perhaps, too much voluntarism has been one of the most outstanding features of advocations for BT in DCs. Meanwhile, precious time is being wasted.

In principle, there is nothing wrong with a science-push entry, particularly in a science-driven industry like BT, provided that the incentives, markets, capabilities and institutions are in place and work effectively so as to meet social needs and reach consumers competitively. However, the existence of externalities, indivisibilities, and like market failures involve the need for industrial policy. But industrial policy cannot do without the necessary capabilities and conducive institutions.

The main global trends in BT are discussed in section II. Then entry into BT is dealt with in section III. Section IV is devoted to various specific industrial-policy issues focusing on the case of semi-industrial countries. The last section offers some closing remarks.

II. The Global Setting

Some of the most relevant global trends and factors affecting BT industrial policies in DCs will now be reviewed. They are: (a) scientific and technological uncertainties; (b) relative competitiveness; (c) timing of introduction and rate of diffusion; (d) routinization of the basic techniques; (e) threshold barriers and shifting manufacturing frontier; (f) company strategy; (g) national policies; (h) trade reversals; (i) scope/scale trade-offs; (j) privatization of scientific knowledge; (k) industrial property regime; and (l) need focusing.

(a) Scientific and technological uncertainties

BT's future is highly uncertain. Because the knowledge base is growing at a faster rate than the use of such knowledge in practical applications, the technological and industrial trajectory of BT is not yet quite clear, even within the not too distant future.

In the scientific sphere not enough is known yet about things such as the relationships between protein structure and functions, the mechanisms of pathogenicity in plants, and drug delivery methods. However, if feasible technical solutions and profitable economic outlets are found, an ever increasing number of radical technological and commercial breakthroughs will certainly take place. This may lead, among other things, to a shift away from anti-cancer chemotherapies and agro-chemicals, thus bringing about major shifts in market structure. But this is highly unlikely to happen before the turn of the century.

(b) Relative competitiveness

Examples of BT's superiority abound. For instance, BT methods for protein manufacturing are far superior to those relying on extraction from vast amounts of animal tissue or the random screening of organic compounds. However, the relative competitiveness of BT products and processes still remains to be demonstrated, except in the few cases where it has given birth to entirely new products (like monocolonal antibodies – Mabs) or has overcome absolute physical and/or cost limits to input availability (to produce insulin, for instance). High costs related to research, stringent process and quality assurance requirements, handling, delivery systems etc. have so far offset BT's inherent advantages. Sharp changes in relative prices may improve BT competitiveness in some applications and encourage efforts in areas such as energy and commodity chemicals. Technological mastery plus the diminishing quasi-monopoly power in established pharmaceuticals and agro-chemicals will gradually offset the initial handicap of BT products.

(c) Timing of introduction and rate of diffusion

These variables differ widely across sectors. The diffusion rate is the highest in drugs, followed by chemical and agricultural applications, with the rest far behind. Within drugs, the diagnostics sector is more advanced than therapeutics; and therapeutics, in turn, is more advanced than preventive applications. These contrasts follow a complex and uncertain interplay among the state and evolution of the knowledge base, policy priorities, the role of the regulatory environment and public opinion, the relative competitiveness of BT processes and products, the interplay of competitive forces and the status of industrial property rights.

Cross-industry diffusion rates depend much on industry-specific variables such as unit product value, R&D thresholds and payback periods. Drugs, a highly R&D-intensive industry, will keep a head-start in BT as long as the efforts required for scientific breakthroughs, and engineering constraints, are not made trivial by technical progress. As for the rest of potential BT user industries, the key largely lies with technological mastery. The building of *savoir faire maison* is going to take a long time in most BT user industries while the basic techniques are routinized and intermediate supplier-networks developed.

The timing of introduction and pace of diffusion are also influenced by the policy environment. Thus, for instance, in 1989 over $245 billion were poured by OECD countries into import quotas, acreage set-asides, export subsidies and other policies, making agriculture the most manipulated industry of all (*The Economist*, 1990). This affects the timing of introduction and pace of diffusion of BT innovations (like bGH), since these would deprive cosy subsidies of justification. US subsidies that encourage more research into sugar or petroleum substitutes than warranted by market prices work in the opposite direction (*Fortune*, 1990, p. 57).

(d) Routinization of the basic techniques; application-specificity of engineering and manufacturing know-how

The routinization of the basic scientific techniques, coupled with the growing application-specificity of BT engineering and manufacturing know-how, cause BT to be absorbed into the various user sectors rather than to evolve as a readily identifiable industry – except for the intermediate input and instrument segments. The acquisition of core in-house R&D BT capabilities by large firms in many industries strengthens this trend (Toyota being the latest reported entrant) (*Bio-Technology*, 1990, p. 802). As BT matures, so does the growing differentiation of entry barriers relating to sector-specific engineering, manufacturing, marketing, regulatory standards, routines and practices. As a result of this, the current science-led stage will give room to a more market-driven stage.

Over what remains of the twentieth century, the structure of the BT 'industry' will probably become well defined. In the US, it is likely to take a multiple, application-sector-focused and hub-like shape, centred around a rather limited number of large firms playing the part of nexus among hosts of research boutiques, research institutions and dedicated BT firms serving niche markets, through a complicated network of financial and technological arrangements. In the EC and Japan the structure will be less diversified.

(e) Threshold barriers and shifting manufacturing frontier

Because of competition from conventional products, scientific uncertainties, intense R&D rivalry and evolving manufacturing practices, reaching the market with a specific product does not guarantee the recovery of the substantial sunk R&D investments involved. This is why risk-sharing through subcontracting, partnerships or subsidies has become inescapable even for the largest players. Although it is true that BT has brought about a compression between the different stages that go from basic scientific discoveries to actual applications, exaggerating the existence of short-cuts and quick 'fixes' pays lip service to the interests of DCs contemplating their entry into the industry.

(f) Company strategy

Strategic partnering with a large multinational appears so far to be the only way in which new entrants can hope to get into mainstream BT markets. In mutual partnerships, both start-up companies and multinationals have valuable assets to offer. The former provide their ability to leverage knowledge from universities, hire university facilities on a part-time basis and motivate contributions by scientists or entrepreneurs through stock ownership and other economic incentives. The latter contribute with their R&D financing muscle; regulation-related experience and resources; scale-up capacity; established marketing networks; and diversity of product lines that make it possible to reap economies of scope. Often, startups have a high price to pay when they cannot afford but to get into this kind of arrangement, i.e. relinquishing control on their scientific and technological developments. Except in niche and highly specialized market segments, alternatives to this are becoming less and less feasible.

Although multinationals can strongly affect the timing of introduction and pace of diffusion of BTs, they cannot suppress them; nor are they likely to try to do so in order to protect their markets for conventional agro-chemical and pharmaceutical products. For one thing, a good deal of their patents protecting these products are expiring so that profit margins are diminishing. For another, public opinion and pressure groups are creating an atmosphere hardly conducive to continuing to rely on conventional products. Thus, although the intrinsic potential superiority of the BT route remains to be expressed in the economic arena, multinationals are definitely open to the prospect of using it to recreate their weakening quasi-monopoly power.

(g) National policies

ICs are explicitly applying infant-industry policies in BT. For instance, the EC has recently lifted its opposition to proposed Belgian government subsidies to commercial R&D on recombinant products, on grounds of the innovative nature of genetic engineering and associated risks. This is in addition to things such as the 3rd EC Framework Programme (1990/94), recently approved by the Council of Ministers, that will provide $200 million for BT R&D. The US provides subsidies to (tax exempted) schemes such as Research and Development Limited Partnerships (RDLPs) and tax preferences to patent royalty income.

ICs are also targeting support of scale-up efforts. The so-called 'downstream processing club' in the UK involves two research institutes and various firms in the search for improved separation and purification of products from bioreactors. Direct support to scale-up is considered one of the most relevant policy issues in the US. Japan paid attention to scale-up-related problems very early in the development of its own BT industry.

(h) Trade reversals

Cases such as those of sugar and vanilla substitutes show that BT is aggravating the impact of trade reversals originating in automation of labour-intensive processes. Further examples: the plant *shikonin* (grown in China and the Republic of Korea) which, thanks to its medical properties, sells at $4,500 per kilo, is now being produced in bulk through tissue-culture techniques by Mitsui in Japan. Similar is the case with products such as pyrethrin, codeine and quinine. However, industrial use of the knowledge base is often kept on tight hold because of economic and social uncertainties. This cushions the actual impact on DCs.

(i) Scope/scale trade-offs

BT poses the need to master skills such as the ability to manage multidisciplinary R&D teams, and taking prompt advantage of synergies and cross-fertilization in scientific and technical knowledge in order to exploit spin-off potentials. Particularly when the time comes to scale up BT processes, trade-offs arise between reaping economies of scope in R&D and exploiting economies of scale in specialized manufacturing. Few firms can have it both ways. In DCs, lack of markets and interactions induce the first route at the cost of delaying actual entry into the market. But this undermines the economic prospects of the ventures by preventing the timely recovery of R&D investments (see further examples below).

(j) Privatization of scientific knowledge

Basic scientific knowledge is no longer flowing as freely as it used to. Nowadays, when scientists are on the verge of a breakthrough the first thing they are advised is not to publish or disclose it in any way, but to reserve property rights through patenting. Their activity affects stock market quotations directly, which indicates the extreme sensitivity of BT business to shifts in the scientific frontier.

(k) Industrial property regime

The strengthening of industrial property rights is intended to offset diminishing imitation time-lags. There is a conflict of interest between IC-based enterprises that want to maximize global returns accruing to their R&D investments and DC firms trying to gain breathing space for their imitative activities. To make things worse, only very few hold indisputable or undisputed rights on BT patents. But the key to entry into BT resides ever less in getting access to basic knowledge and ever more in knowing how to apply it industrially. Herein lies the main challenge ahead for DCs.

(l) Need focusing

BT's trajectory has so far been focused on the needs of the populations of OECD countries and, within this, on the highest value-added products. Two-thirds of drug R&D in the US goes to applications catering to the needs of the oldest segment of the population, while less than 3% goes to tropical-disease prevention or cure. Meanwhile, the rate of infant mortality in DCs is assessed at 20%, while hundreds of millions are infected by parasitic organisms.

III. Developing Countries' Market Entry

In discussing DCs' entry into BT and related policy issues, the first thing that comes to mind is market failure. Acute imperfections in the markets for factors and information prevent BT developments from reaching those market segments where they are needed most. This poses formidable challenges to policy-makers.

To date, most BT developments are sharply at odds with views that suggest that BT is particularly suitable to DCs – because of what it promises, its allegedly low entry barriers, and its assumed appropriateness or amenability to be used for leapfrogging. However, while BT's birth is still being laboured, basic techniques are becoming routine, the technological trajectory is becoming increasingly userspecific and imitation costs and time-lags are being shortened. All this may facilitate DCs' market entry, provided that scaling-up and downstream processing problems are addressed appropriately. DCs' genetic endowment is a purely static advantage. It will be irremediably lost unless its value is enhanced through S&T efforts. Not even the shrewdest protective legal devices will do in their place.

Although most DCs' (like ICs') entries into BT are supply-led, there are variations. Sometimes the push from science is stronger than the pull coming from industry or vice versa, while strong market-driven elements can be identified in some cases.

Cuba is a good example of a science-driven entry into – largely health-oriented – BT, mainly at the R&D stage. Although some production capacity has developed, it cannot reach world markets because of allegedly deficient quality-assurance guarantees (so far Cuba is only serving some Third World markets based on concessionary assistance and science and technology co-operation deals). Its cost-competitiveness is unknown. The Centre for Biological Research (CIB), set up in 1982, produces its own restriction enzymes and does research on the synthesis of oligonucleotides, the cloning and expression of a number of other genes, and the production of Mabs for diagnostic purposes.

Cuba's entry into BT pursued social ends; i.e., the interest in interferon was prompted by the outbreak of dengue hemorrhagic fever affecting some 300,000 people in the late 1980s. But there was also a science-push drive: first-rate bioscientists were available and it was thought that BT suited Cuba because of its research-intensive nature (which applies to entry into research rather than into manufacturing). If Cuba is to take steps to get closer to the world market, substantive efforts will have to be made to set up cost-efficient and world quality process, product, and production engineering standards as well as marketing and distribution channels.

Argentina's entry into BT shows strong industry-push elements. It is based on a small though rather dynamic industrial BT establishment drawing on the remainders of a world-class biology science base. There are a few BT firms working in the field of diagnostics, vaccines and micro-propagation led by two small pioneer firms mainly active in human health. The predicament facing one of these firms is typical of a DC milieu (i.e., external dis-economies and the need for expensive in-house efforts).

In order to enter the rDNA route, a series of related techniques such as cell culture, protein purification, Mab production, fermentation, etc. had to be learned. But their mastery would not have made sense in order to produce just one product: a steady drive towards exploiting scope economies plus a lack of out-sourcing networks led to a steady growth in the size of an initially modest project. Size escalations and start-up delays followed. What first looked like 'short-cuts', drawing on imitation and extensive use of freely available information, later turned into unexpected bottlenecks and difficulties requiring a good deal of unforeseen experimental work and innovative efforts to learn a wide range of basic techniques and to apply them effectively. The start-up of the lab, isolation of the gene, its expression and optimization, added up to 6 years before commercial production. The initial budget grew ten times (Katz and Bercovich, 1988). Little time was saved compared with what it takes a dedicated BT firm in an IC, although the investment was significantly lower because it relied on reproducing a process already known. Although the project was technically feasible, its economic rationale remains to be demonstrated. No industrial policy framework was available to support this effort.

Much stronger and more effective demand-pull elements are found in Brazil. The elements behind the rationale for the Alcohol Programme were energy dependence, a very high level of photosynthetic efficiency, and an expected price of a barrel of petroleum at over $40. Brazil's headstart in the field of ethanol from sugar-cane relied on natural advantages and upon the mastery of all skills and capabilities needed to turn out complete package deals, including project design, execution and start-up, process know-how, machinery construction, training, technical assistance and planning of integrated agro-industrial operations. The Programme sought to control natural processes rather than to engineer them. Hence, it relied largely on known fermentation-related process control engineering, scaling-up and mass production rather than on the manipulation of genetic information. However, the Programme (which is now re-entering a more favourable phase), along with the exploitation of a variety of biomass sources, created a large and avid market for BT breakthroughs (Sercovich, 1986).

Brazil's headstart in traditional BT has spun-off what has now become an incipient and dynamic development of frontier BT. These efforts are being led largely by academic research scientists and by increasing numbers of innovative start-ups. University–industry links are being forged through initiatives like Bio-Rio, a science park that will offer an incubator facility, central labs for sequencing and synthesis of nucleotides, rDNA experiments and scale-up, administrative support and technical services.

While in Latin America the weak link is usually industry, in developing South East Asian countries it is the domestic science base. The Republic of Korea, Singapore, Taiwan Province of China and Thailand show comparatively stronger market-driven orientations. They also have more explicit and focused industrial policies towards BT, including supply of credits, grants, risk capital and support for skill-formation and process and product development. Thailand pays relatively more attention to agriculture and the other countries to health-related applications. In Singapore, Taiwan Province of China and Thailand, start-ups play an important role. The Republic of Korea relies much on chaebols, i.e. large conglomerates that devote substantial resources to BT R&D. South East Asian countries offset the relative weakness of their science base by drawing directly on ICs' scientific establishment through their expatriates and by setting up BT research firms there – the Republic of Korea's Samsung and Lucky-Goldstar have done so in the US (Yuan, 1988). And the circuit goes both ways. Glaxo is setting up a $50 million research joint-venture with Singapore's Institute of Molecular and Cellular Biology (IMCB). Not accidentally, all three senior scientists involved in the IMCB are, or have been, associated with major research institutions in the US and Europe (*Genetic Engineering News*, 1989, p. 26).

In conclusion: (a) demand-driven elements appear to have a stronger presence in South East Asia than in Latin America, where supply-led elements tend to prevail; (b) within the supply-led experiences, science-push forces are particularly strong, most of the action taking place at university research centres or in research-oriented firms; and (c) there is a pervasive lack of skills and capabilities to bring scientific output into industrial use. The scope for LDC firms to continue to take advantage of shortening imitation time and cost lags is at stake in bilateral and multilateral TRIP (GATT)-related negotiations currently underway. A weak industrial policy content is particularly noticeable in the Latin American experience.

IV. Industrial Policy Issues

BT poses plenty of room for controversy and doubts, for it challenges a good deal of the conventional wisdom regarding issues such as the role of basic science in industrial progress, the economics and management of R&D efforts, the locus and focus of technical change, industrial property rights and biosafety-related issues. However, all this ought not to delay industrial policy action anymore.

The science-push drive fails to work in some cases, like in vaccines, where price competition allegedly discourages leading firms from engaging in development and manufacturing. This case illustrates dramatically the critical importance of threshold barriers to DCs' entry. Plainly, as long as technological and manufacturing barriers are not overcome, a number of vaccines that can be produced today on the basis of existing scientific knowledge just will not reach those who need them. Because IC markets do not justify their commercial development, they remain expensive, and because they are expensive they are beyond the reach of those who need them most.

The progressive routinization of the basic techniques makes it easier for user industries to appropriate the know-how concerned. DCs have the possibility to undertake such appropriation directly, in connection with applications most relevant to them (be it in agriculture, food, health care, mining, waste disposal or whatever).

This prospect is not favoured at all by the increasing privatization of scientific knowledge in ICs. However, this problem concerns particularly the very cutting edge of the scientific frontier. Short of it, DCs have a lot of room to take advantage of the already routine breakthroughs (like gene-splicing engineering).

One of the main promises BT brings with it is that of letting DCs wean themselves from economic dependence on commodity prices. Australia has focused on this problem as the main target of its policy in BT. From this angle, Australia's approach is relevant to most DCs (Freeman, 1989, p. 14). However, such a promise must be looked at with a great deal of caution. The route to it may be hazardous.

DCs remain relatively backward, despite all their potential for catching up, because they lack many or all of the ingredients that concur in forming the social capability required to realise such potential. There should be no illusions as to BT being an exception in this regard. Many DCs can put together a group of first-rate scientists and even endow them, at the cost of great sacrifices, with the resources necessary to undertake high-quality research. But to expect to be able to reach the world market on this basis is an illusion. As Japan, and then the Republic of Korea, Hong Kong, Singapore and Taiwan Province of China have shown, the key to effectively exploiting the leapfrogging potential lies not just in the mastery of the scientific underpinnings of a technology, but rather, in the mastery of the engineering, industrial and commercial skills and capabilities that make it possible to reach the market competitively. Although less successful, Brazil and Mexico have been trying to apply the same lesson. Science-intensiveness does not make matters any easier – rather the opposite.

The case of idiosyncratic, DC-specific, needs for which BT applications may be sought, as well as all those instances where the market fails to operate efficiently (like in vaccines or in bGH), merit a special consideration of the scope for government intervention. But, no matter how much or how little the government intervenes, the fact still remains that entry into BT cannot be seriously considered if enough attention is not paid to things such as skills to be mastered, resources to be commanded, products to be manufactured, organizational modes and manufacturing standards to be adopted and markets to be served, right from the lab throughout all stages up to the distribution to the final consumer.

The above does not mean – particularly after allowing for differences among countries – that DCs should focus on 'low-end' applications, most of which are still to be developed. It simply indicates the need for paying enough attention to bottlenecks and constraints to the 'high-end' applications which are sometimes recommended.

Entry into high-level BT research can render extremely valuable services because, among other things, it makes it possible to keep an eye on what is going on in the scientific frontier and, eventually, take advantage of it as a possible quick follower. However, entry into the research stage without having much chance to proceed forward along the innovative chain, entails the risk of having the results industrialized elsewhere and, what is even worse, of subsidizing ICs' research endeavours.

Over and above the need to bridge the gaps between scientific breakthroughs and technological design, between technological design and engineering development and between engineering development and manufacturing practice, there are also requisites regarding the necessary interaction among the diverse agents of the innovative process. The Brazilian experience in ethanol is a good illustration of the role of the systemic and synergistic aspects in BT development. But only a few DCs can afford to engage in an effort on such a comprehensive scale.

Some 20 to 30 years will elapse before BT becomes a widely utilized technology affecting many industrial sectors. How can DCs take better advantage of it over this period?

The intensity of current international competitive rivalry and the fact that the US, the leading country in the field, is on the defensive and trying to offset its eroding competitive power, is a rather unfortunate coincidence for DCs endeavouring to enter BT. Conditions for access to technological know-how are now harder than they used to be when a lot of knowledge and information regarding manufacturing processes was transferred on a commercial basis. Today, this kind of transfer to DCs has become rare. The rapidly shifting scientific, technological and industrial frontiers in BT accentuate the risks and uncertainties linked to DC moves.

For instance, initial price quotations for BT products are very high since the firms concerned intend to recover R&D costs as quickly as possible. But prices may go down substantially any time. This makes it rather tricky for DC firms considering whether to get into the BT business to undertake a realistic assessment of future returns (even though their own R&D costs may be substantially lower, thanks to imitator's advantages). Another difficulty lies in the sparsity of engineering-cost estimates, since most relevant equipment for advanced BT applications is currently being made to order.

The potential success of attempts at entering BT depend, among other things, on the previous-experience profile at the firm and country levels; inter-organizational synergies within the private sector and between it and the public sector; availability of risk capital; innovation financing; linkages between industry and the scientific and technological system; and application-sector-specific scale-up skills and capabilities.

Although DCs may have little chance of entering directly into high value-added product lines involving heavy R&D expenses, they do have certain indirect strategic routes for taking effective economic and social advantage of advanced BT and building up the experience necessary to enter increasingly higher value-added products. Such routes include applications regarding: (i) plagues and idiosyncratic diseases; (ii) improvement in the competitiveness of traditional industrial sectors (agriculture, biomass, food and drinks, forestry, textiles, mining, etc.) by enhancing existing product quality and process efficiency; and (iii) developing new products based on traditional industrial sectors aimed at niche markets.

But it would be absolutely illusory to attempt entering commercial BT without paying enough attention to the mastery of effective downstream processing technologies through joint work between chemical engineers and biochemists. The lack of bioprocess engineering skills may effectively block scale-up efforts, particularly at the purification stage (the major cost item). The ability to undertake effective scale-up is a major entry barrier into most commercial BT segments. Substantial lead times are involved. Genetic engineering has permitted mass production of proteins and lower fermentation costs for products such as enzymes and amino-acids. But it does not substitute for more traditional engineering disciplines. The choice of techniques (e.g. regarding the optimum expression medium) is still another important challenge to engineering developments involved in scale-up efforts.

The rich variety of agents of BT change in the world market provides plenty of room for identifying and resorting to sources of international scientific and technical co-operation. Many IC-based BT start-ups are eager to enter into technology-transfer agreements with DC-based firms. However, it is necessary to proceed with caution since, in most cases, their technologies are still at an experimental stage. On the other hand, examples of DCs' excellence in BT research abound. There are also many instances of successful applications of the outputs of such research (like Zimbabwe's DNA probes for salmonella; Argentina's diagnostic test for the Chagas disease; and Colombia's malaria vaccines) (Eisner, 1988).

As pointed out, Singapore, along with other South Asian countries and Spain, has pursued a shrewd strategy that consists of taking advantage of expatriate scientists and engaging in joint-research venturs in ICs. Zimbabwe, for instance, takes advantage of expatriate scientists working in France in the area of DNA probes for salmonella. This work is of global interest as the disease causes 3.5 million deaths each year in children with diarrhoea (*The Economist*, 1990a, p. 81).

But joint-research ventures do not necessarily work to DC's advantage. Some agreements may allow IC-based corporations to use DCs' research skills and capabilities as a source of cheap inventive labour whose output is subsequently processed industrially and commercially back in the IC (Thayer, 1989, p. 7. and *Chemical and Engineering*

News, 1989, p. 14). The Chinese are involved in this kind of joint-research venture while acquiring, at the same time, turn-key, prefabricated BT facilities from a major multinational to manufacture recombinant hepatitis B vaccines. This black-box-type transfer includes highly sophisticated hardware items (such as ultra-centrifugation process equipment that brings into play forces hundreds of thousands of times as powerful as gravity) (*Wall Street Journal*, 1989).

V. Concluding Remarks

One of the basic dilemmas a DC faces in BT is how to enter it at the right time, and how to avoid pursuing wrong leads and dead ends. Getting into BT at a point too far removed from the market or too dependent on price-sensitive products in highly competitive and risky markets may not be a sensible approach.

DCs need to understand the dynamics of BT change in order to identify technology and market trends and valid interlocutors (universities, research boutiques, dedicated BT firms, or multinational corporations) according to specific needs. This, in turn, requires a clear assessment of the nature of these different actors, their relationship to each other, and their respective strategies and likely trajectories.

It is also essential for DCs to understand the nature of the most important factors that affect the timing of introduction and rate of diffusion of BT, such as company strategies; scientific, technological and engineering bottlenecks and uncertainties; barriers to entry and threshold factors and the relative competitiveness of BT products and processes.

To bridge the gap between the rapid development of the scientific frontier and the lagging evolution of the technological and manufacturing frontiers will take a great deal of time and resources. An increasing number of entrants at the R&D stage can be anticipated. But it is not so certain that the state of the art in manufacturing will catch up any time soon with the acquisition of applied scientific skills at the enterprise level. Herein lies a vital breathing space for DCs.

However, the inability to supply products and services at competitive prices (net of infant industry learning-related costs and external dis-economies) downgrades the capacity to generate wealth. No matter how creative the efforts involved might be, this kind of situation is likely to lead to a dead end. High value-added products make it possible to pass on high costs of research, but for now they do not appear to be the solution for DCs attempting to enter BT commercially.

The 1970s witnessed the birth of BT industrial applications. During the 1980s MNCs cautiously followed events, becoming more and more involved and thus getting ready to fully enter the BT industry. During the 1990s they are likely to impress thei particular mark upon future developments.

Once the basic BT techniques become routine, one of the main questions to be addressed is what to do with them (new proteins or life forms can be created without a clear purpose). The answer to this question cannot be pre-fabricated. It can only result from a learning process whereby the accumulation of scientific, technological and manufacturing skills and capabilities interacts with social needs and market realities.

Among other things, this process entails, on the one hand, carrying out basic and applied research on a continuous basis and, on the other, setting up the engineering capability that is needed to translate the resulting insights into competitive products. This process will be more and more influenced by the increasing absorption of BT by user industries, whereby its trajectory will be progressively assimilated by that of those industries.

The above is precisely what, once again, the Japanese appear to have understood very early. In their two-tier strategy, the first stage (1981–88) consisted of achieving the mastery of the scientific underpinnings and practical use of the basic techniques of BT. For this, they have taken full advantage of research links with the best centres of excellence in the world. The second stage (1988 onwards), which started while the first was still in progress, consists of acquiring the necessary manufacturing experience through licences – and then starting to enter the real game as innovators, forging ahead both at the scientific, technological and commercial levels (Masuda, 1989).

International technical co-operation has an important catalytic role to play. This includes, first, supporting the setting up of information networks. In the second place, it concerns the building up and strengthening of domestic scientific and technological capabilities. This comprises areas such as bioprocess-engineering skill formation, experimental development and scale-up efforts, setting up and upgrading standards of manufacturing, quality, and process and product safety, and working out of industrial policy guidelines. Thirdly, it consists of assisting in the transfer and adaptation of technology; and, fourthly, supporting the development of new products and processes.

Initiatives such as PRATAB – Programme of Policy Research and Technical Assistance in Biotechnology (see Sercovich and Leopold, 1990), would help in tackling an urgent need to avoid duplications, create synergies and improve the use of resources.

PRATAB is intended to perform as a scanning and early warning system for the benefit of DCs through the execution and support of technical assistance and policy research in BT, based on the articulation of the so-far scattered efforts made by governments and international organizations. A network of data banks would be set up and consulting services to DC governments and organizations would be provided. PRATAB's sponsorship is to come from governments and international sponsoring agencies. The programme would establish a network of researchers and policy makers from both DCs and ICs so as to facilitate their reciprocal consultations on a periodical basis. Its financing would result from sums granted by the different sponsoring agencies to specific research, consultant and technical-assistance tasks in the context of their on-going activities so that overheads would be kept at a bare minimum.

NOTE

[1] For an in-depth treatment of many of the points made in this chapter, see Sercovich and Leopold (1990).

REFERENCES

Bio/Technology (1990), Vol. 8, September, p. 802.
Chemical and Engineering News (1989), Cell Technology in Chinese Joint Venture, 3 April.
Eisner, R. (1988), *Genetic Engineering News*, July/August.
Fortune (1990), 8 Oct.
Freeman, K. (1989), Aussies in US, *Genetic Engineering News*, February.
Genetic Engineering News (1989), June.
Katz, J. and Bercovich, N. (1988), Biotecnología e Industria Farmacéutica, CEPAL, Buenos Aires.
Masuda, M. (1989), Bio-Industry Policy Reviewed in Mid-Term, *Business JAPAN*, July.
Sercovich, F. (1986), The political economy of biomass in Brazil – the case of ethanol, in Jacobsson et al., *The Biotechnological Challenge*, Cambridge University Press, Cambridge.
Sercovich, F. and Leopold, M. (1990), *Developing Countries and the 'New' Biotechnology: Market Entry and Related Industrial Policy Issues*. Manuscript Series, IDRC, Ottawa.
Thayer, A. M. (1989), 'US firm, Chinese set up biotech venture', *Chemical and Engineering News*, 20 March.
The Economist (1990), *Special Report on World Trade*, 22 September.
The Economist (1990a), The slow march of technology, 13 January, p. 81.
Yuan, R. T. (1988), *Biotechnology in South Korea, Singapore and Taiwan*, International Trade Administration, Washington, D.C.

Biotechnology and the Third World: The Missing Link Between Research and Applications

Raymond A. Zilinskas

Introduction

This chapter contains the synopsis of a study undertaken to investigate the activities that four major United Nations (UN) agencies are undertaking to help developing countries gain advanced capabilities in biotechnology (Zilinskas, 1987b). First, the author researched the growth of biotechnology in the industrialized countries, particularly its applied aspects. Second, information was collected about relevant projects at the headquarters of the agencies, which included the UN Development Programme (UNDP), the UN Educational, Scientific and Cultural Organization (UNESCO), the UN Industrial Development Organization (UNIDO), and the World Health Organization (WHO). Third, three developing countries selected as case studies (Egypt, Thailand, and Venezuela) were visited to scrutinize the biotechnology- related activities that they had planned or that were underway. Fourth, the development of biotechnology in the case countries was compared with and contrasted to that process in developed countries. Fifth, information derived from the first four tasks was analysed to assess whether the UN-sponsored projects were fulfilling national needs and if they were indeed augmenting and advancing these countries' capabilities in biotechnology. Last, suggestions were made for initiating a cohesive approach towards biotechnology for the developing countries by UN agencies.

Biotechnology in the Industrialized Countries

Beginning about 10 years ago, advanced biotechnology research and development (R&D) has given rise to a rapidly growing

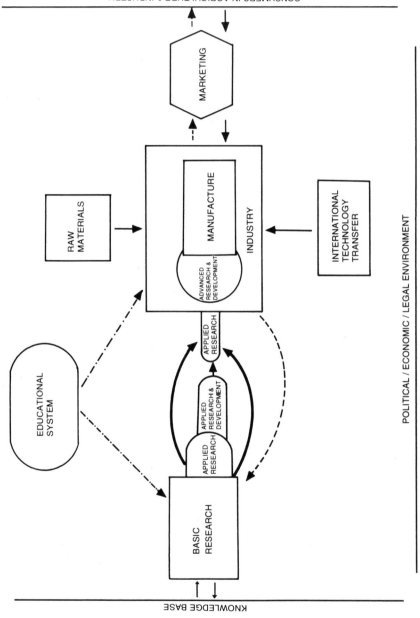

Figure 36.1 The concept development process in industrialized countries.

bioscience-based industry that is now starting to market its first products. With this development, it is possible to chart the progression of events whereby an idea or concept becomes a marketable product (i.e., the concept development process), as well as to identify its important components and the forces that act on it. Furthermore, it is possible to distinguish and evaluate the elements essential for capability building in its research and applications sectors. Hence, the concept development process may be schematically presented (see figure 36.1).

Fundamental to the process is the easily accessed knowledge base. In science, the most important contributions to the knowledge base come from basic research. Only a small fraction of this base has the potential for practical application. That fraction, accessed by innovators, inventors, and entrepreneurs, is further researched in an applied research facility. If the concept is demonstrated as having the potential for meeting a need and of being workable, it enters advanced R&D. In a developed country, advanced R&D typically takes place at one of three types of facilities: national laboratory, defence establishment laboratory, or industrial laboratory. If development is successful, a product or process results. To lay the technical basis for its large-scale manufacture and to determine its economics, the product is scaled up at a pilot plant facility. Pilot plant operations and downstream processing may occasionally be done at development laboratories but more typically take place at the industrial plant. If the feasibility of the product is proved, the industrial plant scales up for the commercial manufacture of the product, and the marketing and sales division makes certain that the product reaches those for whom it is intended. The education system produces the scientists, technicians, and managers that operate the process. And the process functions well in a favourable political–economic–legal environment.

UN-sponsored Activities in Biotechnology

The survey of past, ongoing, and planned biotechnology-related activities sponsored by the UNDP, UNESCO, UNIDO, and WHO (Zilinskas, 1987a) was performed to answer three questions: is there a coherent approach by international organizations (IOs) in the field of biotechnology? What kind of assistance is being provided by IOs to the Third World and what are its impacts? What assistance is not being rendered?

Coherence of IO approaches

Usually UN agencies do not formulate general policies that serve as guidelines for advancing technologies in the Third World because there is no need for such policies. Neither low nor medium technologies require special conditions to flourish; for example, even the complex technologies necessary to build an automobile do not require their users to access the knowledge base or to develop close associations with universities or research institutes. Thus, technologies may be introduced directly into the industrial sector without involving the research component. However, to deploy a science-based technology, a far-reaching, complex environment is required, as demonstrated by the ideal concept development process (fig. 36.1). Wise policies must be formulated and applied to create that environment and to build the infrastructure that is part of it. However, the number of biotechnology-related projects being sponsored by the IOs is very small, as can be witnessed by UNDP supporting less than 10 biotechnology projects out of more than 4600 in 1986, while UNIDO is running fewer than 15 out of a total of more than 1500 projects. The low number of IO biotechnology-related activities probably stems from two factors: few IO managers are familiar with the field, and its practical applications are perceived by knowledgeable IO and national officials to be far off in time. Thus, for the present, biotechnology is of less importance to the IO policy makers than are many low and medium technologies and, consequently, agencies have not formulated a general policy for this field among, or within, UN agencies.

Kinds of assistance provided by IOs

An examination of IO activities indicates that there are three categories of assistance: help to policy makers, support for countries' research and training activities, and provision of backing for the establishment of international networks.

Help to policy makers. UNESCO and UNIDO in particular have taken a 'from the ground floor up' approach to introducing biotechnology to developing countries: i.e., they first seek to familiarize decision makers and their scientific advisers with biotechnology through briefings by scientific missions, the holding of national symposia, and the providing of informative material. This approach appears to have borne fruit: information elicited from interviews conducted with scientists and politicians in developing countries indicates that there is widespread awareness among Third World officials about biotechnology.

Yet, it is not clear whether the widespread appreciation of biotechnology by politicians has translated, or will translate, into national commitments by countries to actually proceed with capability-building measures. Few developing-country governments have taken the hard decisions that would lead to the building of biotechnology capabilities in their countries. Perhaps the distressed economic situations many countries find themselves in prevent governments from taking on new obligations in science and technology, or it could be that other endeavours are perceived as offering better prospects for national development.

Once officials have gained an awareness of biotechnology, they must consider the range of actions needed to deploy biotechnology for designated purposes. UNIDO especially has assisted nations in this regard by setting up fora where the possibilities of biotechnology for the nation in question have been discussed and evaluated. National conferences have been held in Morocco, South Korea, Kuwait, Saudi Arabia, and other countries. However, since all such national symposia or workshops have been held rather recently, their effects cannot be evaluated as yet. Early indications are that national symposia and workshops are worthwhile for IOs to sponsor since they tend to focus national efforts in biotechnology.

Support for capability building in research. Support for capability building in research takes four forms: provision of technical information, training opportunities, funds to purchase major equipment, and funds to purchase chemicals.

(i) Technical information. Scientists in the case countries are usually unable to access technical information easily or quickly. Sometimes information that is sent to them is lost in the mail, a relatively common experience in, for example, Venezuela. Most often, however, there is an invisible, yet real, administrative barrier that prevents the dissemination of information to researchers. These barriers are found in the ministries under whose auspices or authority researchers work. Simply put, huge quantities of general and technical information is routinely sent by IOs to the ministries they liaison with: very little of it is forwarded to anyone outside these ministries. Apparently, the concept of information dissemination is poorly developed in governments of the case countries.

Other information is provided directly by IOs on request; yet few take advantage. For example, free subscriptions to UNIDO's *Genetic Engineering and Biotechnology Monitor*, a quarterly publication that sums up recent developments in biotechnology throughout the world, are extended to anyone requesting it. One would believe that scientists and libraries in developing countries would be clamouring for such a valuable source of information. Instead, there are only about 1200 subscribers, mostly in developed countries. The likely reason for the low number of subscribers to the *Monitor* is that Third World scientists do not know of it.

Even in those few cases where the issuing organization has a programme for the active, direct forwarding of information to scientists, the potential recipients rarely benefit. Thus, for example, the International Network of Biotechnology (INB), through its member governments' diplomatic missions in developing countries, exerts much effort in sending information circulars directly to scientific societies, university departments, institute research units, and individual scientists. Yet, it was only the rare researcher who had heard of the INB.[1]

Scientists in the case countries are unanimous in claiming that they do not receive the technical information they need from IOs, nor are they informed about the possibilities of assistance made available by IOs. The conclusion may be made that the passive method of information dissemination that is usually employed by IOs is ineffective; better, active mechanisms should be developed to ensure that technical information gets to the intended recipients.

(ii) Training opportunities. By providing training opportunities, IOs can augment present capabilities of scientists to perform research and introduce new techniques to research units, enabling them to take on previously impossible projects. Conversely, newly trained researchers will often be unable to apply new skills. Many scientists from developing countries who receive advanced training cannot use it in their home institutions because of lack of resources, facilities, equipment, and supplies. Some respond by allowing their newly gained skills to languish. Others leave their home countries and use their new knowledge in developed-country laboratories.

The question of how one calculates cost–benefit in these situations in unclear. After all, some scientists who receive training will be able to utilize it at home for designated purposes. Others could gainfully employ new knowledge in the future. The expatriate scientist is using his new found skills, albeit not as originally planned. Probably, only a few trainees represent a total loss.

(iii) Funds to purchase major equipment. With the possible exception of WHO (through its Research Capability Strengthening programme), IOs are wary of financing projects when the equipment component adds up to more than 30% of the project's budget. Several possible reasons follow.

First, since major equipment is usually shared, because it remains with the supported research unit after a project ends, and because very few projects would by themselves necessitate its sole use, it could be considered part of the infrastructure. Thus, in the IO's view, the apparatus should be purchased by the institute running the project or by the home government.

Second, it can be difficult for an IO to evaluate requests for major equipment. The need for some can be estimated: for example, the request for a fermentor may be assessed by considering the value of its output over time. Other requests can be problematic, for instance that for a gamma counter. Can the high cost of a gamma counter be justified in terms of its value to the project under consideration? Many times it cannot; in the case countries only a rare project would have need of that highly sophisticated equipment, few researchers would know how to use it, backup support by its manufacturer would be unavailable, and no technician would be able to maintain it.

Third, every IO project manager is aware of equipment that has been unnecessarily purchased with IO funds. For instance, UNIDO's Selected Committee observed in an institute it visited 'a set of about 20 gamma counters laid out on benches behind locked doors' (United Nations Industrial Development Organization, 1983).

Whatever the reason, the amount of major equipment being provided by IOs to research institutions in developing countries is minute when compared with the need.

(iv) Funds to purchase chemicals. Rare chemicals, and other expendable supplies, are vital to research and to bioscience-based industry. Thus, practically all IO-sponsored biotechnology-related research projects include provisions for their purchase. While such funding helps carry individual projects forward, it does nothing to help solve the overriding problem, namely, no dependable manufacturers of rare chemicals exist in the case countries or, with the rare exception, in the Third World. To secure these chemicals, researchers spend scarce hard currency, and even so they will always be in short supply. Clearly, the shortage of rare chemicals limits growth of biotechnology R&D and places what may be a fundamental barrier in the way to the establishing of bioscience-based industry in the Third World.

Most of the surveyed IO activities in biotechnology fall within this category of direct support for capability building in research. There appear to be two reasons why. First, IOs themselves support them. There is the pervasive belief among IO managers that a high-technology industry cannot be built unless its underlying science is firmly established. Since few developing countries have an adequate scientific base for biotechnology, it logically follows that assistance should first be given to lay or strengthen such a base.

Second, bioscientists have successfully lobbied IOs to support biotechnology research projects. There are constant interactions between IO managers and bioscientists. Those from developed and developing countries frequently serve as consultants to UN agencies. Third World scientists are often advisers to their governments in negotiations having high scientific–technical content. Several have become aware of the range of UN-sponsored projects and use that knowledge to seek assistance for national programmes. Programme managers cannot help but be influenced by these people: they are articulate, sensitive, caring, and persuasive. The end result has been that bioscientists, who without doubt firmly believe that biotechnology can and will help the Third World develop, are identified by project managers as presenting the prevalent view among policy makers in developing countries: a perception that is at the least one-sided. By not taking into account differing interests, even if not articulated, projects are designed to boost capability building in research, but they are otherwise too circumscribed to have practical effects.

IOs and international networks. IOs have the important function of acting as catalysts in the setting up, and in the operation, of networks having international reach. Two networks have the longest history: the Consultative Group on International Agricultural Research (CGIAR) and the Microbiological Resource Centers (MIRCENs). The two are quite different: CGIAR, although an informal organization without constitution or charter, fully funds its 13 member institutions, and its research activities are in a general sense coordinated by the network secretariat (located at the World Bank's headquarters in Washington, DC). Its focus is on tropical agriculture and animal diseases prevalent in the tropics (United Nations Development Program 1986a). The 16 MIRCENs are independent units, only loosely connected through networking, and minimally funded by UNESCO. Initially, MIRCENs' interest was focused on research pertaining to gene pools; now its scope is wider, including both agriculture and industry (DaSilva and Taguchi 1986). Despite differences, both networks facilitate communications between widely dispersed units so they may work in unison for achieving general objectives by sharing expertise, undertaking cooperative research, integrating training of scientific personnel, and sharing certain resources. Occasionally, research results from a network have had dramatic effects on the Third World: for instance, the highly productive dwarf rice developed by CGIAR's International Rice Research Institute has been adopted by farmers throughout the world, resulting in a marked increase in production.

Apparently, neither CGIAR nor MIRCENs have been objectively assessed. Nevertheless, that over 30 nations (as well as IOs and foundations) contribute in excess of $180 million per year to support CGIAR would indicate they find it worthwhile. Similarly, scientists who work at institutes that are part of MIRCENs uniformly espouse its value (Colwell 1983, 1986).

New networks that concentrate on biotechnology are in the process of being set up by IOs. Thus, three UN agencies (UNESCO, UNDP, and UNIDO) are helping in the establishment of the Regional Biotechnology Programme for Latin America and the Caribbean (United Nations Development Program, 1986b); UNIDO is setting up the International Centre for Genetic Engineering and Biotechnology (ICGEB) and its network of affiliated centres (Zilinskas, 1987a); and UNU is formulating a network for international cooperation in biotechnology (United Nations University, 1985). The functions of the new networks will be similar to those of the older ones; i.e., they will seek to integrate the work and training programmes of participating institutes, share expertise, pool resources, and perform cooperative research. If all works as planned, the network institutes should be able to achieve more collectively than they would if operating separately. Furthermore, with the backing of a network, an individual institute can take on projects that it otherwise would have to forsake, perhaps because they would be too expensive, large, or difficult.

The preceding analysis is summed up as follows. (*i*) The number of biotechnology-related projects, whether completed, underway, or planned, is small when compared with the totality of projects being undertaken by the UN agencies, and the resources being committed to them are also minute. (*ii*) No general policy in regard to biotechnology has been formulated among the IOs, or indeed within any individual agency. However, even without a policy, or policies, and whichever the executing agency, the activities pertaining to biotechnology are remarkably similar in that they can be grouped under one of three types of activities: either rendering assistance to policy makers (informing them about biotechnology and its promises, and organizing fora to delineate national programmes in biotechnology), or promoting capability building in biotechnology research (supporting research and training projects at the national and international levels), or helping in establishing international biotechnology networks. (*iii*) No instance of wasteful overlap between activities of different IOs is noted. Possibly, the small number of projects being done in a large technological field occasions few chances for overlaps. (*iv*) IO biotechnology projects have created opportunities for productive cooperation between agencies. Specifically, UNDP, UNESCO, and UNIDO cooperate in the Latin American regional programme; and the ICGEB is expected to work closely with FAO, UNESCO, UNIDO, and WHO. Future regional projects involving Africa and Asia are likely to provide additional possibilities for IO cooperation. Collaboration on policy-related issues has barely begun; for example, UN Environment Program (UNEP), UNIDO, and WHO have recently formed a joint working group on biotechnology safety issues.

Major gap in assistance by UN agencies

The major gap in UN-rendered assistance is related to making certain that results from research are applied. In particular, two problem areas are identified.

Advanced research, development, and industry. The scrutiny of UN activities indicates that no resources have been expended on advanced biotechnology research, development, or industry, except perhaps indirectly, and no resources are being allocated for this purpose in projects being planned. Several of the IO-sponsored projects do include measures for the dissemination of the results they generate to industry, usually through a workshop or symposium, and publications. In actuality, these approaches are not likely to work for five reasons. First, information about the holding of workshops or symposia, or of the availability of publications, most often does not reach those who would be interested. Second, the described results may be in a narrow scientific area, of interest to few people in industry. Third, since industries as a rule do not have R&D capability, they cannot take advantage of the presented results. Fourth, even if results could be usable, the adaptive capabilities of industry would most probably be so low that it would be unable to utilize findings without making major new investments in manpower and equipment. Fifth, capital for development, scale-up, and manufacture is difficult to raise in the developing countries.

IOs and research consumers. The acquisition of scientific knowledge and new technology may enable the recipient industry to become more competitive in local or international markets. At times, changing markets create conditions whereby a continuous flow of new technologies is required by industry. To meet these demands, industrial managers have to rely on science and technology providers to respond quickly and appropriately. As the research sectors in the case countries are seemingly unable to do so, indigenous industry habitually turns to foreign technology suppliers for these needs. Early indications are that this pattern will prevail as well in biotechnology. For example, in Thailand it was noted that a local industry was interested in acquiring a technology for processing palm oil. Although appropriate R&D was taking place at a Thai university, the firm chose to buy the technology from an enterprise in the United Kingdom.

Although the major responsibility for correcting problems related to the bridging of the gap between research laboratories and industry must rest with the home government, IOs could make this task easier.

IOs could provide counsel to governments on how changes can best be accomplished, and what steps they can take to encourage indigenous industry to contract with local research establishments for needed research or to develop customized biotechnology. However, no such IO projects are active or planned.

Biotechnology-related Activities in the Case Countries

The 'ideal' concept-development process provides a framework for analysing biotechnology-related activities in the case countries of Egypt, Thailand and Venezuela. The major findings are as follows.

Knowledge base

Practically speaking, severe impediments prevent the Third World researcher from using important information sources. First, as a result of high expense, scarce funds, and at times, a lack of hard currency, only a few libraries in the case countries have book and journal collections adequate to support strong research efforts in biotechnology. These few are located in the capitals; researchers elsewhere in the countries are generally not adequately served by libraries. Second, access to data banks is not available, or is limited, because of lack of funds, poor communications lines, and lack of technical expertise. Third, person-to-person contacts between researchers from developed and developing countries are relatively infrequent because the Third World researchers have few opportunities to travel, while scientists from industrialized countries do not usually travel to developing countries. One may conclude that apart from scientists working in the capital cities, the knowledge base is neither readily nor easily accessible to bioscientists in the case countries.

Basic research

Wide-ranging basic research in the traditional biosciences is proceeding at universities and government institutes in the case countries. Thus, a substantial scientific base exists from which research in biotechnology could expand. Nevertheless, before such research can be undertaken on a larger scale, certain barriers have to be overcome. One is mentioned above: access to the knowledge base must be improved. Second, there is a pervasive shortage or lack of rare chemicals; enzymes and radioisotopes are particularly difficult to procure. The barrier presented by the unavailability of rare chemicals is very serious since some of them, particularly the endonucleases, are the indispensable tools of the modern biotechnologists. Without an assured and adequate supply of these substances, it is not practicable to expand a biotechnology research programme, nor is it possible to lay a basis for a bioscience-based industry.

Research laboratories in the case countries in general lack recent equipment, especially major pieces of hardware such as mass spectrometers, gamma counters, ultracentrifuges, automatic DNA sequencers, and flow cytometers. The relatively small sums of money allocated by governments for research does not allow the purchase of these 'big ticket' items. Without major equipment, researchers performing advanced biotechnology research will sooner or later reach limits that cannot be surmounted. Sophisticated research cannot usually be undertaken.

Even if equipment and instruments are available, other problems may prevent their full utilization. Spare parts are often unavailable, so when equipment breaks down it stays down for lengthy periods of time. The problem of breakdowns is compounded by instrument and repair technicians lacking training, and motivation, to clean and perform preventive maintenance of equipment. This situation is very difficult to correct because it often stems from systemic reasons: low wages, job security despite poor performance, minimum or nonexisting criteria for licensing of technicians, and lack of incentives for superior job performance. The net effect of equipment being down, whether from normal wear or as a result of careless maintenance, is that the productivity of the affected research laboratory goes down. Projects underway are stopped; some may have to be redone. Planned projects are delayed. Researchers spend precious time away from the bench on frustrating tasks, such as trying to deal with bureaucracies to effect repair, negotiating with irritable officials and repair men, and designing temporary fixes.

Scientists and technicians are civil servants in the case countries, compensated according to scales set for government employees. As part of austerity measures, governments have for the last 5 years or so refused pay rises to civil servants. The wages in the public sector have accordingly not kept pace with increases in the cost of living or with the wages in the private sector. Scientists and technicians are poorly compensated when compared with somewhat equivalent employees in the private sector. Furthermore, the possibilities for earning supplemental income are severely restricted for scientists as no research and little development is done by industry.

On sometimes a daily basis, Third World scientists face difficulties rarely encountered by their colleagues in developed countries, such as power shortages or outages, interrupted water supplies, supplies of vital reagents running out with little chance for timely replacement, frequent equipment breakdowns, lengthy delays in communications caused by poor telephone systems, and so on. The net effect of these factors on research productivity is without doubt negative.

Applied research

In the case countries, little applied research is being done at universities. There appear to be two reasons. First, the role of scientists in universities is a traditional one: to teach, to implant a love of learning among students, and to perform basic research. A university scientist would find it trying to take up applied research projects since there is minimal appreciation for such work among colleagues. On a professional level, academics do not interact with industrialists.

Second, with few exceptions, there are no contractors for applied research or consumers for research results in the case countries. As a result, it would make little sense for university researchers to perform applied research since results, even if potentially useful, will not be developed. The exceptions tend to prove the rule. The few applied projects underway in health are being done by groups that have through strong efforts in basic research developed expertise in narrow areas pertaining to various disease agents. The expertise allows the groups to take on projects supported by IOs or funding organizations in developed countries, who may then use the results for their own purposes.

A certain amount of applied research does take place at public research institutes in the three case countries; in Egypt it is quite significant: more than 75% of research carried out by its science and technology organizations is applied (El Nockrashy et al., 1986). Only a small percentage of research results is eventually used because the work is inappropriate or of low quality or there is little association between the research units and potential users of results in agriculture and industry. No applied biotechnology research funded by indigenous industry is done in the case countries.

Advanced research and development

No units capable of advanced biotechnology R&D exist in the case countries. The advanced R&D component so strategically located in the ideal concept development process (see fig. 36.1) is missing in the case countries. As a result, it is difficult, if not impossible, to effect the transfer of knowledge and technology from research units to indigenous industry.

Industry

The industrial unit in the case countries takes one of two forms. In the first, the enterprise is merely a packaging and marketing unit of bulk products produced elsewhere and imported. This is probably the most common form of transnational corporations' subsidiaries. The second form is the indigenous industry, which can have a development capability but is in the main a manufacturing facility. Neither has a research capability. When industry requires a technology, it is imported. The importing industry's development capability, at the maximum, allows the industry to adapt the imported technology to meet its own requirements.

Because scientists in the public institutions are perceived as having little appreciation of real-life problems, and since it is difficult to engage their services because of bureaucratic restrictions, industrialists as a rule have no professional contacts with the research establishment. As a result, industry does not access the knowledge base, cannot assimilate results from basic or applied research, and is incapable of independently performing research to solve problems or to develop new products or processes.

'Technology push' and 'demand pull'

In the ideal scheme of concept development, technology push and demand pull exert forces that act at every point along the concept-development process. Thus, the push of a powerful technology is likely, for better or worse, to result in the delivery to customers of new products and processes. Conversely, the pull generated by customer demands can and does give rise to applied research. However, the concept development process in the case countries is different from the ideal as a result of the separation between the research and the industry–marketing components. Primarily, there is no demand pull for biotechnology research since there are no consumers of research results. Furthermore, it is probable that demand pull is less of a force in developing countries than in the industrialized world. Consumers in the Third World are apparently considered only marginally important in economic terms by research directors in the industrialized countries, as is demonstrated by the finding of the Science Policy Research Unit at Sussex University that less than 1% of the research in the developed world has relevance to developing countries (Freemantle, 1983). Therefore, products and processes developed and marketed for consumers in the industrialized world are also marketed willy-nilly in developing countries, without much regard for their populations' wants or needs.

Nearly all biotechnology-related activities in the case countries are designed to increase capabilities in research; practically none is designated to increase capabilities in development or industry. The relatively high level of support for the biotechnology research establish- ment in these countries results from bioscientists having successfully organized themselves to form relatively powerful political pressure groups to lobby for the support they want. Close, continuous relationships exists between scientists and the policy makers who deal with scientific–technical matters in governments. This 'normal' political activity in democracies cannot be faulted, except for one matter, the bioscientists appear myopic in how they neglect establishing relationships with the industrialists. Although scientists acknowledge and deplore the lack of communications between themselves and the industrialists, neither they as individuals nor their interest groups have made meaningful attempts to bridge the gap. As a result, there is a discontinuity between the research establishment and the industrial–marketing components. The prevalent sentiment among scientists appears to be that they are not interested in helping turn their research into profit-making enterprises for others and that their governments, rather than themselves, should take the necessary steps whereby their resarch findings are applied. The attitude of bioscientists also partially explains why there are no consumers for bioscience R&D in the case countries.

Efforts by bioscientists are leading to an increase in their countries' research capabilities; possibly the research establishments' productivity will go up, and more remarkable results could be generated from improved research units. However, even if research productivity increases, its results are not likely to be applied by industry or agriculture. As a consequence, economic growth will not be stimulated, national self-reliance will not be furthered, and no particular region or sector of industry or agriculture will benefit. It is possible that if scientists in the case countries maintain their present course of action, they may in the end weaken their own position since the public and its representatives could come to the realization that the bioscientists, while ready and able to draw scarce funds from governmental coffers to feather their own nests, do little of practical worth for their home nations. The disparity between the great expectations of biotechnology and the actual lack of achievements could in the future cause a government to look less favourably on proposals to support biotechnology and bioscientists.

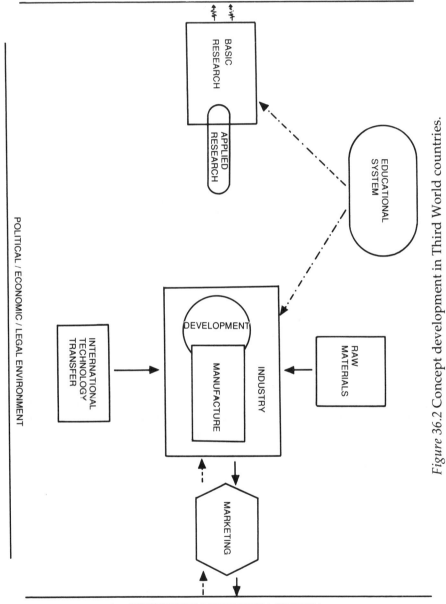

Figure 36.2 Concept development in Third World countries.

The Advance of Biotechnology in Case Countries versus Developed Countries

Findings from the foregoing may be used to construct a scheme of the concept-development process in the case countries (see fig. 36.2). The process is quite different from the ideal depicted in fig. 36.1.

The first difference is that accessing the knowledge base by researchers in the case countries is often difficult and time-consuming and may be expensive.

Second, in contrast with industrialized countries, where a proportion of basic research is performed to solve fundamental problems that may have arisen in applied research or in development, in the case countries practically all is basic research in its purest sense. In other words, results from basic research becomes part of the knowledge base; little, if any, is applied or used for problem solving.

Third, an important difference partains to applied research. In contradistinction to industrialized countries, where applied research is strongly supported whether in universities or in industry, in the case countries it is either weak buds of strong basic research units in universities or it is performed in public institutions. No applied research takes place in industry.

Fourth, a major difference is the discontinuity between the research and the industrial–marketing components. In the ideal process, continuum is provided by the advanced R&D units that are either part of the research component or, more commonly, part of industry. These units do not exist in the case countries.

A fifth difference is that a technology, usually imported, is injected directly into the industrial component without in any way involving the research sector.

Sixth, industrial units prevalent in the case countries are one of two types: either a subsidiary of a multinational corporation or an indigenous industry. Both import technologies as need arises. In contradistinction to industry in developed countries, neither possesses advanced R&D capabilities.

Relevance of UN-sponsored Activities to Third World Development

IO activities may be assessed in terms of whether they help to remove the barriers present in the concept-development process of the case countries, as depicted in fig. 36.2. The following is a listing of these barriers: each is followed by a capsule appraisal indicating whether or not UN activities are planned or underway to alleviate or remove it.

Fundamental barrier: there is a lack of advanced research and development units, whether in the research or industrial sectors

There are no active or planned IO activities to remove this barrier to biotechnology.

Serious barrier: there are no consumers of research results

There are no active or planned IO activities to remove this barrier to biotechnology.

Serious barrier: applied research units in the industrial sector do not exist

There are no IO activities designed to encourage industry to set up biotechnology applied research units.

Serious barrier: there is a shortage of major equipment in the research sector

There are some IO projects that include funds for the purchase of major equipment. However, this assistance does not come close to fulfilling the need of the Third World for this kind of equipment.

Serious barrier: rare chemicals, especially enzymes, are in short supply

Most IO projects include funds for the purchase of chemicals on an ad hoc basic. However, no IO activity is underway, or is planned, that deals systematically with this barrier.

Serious barrier: accessing the knowledge base is problematic

The different means of communication that enable researchers to acess the knowledge base are controlled by governments and are therefore all outside the purview of IOs. However, there is another aspect to this problem. IOs generate information useful to researchers but have taken few, if any, practical steps to ensure it reaches those who can use it.

Barrier: in most countries there are inadequate instrument-repair facilities and there is a shortage of spare parts

These are systemic problems that governments are best placed to alleviate. No IO activity is active or planned that deals with this issue.

Barrier: the poor working conditions under which many scientists work hinder the perjormance of research

These are systemic problems that governments are best placed to alleviate. If and when IO-supported projects are likely to be negatively affected by these conditions, the responsible IO can, and should, suggest corrective actions based on good management practices.

A Cohesive Approach by UN agencies

For reasons discussed, IO biotechnology projects are almost entirely aimed at increasing capabilities in research, while little effort is given to making certain research results are applied. Unless a more balanced approach is taken, i.e., the applications side should receive at least equal attention to that given to the research side, biotechnology for the foreseeable future is not likely to be deployed for economic development or to help solve pressing problems facing the Third World. IOs should take the lead to prevent such a negative outcome.

The first step in designing a coherent action plan for making certain that biotechnology is deployed for the benefit of the developing countries would be to hold a workshop for IO managers. One of the more policy-oriented IOs, perhaps the UNU or the UN centre for Science and Technology for Development, should organize a workshop comprising managers responsible for funding and executing biotechnology-related projects by major UN agencies. The workshop would have four objectives: (*i*) to develop a broad consensus among the UN agencies on concepts, ideas, and issues pertaining to capability building in biotechnology research and applications by developing countries; (*ii*) to identify key policy and institutional constraints that prevent capability building in biotechnology by developing countries, especially in their health delivery, industrial, and agricultural sectors; (*iii*) to identify promising approaches that may be taken by the UN agencies to help developing countries overcome these constraints and to otherwise facilitate capability building in biotechnology; and (*iv*) to delineate in general terms areas of responsibilities for the various UN agencies in future efforts to overcome constraints and to facilitate capability building.

Once UN programme managers have a good perspective of the problems hindering the advancement of biotechnology in the Third World and how to deal with them, follow-up corrective measures can be taken. An initial measure could be to make certain that both industrialists and researchers participate in UN-sponsored projects, from inception to completion. In particular, projects should contain concrete mechanisms for applying research results – the line from research to application would have to be clear and unbroken.

NOTE

[1] The INB is jointly coordinated by England and France. In addition, Canada, the Federal Republic of Germany, and Japan belong to it. The INB provides financial aid and training opportunities in biotechnology for Third World students at universities in member countries.

REFERENCES

Colwell, R. R. 1983. A world network for environmental, applied, and biotechnological research. *Am. Soc. Microbiol. News.* **49**(2): 72–73.
Colwell, R. R. 1986. Microbiological resource centers. *Science* (Washington, DC), **233**: 401.
DaSilva, E. J. and Taguchi, H. 1986. MIRCENs. Mechanism for international co-operation in applied microbiology and biotechnology. *MIRCEN J. Appl. Microbiol. Biotechnol.* **2**(1): 27–40.
El Nockrashy, A. S., Galal, O. and Davenport, J. 1986. Applied science and technology: an Egyptian–American cooperative program. National Research Council, Washington, DC.
Freemantle, M. 1983. The poor world needs chemists. *New Sci.* **98**: 226–229.
United Nations Development Program. 1986a. Summaries of global and interregional projects. UNDP Division for Global and Interregional projects, May.
United Nations Development Program. 1986b. Programa Regional de Biotecnologia PNUD/UNESCO/ONUDI Para América Latina y el Caribe. Projects RLA/83/003 and RLA/83/009.
United Nations Industrial Development Organization. 1983. Report of the selected committee. UNIDO document ID/WG.397/1.
United Nations University. 1985. Harnessing genes for development. *UNU Focus*, No. 1, August. p. 1.
Zilinskas, R. A. 1987a. The International Centre for Genetic Engineering and Biotechnology. *Technol. Soc.* **9**: 47–61.
Zilinskas, R. A. 1987b. Biotechnology for the developing countries: the role of selected United Nations agencies. A report to the International Cell Research Organization and the United Nations University, 1 September.

Index

Note : References to figures are in italic; references to tables are in bold.